Engineering Cyber-Physical Systems and Critical Infrastructures

Volume 14

Series Editor

Fatos Xhafa, Departament de Ciències de la Computació, Technical University of Catalonia, Barcelona, Spain

The aim of this book series is to present state of the art studies, research and best engineering practices, real-world applications and real-world case studies for the risks, security, and reliability of critical infrastructure systems and Cyber-Physical Systems. Volumes of this book series will cover modelling, analysis, frameworks, digital twin simulations of risks, failures and vulnerabilities of cyber critical infrastructures as well as will provide ICT approaches to ensure protection and avoid disruption of vital fields such as economy, utility supplies networks, telecommunications, transports, etc. in the everyday life of citizens. The intertwine of cyber and real nature of critical infrastructures will be analyzed and challenges of risks, security, and reliability of critical infrastructure systems will be revealed. Computational intelligence provided by sensing and processing through the whole spectrum of Cloud-to-thing continuum technologies will be the basis for real-time detection of risks, threats, anomalies, etc. in cyber critical infrastructures and will prompt for human and automated protection actions. Finally, studies and recommendations to policy makers, managers, local and governmental administrations and global international organizations will be sought.

Alvaro Rocha · Fernando Moreira ·
Sachi Nandan Mohanty · Shu Hu
Editors

Integrating Advanced Technologies for Enhanced Security and Efficiency

Springer

Editors
Alvaro Rocha
ISEG
University of Lisbon
Lisboa, Portugal

Fernando Moreira
Department of Science and Technology
Portucalense University
Porto, Portugal

Sachi Nandan Mohanty
School of Computer Science
and Engineering
VIT-AP University
Amaravati, Andhra Pradesh, India

Shu Hu
School of Applied and Creative Computing
Purdue University
West Lafayette, IN, USA

ISSN 2731-5002 ISSN 2731-5010 (electronic)
Engineering Cyber-Physical Systems and Critical Infrastructures
ISBN 978-3-031-91797-4 ISBN 978-3-031-91798-1 (eBook)
https://doi.org/10.1007/978-3-031-91798-1

This Springer imprint is published by the registered company Springer Nature Switzerland AG
The registered company address is: Gewerbestrasse 11, 6330 Cham, Switzerland

Preface

The rapid evolution of technology has fundamentally transformed the way we live, work, and interact. From artificial intelligence and block chain to the Internet of Things (IoT), advanced technologies are reshaping industries and creating unprecedented opportunities. Alongside these advancements, however, come complex challenges in ensuring security and efficiency—two pillars essential to the sustained progress and adoption of these innovations.

This book, *Integrating Advanced Technologies for Enhanced Security and Efficiency*, explores the intricate interplay between cutting-edge technologies and their role in addressing contemporary challenges. It serves as a roadmap for technologists, policymakers, and industry leaders seeking to harness these advancements responsibly and effectively.

Our journey begins by examining the core principles behind these transformative technologies, delving into their capabilities and potential. We then shift focus to the practical challenges that accompany their integration—ranging from cybersecurity threats to operational inefficiencies. By highlighting real-world applications, we aim to illustrate not only the "what" but also the "how" of achieving seamless integration for maximum impact.

We are at a pivotal moment where the responsible integration of advanced technologies can yield profound benefits, from securing critical infrastructures to streamlining global supply chains. However, achieving these outcomes demands a collaborative approach that bridges disciplines, industries, and geographies.

This book is intended to spark dialogue, inspire innovation, and equip readers with the knowledge and tools needed to navigate the complexities of technological integration. We hope it will serve as a valuable guide for those committed to shaping a future that is both secure and efficient.

We extend our deepest gratitude to the authors, reviewers, and contributors who made this volume possible. Their dedication and expertise are the foundation of this work. Finally, we thank our readers for embarking on this journey with us and for their commitment to advancing the frontiers of knowledge in this critical field.

Lisboa, Portugal	Alvaro Rocha
Porto, Portugal	Fernando Moreira
Amaravati, India	Sachi Nandan Mohanty
West Lafayette, USA	Shu Hu

Contents

AI-Powered Threat Detection and Prevention

Cyber Threat Detection and Classification Using Machine Learning Techniques

Bhagirathi Nayak, Sunil Mishra, Pritidhara Hota, and Rajesh Kumar Ojha

Abstract The foremost military and security issues of the twenty-first century have been and will persistently be influenced by the strategic dynamics of cyberspace and terrorism invalid input. Nevertheless, multiple conflicting definitions exist for both cyberspace and terrorism, and a globally accepted definition for numerous cyber-related behaviours (such as cyber-terrorism, cyber-warfare, and cyber-crime) is lacking. Cyber-terrorism is frequently characterized as the "intersection of terrorism and cyberspace," resulting in extensive interpretation and ambiguity. Cybercriminals are evolving their methods over time to circumvent security measures. Traditional methods are inadequate for identifying zero-day attacks and advanced threats. Various ML algorithms have been created to identify cybercrimes and counteract cyber threats. This research aims to evaluate many prevalent machine-learning approaches to identify significant cyber risks in cyberspace. Five principal machine learning algorithms are mostly used: Random Forest, Naïve Bayes, Decision Tree, Logistic Regression and KNN. We have conducted a concise examination to assess the efficacy of these ML algorithms in Cyberattack detection and classification also used an ensemble technique adaboost utilizing commonly employed and benchmark datasets of the NSL-KDD dataset.

Keywords Machine learning (ML) · Cyberattack · Cyber-terrorism · KDD dataset

B. Nayak (✉) · P. Hota
Sri Sri University, Cuttack, Odisha, India
e-mail: bhagirathi.n@srisriuniversity.edu.in

S. Mishra
ASTHA Institute of Management, Bhubaneswar, Odisha, India

R. K. Ojha
Silicon University, Bhubaneswar, Odisha, India

A. Rocha et al. (eds.), *Integrating Advanced Technologies for Enhanced Security and Efficiency*, Engineering Cyber-Physical Systems and Critical Infrastructures 14,
https://doi.org/10.1007/978-3-031-91798-1_1

1 Introduction

According to a recent analysis by Check Point Research (CPR), global cyber threats surged by 30% in the second quarter of 2024 relative to the same timeframe in the previous year. Organizations encountered an average of 1,636 attacks weekly, underscoring the increasing complexity of cyber criminals and the pressing necessity for comprehensive security measures. India witnessed a significant surge, with a 46% year-over-year escalation in cyber attacks. Indian organizations experienced an average of 3,201 attacks weekly, ranking second in the Asia–Pacific region, following Taiwan. The education and research sector was the most attacked worldwide, experiencing a 53% rise in attacks compared to the second quarter of 2023, averaging 3,341 weekly attacks per organization [1]. In response to the increasing variety of cyber threats, cybersecurity has implemented significant upgrades to combat cybercrimes. Cybersecurity encompasses a collection of technologies, specialists, and processes designed to implement protective measures against cybercriminals in cyberspace. There are two primary methodologies in cybersecurity: traditional cybersecurity and automated cybersecurity. Conventional cybersecurity has several drawbacks that exacerbate cybercrimes, including unqualified people, inadequate system resource configuration, and restricted access to reliable data [2]. The future of cybersecurity is centred on automation. Advanced and automated cybersecurity techniques are in high demand. They can adjust to the evolving scene of cybercrime by using their experience to detect novel polymorphic cyberattacks [3]. A cyber threat is an attempt to unlawfully acquire information, breach integrity protocols, or damage computing devices or networks. Phishing, spyware, IoT device assaults, denial of service attacks, spam, network or mobile device intrusions, financial fraud, and ransomware are just a few examples of cyber threats [4]. Malware, intrusion, and spam detection are covered in this study. The most effective and crucial strategy for thwarting cyber threats and overcoming the limitations of conventional security solutions is ML. Notwithstanding its numerous advantages, ML approaches have inherent limits and limitations. The intriguing characteristic of ML techniques is that they do not require explicit programming, as they can autonomously learn from experience to produce outcomes [5]. Due to the numerous advantages of machine learning techniques, their application is broadening across various domains, including cybersecurity, medical science, education, intrusion detection, spam detection, and malware detection. Nearly all renowned ML methodologies have been utilized to identify and categorize various cyber risks. Frequently employed machine learning methodologies [6]. We have evaluated the Random Forest, Naïve Bayes, Decision Tree, Logistic Regression and KNN as well as ensemble methodology for this study. We have presented a comparison of ML approaches utilizing commonly employed and benchmark datasets, which are freely available. Total data 125,973.

2 Related Work

Recent advancements in digital technology have led to a rise in cyberattacks, necessitating the development of more sophisticated security solutions. Machine learning (ML) has become an essential instrument for detecting and classifying cyberattacks by recognizing patterns and abnormalities in extensive datasets. This literature review examines developments in cyberattack detection through machine learning, emphasizing approaches, models, problems, and future directions. Machine learning algorithms exhibit considerable capability in detecting cyberattacks through the analysis of extensive network data. Diverse machine learning methodologies, encompassing supervised, unsupervised, and reinforcement learning, have been employed to identify various cyberattack types, including Distributed Denial of Service (DDoS), malware, phishing, and advanced persistent threats (APTs). Supervised learning algorithms, including SVM, RF, and ANN, are frequently utilized on labeled datasets to categorize attack types according to established signatures or behaviours [7].

Unsupervised learning methods such as clustering (e.g., k-Means, DBSCAN) are employed to identify anomalies in network traffic, signalling potential zero-day attacks in the absence of labeled data [8]. Conversely, reinforcement learning emphasizes dynamic settings, enabling systems to adjust to changing threats over time [9]. Supervised machine learning methods depend on annotated datasets to identify recognized attack patterns. Support Vector Machines and Random Forest are widely utilized for intrusion detection and classification. Support Vector Machine (SVM) models have exhibited effective performance in categorizing assaults, including Distributed Denial of Service (DDoS) and port scans, particularly when integrated with feature selection methods to minimize false positives [10].

Random Forest, a robust ensemble learning method, has been effectively utilized in detecting both binary and multi-class intrusions. A study demonstrated that Random Forest (RF) operates effectively in contexts with extensive datasets, rendering it appropriate for real-time detection. Supervised models sometimes necessitate retraining to maintain efficacy, especially as novel attack patterns arise [11]. Unsupervised machine learning techniques are extensively employed for identifying anomalies in network traffic, potentially signalling novel, unidentified assaults. These strategies are especially advantageous when labeled attack data is lacking or inadequate. Clustering algorithms, such as k-means and DBSCAN, facilitate the ID of anomalies in network data, providing viable methods for identifying zero-day attacks [12].

Autoencoders based on deep learning, an unsupervised neural network architecture, have recently attracted interest for their efficacy in identifying abnormalities in cybersecurity. Introduced a framework of Deep Learning (DL) employing autoencoders for anomaly exposure in Industrial Internet of Things (IIoT) settings. Although effective in identifying anomalies, unsupervised models sometimes exhibit elevated false-positive rates, requiring additional refining to minimize detection noise [13]. To address the constraints of singular learning methodologies, numerous researchers have suggested hybrid models that integrate supervised, unsupervised, and deep learning strategies. Hybrid models utilize the advantages of many methods to enhance detection precision and minimize false positives. A hybrid methodology that integrates SVM and k-means clustering to facilitate real-time intrusion detection while minimizing false positives. The integration of anomaly detection with signature-based techniques improves the system's ability to identify both known and undiscovered threats [14].

Notwithstanding the efficacy of machine learning models in identifying and categorizing cyberattacks, numerous obstacles remain. A significant difficulty is the absence of high-quality, current, and labeled datasets. Numerous supervised learning methods necessitate extensive labeled datasets, and the lack of datasets including authentic cyberattack data hinders the advancement of resilient models [15]. Furthermore, machine learning models are susceptible to adversarial assaults, in which attackers create inputs intended to mislead machine learning algorithms into incorrectly categorizing innocuous actions as harmful or the opposite. The scalability of machine learning models for high-velocity and extensive networks continues to pose a difficulty. Although deep learning models can handle extensive data, they necessitate considerable processing resources, which may not always be practical in real-time intrusion detection settings [16, 17]. Furthermore, minimizing false positives is a crucial concern, as an elevated false-positive rate might inundate security analysts and result in overlooked alerts. The domain of machine learning-based cyberattack detection is advancing swiftly, and further research may concentrate on tackling the aforementioned issues. A possible avenue is the enhancement of resilient adversarial defences to safeguard ML models from evasion attempts. Methods such as adversarial training, which involves training models using adversarial examples, are being investigated to enhance model resilience.

Moreover, transfer learning and few-shot learning methodologies present viable answers to the challenge of inadequate labeled data. Transfer learning facilitates the utilization of models developed in one domain for another, diminishing the dependence on domain-specific labeled data. Few-shot learning emphasizes the training of models using a limited number of examples, which may be advantageous for identifying new risks without necessitating substantial retraining [18, 19]. Machine learning has become an effective instrument for the detection and categorization of cyberattacks, providing novel methods to address the growing complexity of cyber threats. Although supervised, unsupervised, and hybrid methodologies offer distinct benefits, other obstacles, including data quality, model robustness, and scalability, have to be resolved. Future research will likely concentrate on addressing these constraints, prioritizing adversarial resilience and effective real-time detection.

3 Methodology

The methodology for detecting and classifying cyberattacks by machine learning is a multi-phase process encompassing data collection, preprocessing, feature engineering, model selection, training, evaluation, and deployment. Every stage is essential for constructing a system capable of accurately detecting both recognized and unrecognized cyber threats. As cyberattacks progress, forthcoming methodologies may integrate sophisticated techniques such as deep learning and reinforcement learning to respond to evolving threats in real-time contexts. The approach for identifying and categorizing cyberattacks by ML encompasses systematic procedures such as data gathering, data preprocessing, feature engineering, model selection, model training, evaluation, and deployment. These measures assurance that machine learning models can proficiently detect cyber risks and categorize diverse attack types with elevated precision. This section delineates the essential elements of an ML methodology for the detection and classification of cyberattacks.

The initial phase in creating an ML-based cyberattack recognition system involves the aggregation of an extensive dataset comprising network traffic, system logs, or other cybersecurity-related information. We have collected 125,973 data including intrusion detection systems and publically accessible datasets such as NSL-KDD. These datasets encompass both malware, ransomware, and phishing attacks, to provide a comprehensive training set. We have designed a model to execute the finding and classification of cyberattacks using ML algorithms, and also compare the accuracy of the algorithms.

Cyberattacks Classification Model

The above model is designed to execute the Cyberattack data classifications. Here we can observe that after preprocessing the data, it is executed one by one ML algorithms. After that apply the model and measure the output performance of preprocessing tasks. The process involves importing data from various sources, cleaning and transforming it to ensure quality, and preparing it for model training. By following these steps, we can ensure that our data is in the best possible shape for building accurate and robust ML models with different performances. At last, the above model has been designed to run the ensemble techniques with AdaBoost algorithms (Fig. 1).

4 Data Analysis and Findings

The Cyberattacks Classification Model illustrates the consequences of model processing in ML, encompassing the evaluation of model efficacy, interpretation of classification, and comparison of the accuracy based on these algorithms. This step is crucial for evaluating the model's capacity to generalize to novel data and its conformity with the project's goals. The subsequent part presents an analysis of

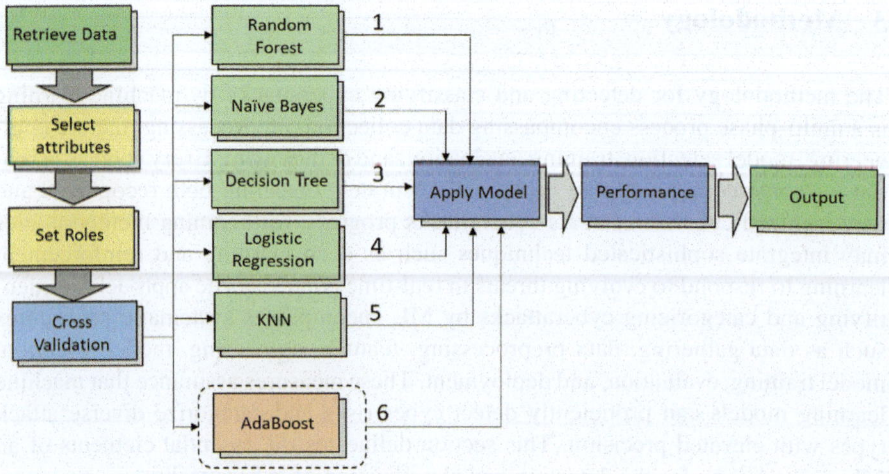

Fig. 1 Cyberattacks classification model

the procedures involved in managing outcomes in ML. We have also measured the Accuracy, Precision and Recall of all classifier algorithms.

Accuracy: The formula for accuracy is used to measure the performance of a classification model. It is calculated as the ratio of correctly predicted observations to the total number of observations.

$$\text{Accuracy} = \frac{\text{Total Number of Predictions}}{\text{Number of Correct Predictions}} = \frac{TP + TN}{TP + TN + FP + FN} \quad (1)$$

where:

TP (True Positives): Accurately identified positive instances.

TN (True Negatives): Accurately identified negative instances.

FP (False Positives): Erroneously identified positive instances (false alarms).

FN (False Negatives): Incorrectly identified negative instances (missed detections).

This formula quantifies the model's total efficacy. Nonetheless, accuracy alone may be inadequate when addressing imbalanced datasets, necessitating the consideration of additional metrics such as precision and recall.

Precision: The precision formula assesses the accuracy of positive predictions generated by a classification model. It is the proportion of accurately anticipated positive instances to the total expected positive instances.

$$\text{Precision} = \frac{\text{TP}}{\text{TP} + \text{FP}} \tag{2}$$

where:

TP (True Positives): Accurately identified positive instances.

FP (False Positives): Erroneously identified positive instances (false alarms).

Since precision indicates how well the model minimizes these incorrect positive predictions, it is especially important in situations where the cost of false positives is high.

Recall: The recall formula, sometimes referred to as sensitivity or true positive rate, quantifies a model's capacity to accurately detect all positive cases. It is the proportion of accurately predicted positive instances to the total actual positive instances.

$$\text{Recall} = \frac{\text{TP}}{\text{TP} + \text{FN}} \tag{3}$$

where:

TP (True Positives): Accurately identified positive instances.

FN (False Negatives): Positive instances that the model erroneously classified as negative.

Recall is crucial when the objective is to reduce false negatives, as in medical diagnostics or fraud detection, because overlooking a positive case can have severe repercussions.

4.1 Findings

Naïve Bayes

We can observe from the above Fig. 2, that the Cyberattacks classification got an anomaly of 61,042 numbers and the normal of 64,931 numbers out of a total of 125,973 numbers NS-KDD data using Naïve Bayes Classification.

It can be observed that the Accuracy is 90.33%, Precision is 88.05% and Recall is 91.67 using Naïve Bayes. We can also observe from above Fig. 3 that the AUC is 0.9 which is near 1.0 means the model is a perfect fit.

Decision Tree

We can observe from the above Fig. 4, that the Cyberattacks classification got an anomaly of 57,464 numbers and the normal of 68,509 numbers out of a total of 125,973 numbers NS-KDD data using Decision Tree.

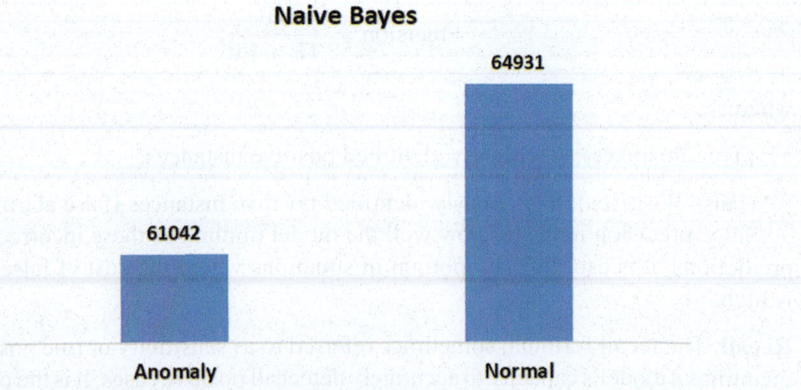

Fig. 2 Cyberattacks classification using Naïve Bayes

Accuracy: 90.33%, Precision: 88.05%, Recall: 91.67%

	true normal	true anomaly	class precision
pred. normal	60047	4884	92.48%
pred. anomaly	7296	53746	88.05%
class recall	89.17%	91.67%	

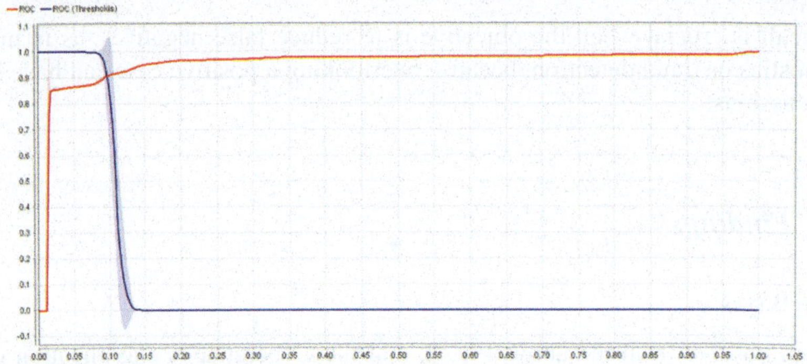

Fig. 3 The AUC of the Naïve Bayes

It can be observed that the Accuracy is 97.09%, Precision is 97.82% and Recall is 95.87% using Decision Tree. We can also observe from above Fig. 5 that the AUC is 0.9 which is near 1.0 means the model is a perfect fit.

Random Forest

We can observe from the above Fig. 6, that the Cyberattacks classification got an anomaly of 57,716 numbers and the normal of 68,257 numbers out of a total of 125,973 numbers NS-KDD data using Random Forest.

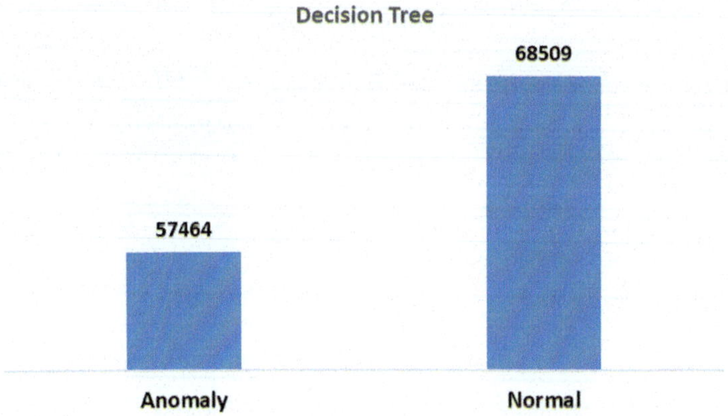

Fig. 4 Cyberattacks classification using decision tree

Accuracy: 97.09%, Precision: 97.82%, Recall: 95.87%

	true normal	true anomaly	class precision
pred. normal	66090	2419	96.47%
pred. anomaly	1253	56211	97.82%
class recall	98.14%	95.87%	

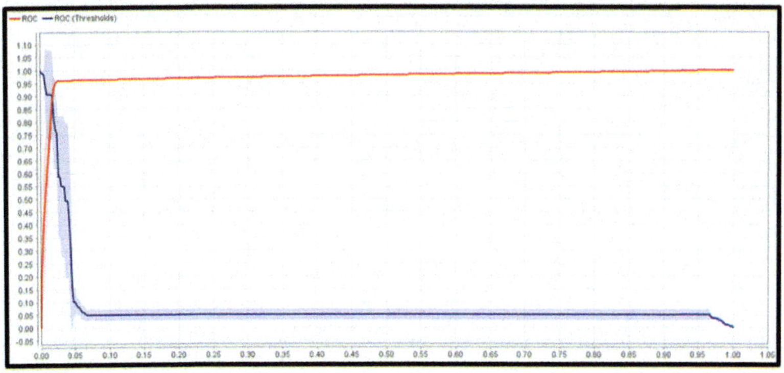

Fig. 5 The AUC of the decision tree

It can be observed that the Accuracy is 99.20%, Precision is 99.92% and Recall is 98.36% using Random Forest. We can also observe from the above Fig. 7 that the AUC is 1.0 which means the model is a perfect fit.

Logistic Regression

We can observe from the above Fig. 8, that the Cyberattacks classification got an anomaly of 57,917 numbers and the normal of 68,056 numbers out of a total of 125,973 numbers NS-KDD data using Logistic Regression.

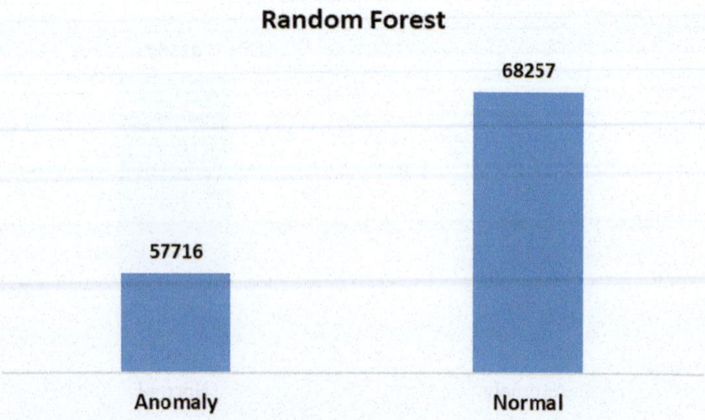

Fig. 6 Cyberattacks classification using random forest

Accuracy: 99.20%, Precision: 99.92%, Recall: 98.36%

	true normal	true anomaly	class precision
pred. normal	67294	963	98.59%
pred. anomaly	49	57667	99.92%
class recall	99.93%	98.36%	

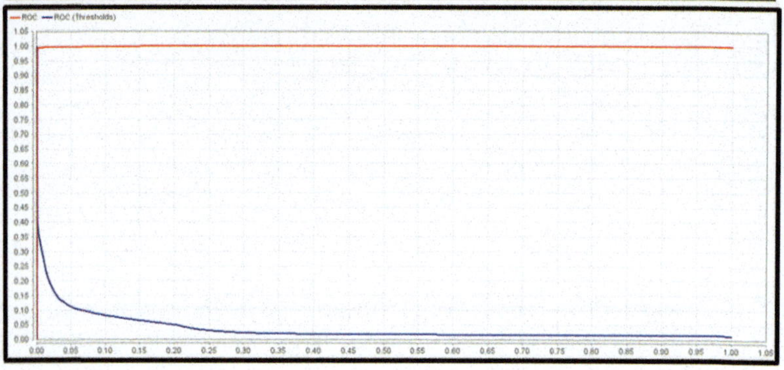

Fig. 7 The AUC of the random forest

We can observe that the Accuracy is 97.44%, Precision is 97.84% and Recall is 96.65% using Logistic Regression classification. We can also observe from the above Fig. 9 that the AUC is 0.94 which is near 1.0 means the model is a perfect fit.

k-NN

We can observe from the Fig. 10, that the Cyberattacks classification got an anomaly of 58,745 numbers and the normal of 67,228 numbers out of a total of 125,973 numbers NS-KDD data using the k-NN classifier.

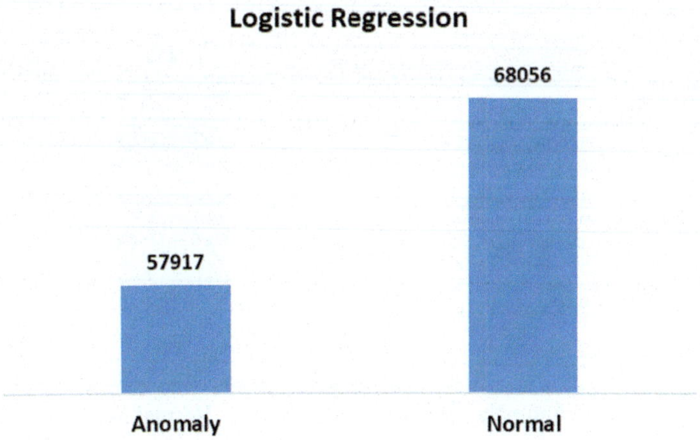

Fig. 8 Cyberattacks classification using logistic regression

Accuracy: 97.44%, Precision: 97.84%, Recall: 96.65%

	true normal	true anomaly	class precision
pred. normal	66090	1966	97.11%
pred. anomaly	1253	56664	97.84%
class recall	98.14%	96.65%	

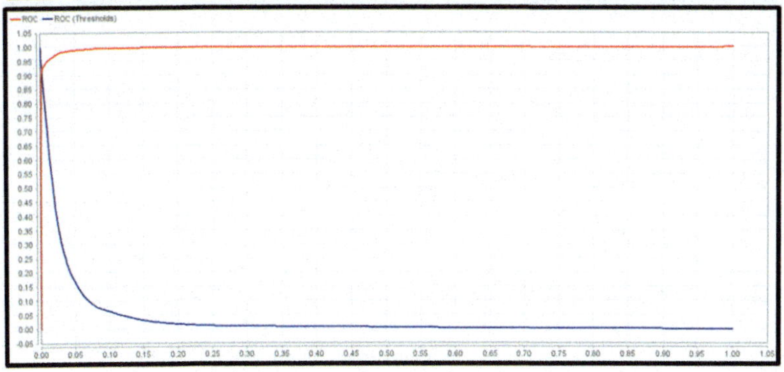

Fig. 9 The AUC of the logistic regression

It can be observed that the Accuracy is 99.57%, Precision is 99.44% and Recall is 99.63% using k-NN classification. We can also observe from the above Fig. 11 that the AUC is 1.0 which means the model is a perfect fit.

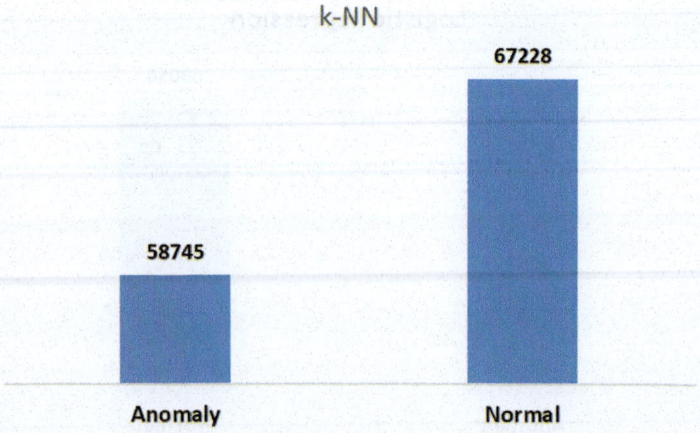

Fig. 10 Cyberattacks classification using k-NN

Accuracy: 99.57%, Precision: 99.44%, Recall: 99.63%

	true normal	true anomaly	class precision
pred. normal	67014	214	99.68%
pred. anomaly	329	58416	99.44%
class recall	99.51%	99.63%	

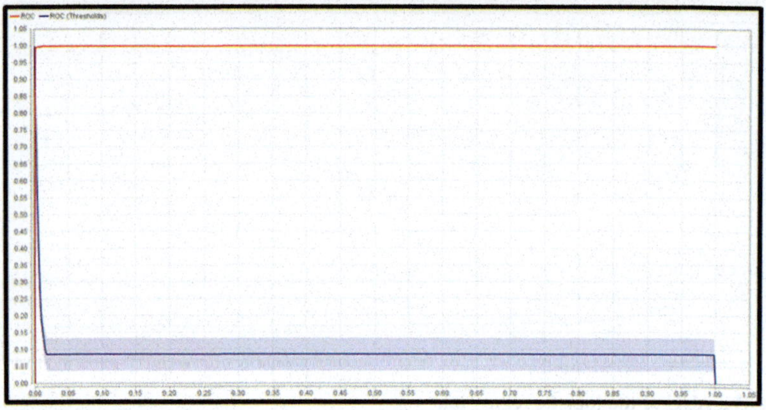

Fig. 11 The AUC of the k-NN

AdaBoost

We can observe from the above Fig. 12, that the Cyberattacks classification got an anomaly of 58,682 numbers and the normal of 67,291 numbers out of a total of 125,973 numbers NS-KDD data using the k-NN classifier.

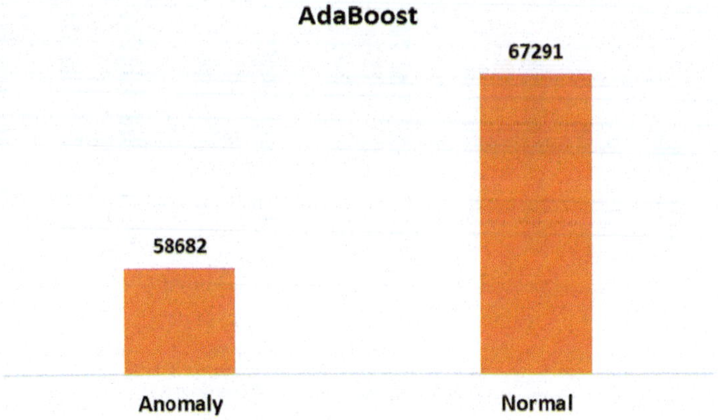

Fig. 12 Cyberattacks classification using AdaBoost

It can be observed that the Accuracy is 99.69%, Precision is 99.62% and Recall is 99.71% using the AdaBoost Ensemble algorithm. We can also observe from the above Fig. 13 that the AUC is 1.0 which means the model is a perfect fit.

Accuracy: 99.69%, Precision: 99.62%, Recall: 99.71%

	true normal	true anomaly	class precision
pred. normal	67119	172	99.74%
pred. anomaly	224	58458	99.62%
class recall	99.67%	99.71%	

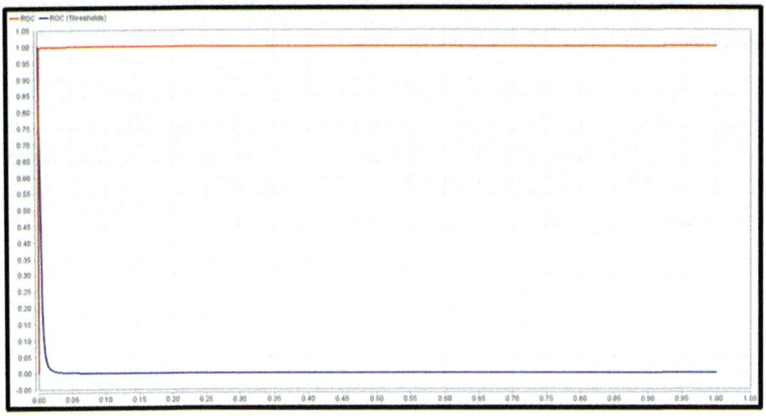

Fig. 13 The AUC of the AdaBoost

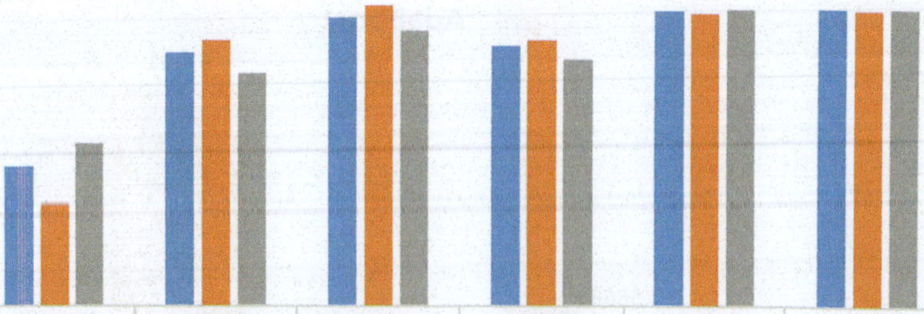

	Naïve Bayes	Decision Tree	Random Forest	Logistic Regression	KNN	AdaBoost
Accuracy	90.33%	97.09%	99.20%	97.44%	99.57%	99.69%
Precision	88.05%	97.82%	99.92%	97.84%	99.44%	99.62%
Recall	91.67%	95.87%	98.36%	96.65%	99.63%	99.71%

Fig. 14 Comparison of different ML algorithms

Comparison of all ML algorithms

According to Fig. 14, we found that the k-NN classifier has the highest accuracy 99.57% and the Naïve Bayes classifier has the lowest accuracy 90.33% within the comparison of five classifier ML algorithms. The other classifier's accuracy, Decision Tree is 97.09%, Random Forest is 99.20%, and Logistic regression is 97.44%. Whereas the Random Forest has the highest precision is 99.92%, the Naïve Bayes classifier has the lowest precision is 88.05% and the other classifier's precision of has Decision Tree is 97.82%, the Logistic regression is 97.84%, the k-NN is 99.44%. Apart from these classifiers, the AdaBoost ensemble algorithm got the highest accuracy at 99.69%, the Precision is 99.62% and the Recall is 99.71%.

Comparison of Cyberattack Classification

According to Fig. 15, we found that the k-NN classifier has classified an anomaly of 58,745 and the normal is 67228 out of 125,973 numbers of data. The Logistic Regression has classified an anomaly of 57,464 and the normal is 68509. The Naïve Bayes classifier has the lowest accuracy 90.33% within the comparison of five classifier ML algorithms.

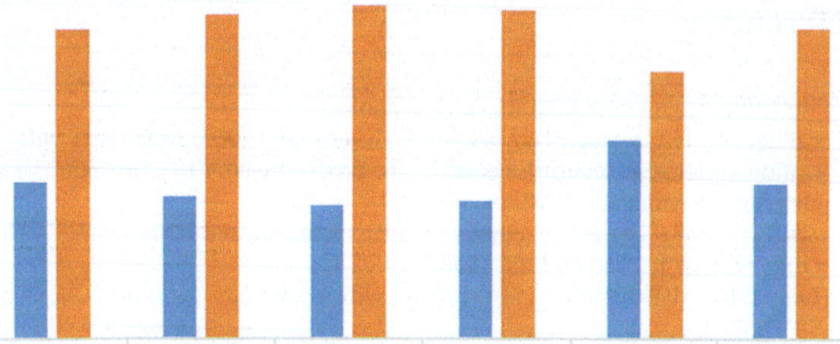

	k-NN	Logistic Regresion	Decision Tree	Random Forest	Naïve Bayes	AdaBoost
Anomaly	58745	57917	57464	57716	61042	58682
Normal	67228	68056	68509	68257	64931	67291

Fig. 15 Comparison of different cyberattack classification

5 Conclusion

In summary, the categorization of cyberattacks by the classifier algorithm of ML has become an essential strategy for improving cybersecurity systems against progressively advanced threats. When adequately trained and refined, machine learning models offer substantial benefits in the detection, classification, and response to both known and new cyberattacks. By utilizing NS-KDD datasets, our models can discern experiments such as data imbalance, feature selection, and dynamic attack strategies are require the implementation of advanced approaches. We have used five classifier algorithms and the AdaBoost ensemble algorithms for the classification of cyberattacks and finally, we got the k-NN algorithm got the best result as per our robust model. Apart from these classifier algorithms, we also got the best result in an ensemble technique. So we can conclude as per the above-discussed research algorithms, that the AdaBoost ensemble technique is the best for its performance of classification (Accuracy, Precision, Recall). In this study, we have proved the accurate classification of the cyberattack dataset with the AdaBoost algorithm of an accuracy 99.69% and the anomaly and normal of 58,682 and 67,291 out of 1,259,273 data.

Future research and development should prioritize the creation of more adaptive algorithms that can detect developing cyber risks in real-time. Integrating ML techniques into cybersecurity frameworks enables enterprises to enhance their defences against various assaults, hence providing greater system reliability, security, and resilience.

References

1. https://timesofindia.indiatimes.com/technology
2. Farahmand F, Navathe SB, Enslow PH, Sharp GP (2003) Managing vulnerabilities of information systems to security incidents. In: Proceedings of the 5th international conference on electronic commerce. ACM, pp 348–354
3. Szor P (2005) The art of computer virus research and defense: ART COMP VIRUSRES DEFENSE, p 1. Pearson Education
4. Jump M (2019) Fighting cyberthreats with technology solutions. Biomed Instrum Technol 53(1):38–43
5. Michie D, Spiegelhalter DJ, Taylor C (1994) Machine learning. In: Neural and statistical classification, p 13
6. Iyer SS, Rajagopal S (2020) Applications of machine learning in cyber security domain. In: Handbook of research on machine and deep learning applications for cyber security. IGI Global, pp 64–82
7. Amanullah MA, Ahamad M, Rahman MS (2020) Machine learning models for DDoS attack detection and mitigation in SDN environment. IEEE Access 8:161517–161537
8. Sethi T, Kantardzic M (2020) Handling adversaries in cyber security using reinforcement learning. IEEE Access 8:21692–21709
9. Niyaz Q, Sun W, Javaid AY (2016) A deep learning-based DDoS detection system in software-defined networking (SDN). In: ICCS 2016
10. Shirazi SM, Khosravi MR, Alipour M, Ghaffarian SM (2019) Intrusion detection using support vector machine. J Inf Secur Appl 47:182–194
11. Buczak AL, Guven E (2016) A survey of data mining and machine learning methods for cybersecurity intrusion detection. IEEE Commun Surv Tutor 18(2):1153–1176
12. Ahmed M, Mahmood AN, Hu J (2016) A survey of network anomaly detection techniques. J Netw Comput Appl 60:19–31
13. Lyu X, Zeng Z, Yu Y, Yang Q (2020) Anomaly detection for industrial IoT using autoencoders with generative adversarial networks in smart factory. IEEE Trans Ind Inf 16(9):5934–5944
14. Vinayakumar R, Soman KP, Poornachandran P (2019) Applying convolutional neural network for network intrusion detection. In: ICMLA 2019
15. Hindy H, Brosset D, Bayne E, Seeam A, Bellekens X, Tachtatzis C (2020) A taxonomy and survey of intrusion detection system design techniques, network threats and datasets. arXiv preprint arXiv:2003.06979
16. Pattnaik LM, Swain PK, Satpathy S, Panda AN (2023) Cloud DDoS attack detection model with data fusion & machine learning classifiers. EAI Endorsed Trans Scalable Inf Syst 10(6)
17. Satpathy S, Swain PK, Mohanty SN, Basa SS (2024) Enhancing security: federated learning against man-in-the-middle threats with gradient boosting machines and LSTM. In: 2024 IEEE international conference on advanced video and signal based surveillance (AVSS). IEEE, pp 1–8
18. Zhang X, Li Q, Zhang W (2019) Machine learning for network intrusion detection: a survey. IEEE Access 7:68270–68284
19. Li W, Xu Y, Wu Y, Zhang Y (2020) Few-shot learning for cyber-attack detection. In: International conference on computer and communications security (ICCCS)

A Network Intrusion Detection System Using Machine Learning Algorithms

Anannya Barua, Padmasri Turaka, and Saroj Kumar Panigrahy

Abstract The inception of a system that can detect if there is an intrusion into the host system is conceptualized. This is performed using standard machine-learning algorithms. Our focus is to collect information about the network traffic and related domains of the host system and store it in a dataset. The dataset is then used to critically examine the data and find a pattern for the normal systems. These data are then comparatively studied by the algorithms to detect if there is a greater probabilistic chance of having an anomaly in the system. We are using a binary classification model that predicts the possibility of malicious activities on the host server through the collected dataset. In this work, we have integrated a user-interactive web page that takes certain inputs from the user, which are necessary for accurate results. Random forest outperforms all four machine learning algorithms (k nearest neighbor, logistic regression, decision tree, and random forest) on the same dataset, with an accuracy of 99.6%. Hence, for our web page development, we use a random forest algorithm that shows the output according to the inputs given by the users.

Keywords Intrusion detection · Machine learning · kNN · Decision tree

1 Introduction

Cyber risks must be brought to the forefront of today's culture, with digitization spreading to practically every part of our lives. Rapid advances in the Internet and network communication industries have resulted in the creation of massive amounts of data. Thus, it becomes necessary for us to build an intrusion detection system that centrally controls all the ongoing activities. A network intrusion detector (NID) would inspect the network traffic, ensuring confidentiality, integrity, and availability in the vast cloud environment. Typically, all the previous IDSs are built to detect

A. Barua · P. Turaka · S. K. Panigrahy (✉)
School of Computer Science and Engineering, VIT-AP University, Amaravati, Andhra Pradesh, India
e-mail: saroj.panigrahy@vitap.ac.in

A. Rocha et al. (eds.), *Integrating Advanced Technologies for Enhanced Security and Efficiency*, Engineering Cyber-Physical Systems and Critical Infrastructures 14,
https://doi.org/10.1007/978-3-031-91798-1_2

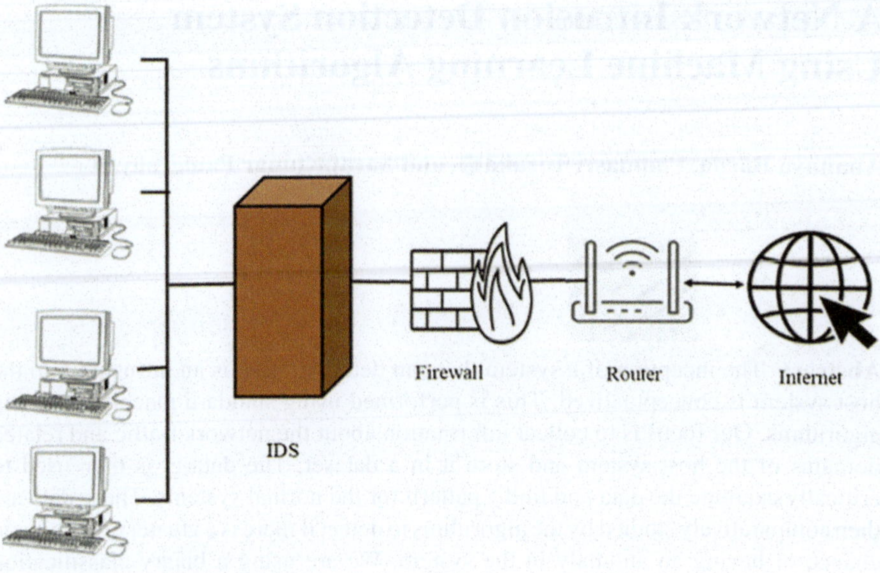

Fig. 1 Intrusion detection system

anomalies and suspicious activities and then alert the system administrators of any policy breach that has been found in a particular area shown in Fig. 1. There are five divisions under which the intrusion detection systems have been subdivided.

1.1 Network Intrusion Detection System (NIDS)

Their sole purpose in the system is to examine the incoming and outgoing network traffic flow from all devices. It compares the information collected from subnets. It becomes important to safeguard the firewalls from possible attacks.

1.2 Host Intrusion Detection Systems (HIDS)

This system runs on networks that have independent hosts or devices. It uses snapshots of existing system files to detect malicious activities by comparing them with previous snapshots of those files. These snapshots help them monitor the incoming and outgoing packets. If any changes are detected in those files, an alert is generated, which would be able to inform the administrator. HIDS will work best on systems that have relatively similar layouts throughout, like in the case of mission-critical systems.

1.3 *Protocol-Based IDS*

This system is designed to protect the system by following and observing the HTTP protocol stream. The agent here would focus more on controlling and interpreting the protocols between the host server and a user or a device. It does that by residing on the front-end side of the server. So, this is how the system protects the web server. It would reside in the interface as a safety net as HPPT is unencrypted, which it would verify before entering the web presentation layer. Only then can HTTPS be used.

1.4 *Application Protocol-Based IDS*

The APIDS monitors and analyzes communication protocols used within the system. The system, also known as the agent, operates within a server group. This is mostly used to monitor the SQL protocol, which interacts with the database on the web server, making it explicit to the middleware.

1.5 *Hybrid IDS*

Hybrid IDS is more efficient than traditional intrusion detection systems because they integrate host or system data with network information to provide a more complete and accurate picture of system occurrences. It is a combined version of two or more IDSs. As a result, it provides us with a comprehensive overview of the framework. Prelude is a hybrid IDS.

1.6 *Intrusion Detection Methods*

Signature-based IDS (SIDS). It seeks to find patterns and compare them to known signals of infiltration. A SIDS uses a database of prior invasions. If activity on your network fits the *Signature* of an attack or breach in the database, the detection system tells your administrator. Because the database is the foundation of a SIDS solution, regular database updates are required, as SIDS can only detect threats that it recognizes. As a result, if your firm is the victim of a previously unknown infiltration technique, no number of database upgrades will protect you. The drawback of signature-based IDS is that it becomes difficult to detect the defect if it does not match the existing patterns of malicious attacks. Hence, this method works best when the attacks are already known rather than when they are not.

Anomaly-based Method. The anomaly-based method is based on the use of ML models to get a more generalized result while predicting the output. It was mainly

Table 1 Various intrusion attack types

Attack category	Description	Examples
DoS [1]	One type of attack called denial of service, aims to stop authorized users from utilizing computer services. DoS attacks can be carried out by overloading the network and stopping a service or connection	Smurf, land, and ping of death (PoD)
U2R [1]	An attack known as "user-to-root" occurs when an attacker first gets access to a regular user account and then uses system flaws to escalate their privileges to root	Buffer-overflow, rootkit, and load module
R2L [1]	An attack known as "remote-to-user" in which an attacker tries to gain unauthorized entry to a local system to transfer packets via a network is known as a remote user attack. The attacker is granted the same rights as a local user would normally have when utilizing the computer in an R2L attack	Imap, Warezclient, and Ftpwrite
Probing [1]	It is a program that automatically scans and monitors network activity or collects network data. The attacker gathers knowledge about the network and discovers flaws that might be leveraged to attack it	Ipsweep, Nmap, and Portsweep

designed to overcome the generalization problem in signature-based IDS. This is achieved by training the models on more data and including the hardware applications and configurations. Here, we would be able to detect unknown malware attacks, as anything that is not found according to the model is declared suspicious. This was necessary because new malware is being developed constantly, so our systems are bound to become more and more sophisticated. In this model, We are using an anomaly-based detection system that uses ML models to detect anomalies in any system using network traffic data. The most common attacks that are found here are shown in Table 1.

1.7 Comparison of IDS with Firewalls

The main difference between a firewall and an IDS is that a firewall stops an attack from happening by restricting access between the networks. Whereas IDS detects an intrusion in the system if it has already happened in the system. Thus, the fundamental difference would be that the firewall looks outwardly for intrusion and prevents it from happening in the first place, but IDS works inside the system and helps find those intrusions if they are already present inside the host system. Both frameworks are needed to ensure the complete security and authenticity of the systems. Various approaches can be used to offer communication security, including access control

[2], optimization-based secure routing strategies [3], agent-based methods [4], and temporal analysis [5].

1.8 Objective

This work proposes an optimized path to find those cyber-attacks using the most suitable algorithms. Our goal is to create an IDS that keeps an eye out for hostile activity, policy violations, and malware on a network. A security information and event management system is used to report and centrally collect any harmful behavior or system intrusions. If an abnormality in the prior report was found, the system would notify the administrator. Our machine learning model will try to predict if there is an anomaly in the system based on the input dataset that is being used. The work is further extended to build a user interface wherein, if the user enters a set of answers or values to the given questions, the algorithm based on the input dataset would be able to detect if there is a higher chance of malware or anomalies in the system based on the most suitable and accurate machine learning algorithm.

2 Related Work

Numerous existing IDSs are already being implemented consistently throughout all major networking systems, and the important component in IDS is to have a robust dataset, which would give us greater accuracy, which would help us reduce false alarms and detect anomalies better. The work [6] is taken as the base paper for our research work. Here they try to compare two of the most accurate ML models, SVM, and Naïve Bayes, to detect anomalies using the dataset NSL-KDD. They give us a comparative study on the accuracy of the model, where we see that SVM outperforms Naïve Bayes.

Another literature [7], tries to improve integrity, confidentiality, and network availability using algorithms. Due to the rapid increase in network traffic and the increasing number of digital users, network traffic is becoming more and more prone to unauthorized and malicious attacks. The paper implements single hybrid and ensemble ML models. Here they deal with seven different datasets, and then the results are discussed for the future scope of the systems. The work [8], in which unusual behaviors are recognized using artificial intelligence that operates as a synthetic analyst simultaneously with NIDS to protect against the threat environment and take relevant steps. In another work [9], better accuracy was found by selecting good features with recursive feature elimination and then classifying them by the decision tree algorithm. In the work [10], the false positives are reduced by preprocessing data with Rough Set Theory and learned by the SVM machine learning algorithm.

Fig. 2 Flow of proposed method

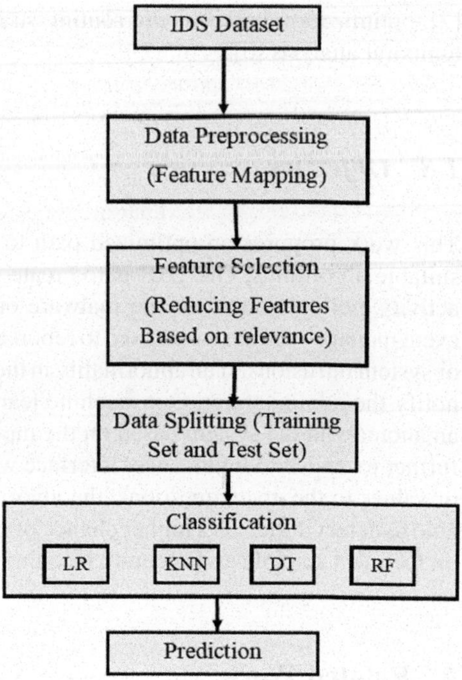

3 Proposed System

In our approach, we try to implement supervised learning techniques and compare their results to be able to detect both known and unknown attacks. We are going to implement all the algorithms in the Python language, which has all the libraries like *sklearn* for successful implementation. The techniques that are to be implemented in this IDS include RF, LR, DT, and KNN. The flow of the work is shown in Fig. 2. The whole experimental work is done in Google Colab[1] in Python, and the Gradio[2] interface is used to implement the user-interactive web page.

3.1 Dataset Used

The Kaggle[3] dataset to be inspected was provided, which included a wide range of intrusions modelled in a military network. It emulated a standard US Air-Force LAN to create an environment for obtaining raw TCP/IP dump data for a network. The

[1] https://research.google.com/colaboratory/.

[2] https://www.gradio.app/.

[3] https://www.kaggle.com/.

LAN was bombarded with several attacks and was focused on a genuine environment. A connection is a series of TCP packets that transport data to and from a source IP address to a target IP address following a predetermined protocol. The packets start and stop at a specific time interval. Every link is additionally classified as either regular or as an attack with a single, distinct attack type. A connection record is made up of roughly 100 bytes. From normal and attack data, 41 quantitative and qualitative features (three qualitative and 38 quantitative features) are derived for each TCP/IP connection. There are two categories in the class variable—Normal and Anomalous. It is divided into two segments—training and testing datasets. The training data consists of 41 columns and 25,193 rows, and the testing has 41 columns and 22,545 rows. The important features of the dataset are—service, Flag, src_bytes, dst_bytes, same srv_rate, diff srv_rate, dst_host srv_count, and dst_host same src_port_rate which were selected to classify the data.

3.2 *The Machine Learning Algorithms*

Logistic Regression (LR). In ML models, LR is a statistical technique for solution prediction. The technique uses the sigmoid function, whose dependable variable 'y' is dichotomous and binary in nature. Using this function, we can determine the link between several distinct variables and their reliable counterparts in the function or situation at hand. The function uses this dependency on one another to plot and even predict the trend of the dependent variable given the independent parameters. The value of the logistic function, or sigmoid function, used in this algorithm ranges between zero and one. The linear regression can be represented as Eq. (1):

$$\hat{y} = b_0 + b_1x_1 + b_2x_2 + \ldots + b_nx_n \tag{1}$$

where $\hat{y}$ represents the predicted variable (dependent variable), b_i is the intercept, and $b_0, b_1, \ldots b_n$ are the coefficients of the independent variables $x_0, x_1, \ldots x_n$ respectively.

Decision Tree (DT). It is a supervised machine-learning algorithm used for both regression and classification problems. It shows the most promising results while solving classification problems is designed to get a classification result at each node based on which node is being split. The internal nodes here represent the features of a dataset, and each of its branches shows the decision rules. The final leaf node, which does not have any branches, shows the final outcome of a decision. The leaf nodes predict the final outcome of the algorithm.

K-Nearest Neighbour (kNN). In some circumstances, kNN is one of the most rudimentary, simplest, and most successful machine learning algorithms available. It is not only simple to grasp but also quick to implement in certain datasets. The algorithm is built on the idea that similar data or test cases are put in a similar class and labeled in a particular category. It is a supervised learning technique, and hence it makes use of the correlation between the k closed items, where k will be predefined, and then, after measuring the similarity between all the items, the k closed items are chosen. From there, the majority prediction is chosen as the class for the test data as well. Both regression and classification issues can be solved with this technique; however, more classification problems are used than regression ones.

Randome Forest (RF). It is a supervised machine learning technique that operates on the notion of majority voting, utilizing an extensive number of decision trees. As a result, increasing the total number of sub-decision trees would result in a more accurate model and improved problem-solving abilities. The various decision trees are constructed from a subset of the specified dataset, allowing for more generalization and avoiding overfitting. The accuracy is significantly higher than that of a standard decision tree since it is based on the ensemble learning idea.

3.3 Software and Libraries Used

The above libraries are used for the implementations of our project, which are being discussed below.

***sklearn* Package**: The goal of Scikit-learn, often known as *sklearn*, is to offer all the machine learning capabilities needed for Python implementation. Other statistical modeling methods via a standardized Python interface are also included, such as clustering, regression, classification, dimensionality manipulation, and selection. Built on top of the NumPy, SciPy, and Matplotlib packages is this library. It is among the most potent and effective Python libraries available for the use of machine learning models. The Scikit-learn library, rather than importing, editing, and summarizing data, focuses on modelling it. Sklearn's most popular collection of models includes the following:

- *Scikitlearn* includes almost all major supervised learning methods, such as LR, SVM, and DT.
- It also includes all of the most common unsupervised learning methods, including clustering and Principal Component Analysis (PCA), as well as unsupervised neural networks.
- Unlabeled data is grouped using clustering Cross-validation is a technique used to assess supervised models' correctness on omitted data.
- Dimensionality Reduction works by lowering the quantity of attributes in the data, which may then be utilized for feature selection, summarization, and visualization.
- Ensembled Methods are utilized to combine the predictions of several supervised models, as its name implies.
- Feature extraction defines characteristics in text and picture data and is utilized to extract features from the data.
- Feature selection is used to find relevant attributes to build supervised models.

Gradio Application. Gradio makes it easy to create an online graphical user interface in a few lines of code, ideal for showcasing model presentations. It is quick, easy to set up, ready to use, and shared as a public connection that anyone may use to run the model on the computer simultaneously and remotely. It uses a variety of media, such as audio, video, and images. It might be used, as is to be expected, in Python code embeddings in addition to ML models.

CatBoost. CatBoost[4] is a decision tree gradient-boosting method. In terms of *scikit-learn* standards, it is a ready-made classifier that would handle categorical features automatically. It is simple to connect with deep learning frameworks, like Google's TensorFlow. It can assist in resolving a variety of issues that businesses currently face by working with a variety of data kinds.

Optuna. Optuna[5] is a software framework for automatic hyperparameter tuning that is specifically made for machine learning. It has an imperative user API that is defined by run. Our define-by-run API allows for high versatility in programs developed with Optuna, and it allows the user to dynamically create search spaces for the hyperparameters.

3.4 Working Methodology

The proposed system is constructed around the data that is being collected from the datasets. The dataset used is taken from Kaggle, but the framework should be effective in all the datasets deployed by the administrator. At first, we import the dataset into the Google Colab Notebook, then we perform the basic data manipulation and visualization to have a clear understanding of it. We omitted the redundant, missing, outliers, and duplicate values at first. Thereafter, encoding and scaling are done to convert the categorical data, which is required for easy processing of the machine learning models. Then, we visualized all the columns or their features using a bar graph and heap map. They allow us to observe the correlation between the attributes and how each contributes to or impacts the production of our output. After careful examination, feature selection is done by selecting those attributes that affect the most in predicting a system's vulnerability. A total of 10 features are being selected. Then we split the dataset into training and testing halves. Finally, this framework is fitted into a machine learning model, and we obtain the accuracy score of both the training and testing datasets. A total of four machine learning models are being deployed: LR, kNN, DT, and RF. RF gives the best accuracy compared to other models. We further tried to extend this work to develop a user-interactive web page. The algorithm used for predicting the result through the web page, that is, the input from the users, would be a random forest here, as it gives us the most precise answers for most of the cases. The web page is built through the Gradio application, which is

[4] https://catboost.ai/.

[5] https://optuna.org/.

Table 2 Accuracy comparison of different classifiers used

Classifier	Accuracy (%)
kNN	98.21
LR	92.31
DT	99.4
RF	99.6

a Python library specifically designed to construct such web pages from scratch for beginners. I have even included a demo authentication page for the web page, as only authorized people should have access to review and test their system's vulnerability by answering certain specially selected questions about their current system's state.

4 Results and Analysis

Performance Metrics. Various performance measures can be used to examine and assess the effectiveness of different strategies. Some of the performance metrics utilized in various research [11] are listed below.

Accuracy. Accuracy is how the IDS detects intrusions and generates true alarms when an incursion is present and identified. The ratio of successfully detected instances to the total number of samples in the dataset determines the accuracy of the IDS [11], which can be shown as Eq. (2).

$$\text{Accuracy} = \frac{\text{Correct Predictions}}{\text{All Predictions}} \tag{2}$$

The accuracies of the algorithms on the testing set are given in Table 2.

Out of these models, RF outperform all of them. Several reasons contribute to the high accuracy of the dataset. This is a binary classification model that allows us to detect an anomaly in any system. Since RF only analyzes a portion of the data rather than the entire set, it performs best on datasets with lots of information. It is therefore more universal than the decision tree. The DT model in the code exhibits 100% accuracy in the training dataset, which could lead to over-fitting. Thus, we use the RF on the webpage we created using the Gradio program, the screenshots are given in the Figs. 3 and 4.

5 Conclusion and Future Scope

With the increase in our involvement with the software industry, intrusions, and cyber go hand in hand. To make our systems more robust and sturdier, we need to constantly update and make appropriate amendments by installing software that can

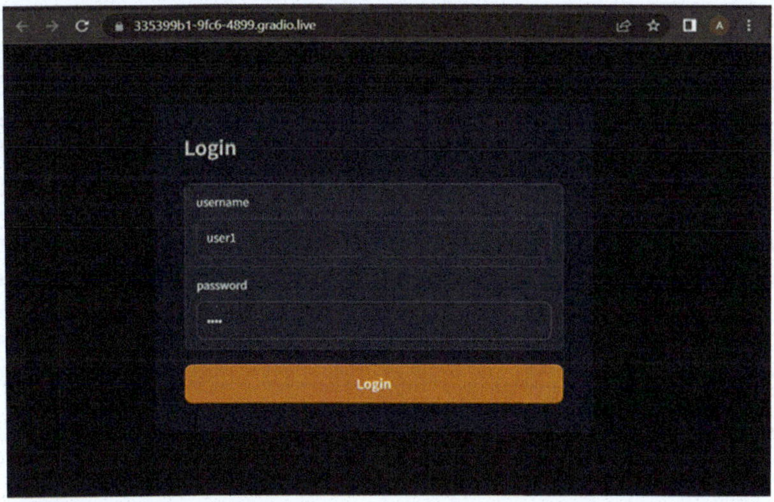

Fig. 3 Our GUI application login page

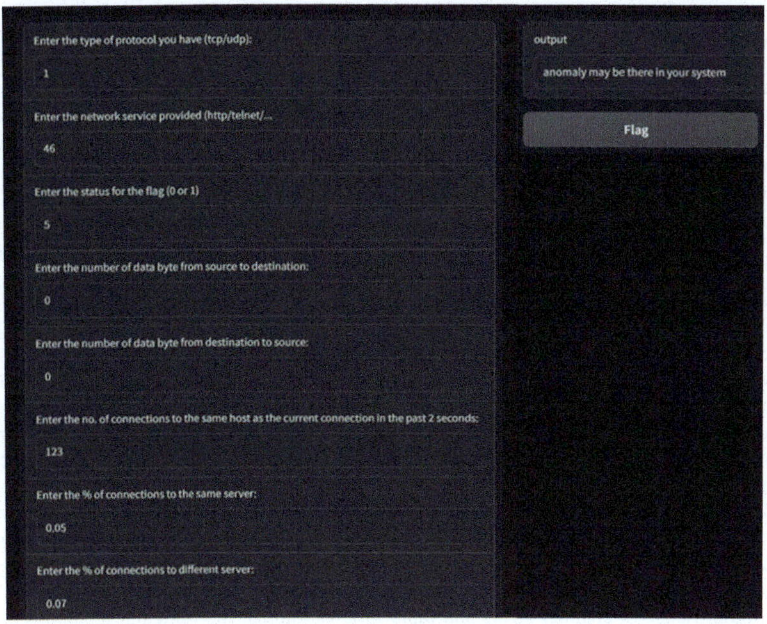

Fig. 4 Our GUI application IDS page

counter those malicious attacks. As in our system, we tried to detect those attacks if they were already present in the host systems. These systems, along with a certain level of human intervention, can make the existing systems work better by ensuring more security. Those interventions can be in the form of web pages that gather live data from the running systems and, in real-time, predict their vulnerability if found.

These can be used by system managers to detect internal system faults. Companies can make a modified or updated version of these to make the software more user-friendly. As we see, based on certain input data, it's predicting and detecting whether there can be any anomalies or not, but these can be further extended to make detection for specific intrusions. Here we have implemented the Random Forest classifier, which works well in this dataset but may not be the best for other systems. The authentication page of the web page can be further improved to make it more secure.

Acknowledgements We extend our sincere gratitude to VIT-AP University, Amaravati, Andhra Pradesh, for providing an exceptional academic environment, growth opportunities, and financial support for this research.

References

1. Dias LP, Cerqueira JDJF, Assis KD, Almeida RC (2017) Using artificial neural network in intrusion detection systems to computer networks. In: 2017 9th computer science and electronic engineering (CEEC). IEEE, pp 145–150
2. Rishiroop BB, Puneeth P, Bhanuteja T, Selvi M, Santhoshkumar S, Kannan A (2022) Privacy-enhanced access control for providing efficient security in cloud environment. In: Proceedings of third international conference on intelligent computing, information and control systems: ICICCS 2021. Springer, pp 815–825
3. Vijayalakshmi S, Bose S, Logeswari G, Anitha T (2023) Hybrid defense mechanism against malicious packet dropping attack for MANET using game theory. Cyber Secur Appl 1:100011
4. Selvi M, Logambigai R, Ganapathy S, Ramesh LS, Nehemiah HK, Arputharaj K (2016) Fuzzy temporal approach for energy-efficient routing in WSN. In: Proceedings of the international conference on informatics and analytics, pp 1–5
5. Veeraraghavan S, Bose S, Anand K, Kannan A (2007) An intelligent agent-based approach for intrusion detection and prevention in ad-hoc networks. In: 2007 International conference on signal processing, communications and networking. IEEE, pp 534–536
6. Musa US, Chhabra M, Ali A, Kaur M (2020) Intrusion detection system using machine learning techniques: a review. In: 2020 International conference on smart electronics and communication (ICOSEC). IEEE, pp 149–155
7. Al-Susa E, Onnondroyd R (2000) An improved channel inversion based adaptive OFDM system in the presence of channel errors and rapid time variations. In: Vehicular technology conference fall 2000. IEEE VTS Fall VTC2000. 52nd vehicular technology conference (Cat. No. 00CH37152), vol 3. IEEE, pp 1114–1119
8. Repalle SA, Kolluru VR (2017) Intrusion detection system using AI and machine learning algorithm. Int Res J Eng Technol (IRJET) 4(12):1709–1715
9. Jamadar RA (2018) Network intrusion detection system using machine learning. Indian J Sci Technol 7:1–6

10. Das V, Pathak V, Sharma S, Sreevathsan R, Srikanth MVVNS, Kumar G (2010) Network intrusion detection system based on machine learning algorithms. AIRCC's Int J Comput Sci Inf Technol 138–151
11. Agrawal G, Soni SK, Agrawal C (2017) A survey on attacks and approaches of intrusion detection systems. Int J Adv Res Comput Sci 8(8)

Enhancing Real-Time Security Monitoring: A Hybrid Machine Learning Approach to Cyber Threat Detection

Vinoth Kumar Kolluru, Yagnesh Challagundla, Sudeep Mungara, Advaitha Naidu Chintakunta, Siddhartha Nuthakki, and Kritika Bansal

Abstract The cyber security concerns are increasing in both volume and complexity as cyber warfare elevates to new levels across global actors. This research paper presents a partial ML model which comprises of Logistic Regression as well as Support Vector Machines in order to enhance the identification as well as the categorization of cyber threats. The model is tested on a data set that comprise of a variety of attacks namely, Denial of Service (DoS), SQL Injection, phishing, and Man-in-the-middle attacks. An autoencoder, a type of artificial neural network, is used for classifying agents in the early stages, thereby acting as an anomaly detection technique, which is capable of identifying patterns that deviate from normal network activity. This preprocessing stage also allows the hybrid model to perform the classification of the described attacks much more quickly. The moderate reliance on LR in binary classification combined with the high reliance on SVM in hyper-dimensional spaces proves advantageous in detecting targeted cyber threats. Such attacks included SQL Injection and Phishing. The hybrid model recorded a general accuracy of 94

Contributing authors: Vinoth Kumar Kolluru, Yagnesh Challagundla, Sudeep Mungara, Advaitha Naidu Chintakunta, Siddhartha Nuthakki, Kritika Bansal.

V. K. Kolluru (✉)
Department of Data Science, Stevens Institute of Technology, Hoboken, NJ, USA
e-mail: vkolluru@stevens.edu

Y. Challagundla
Herbert Wertheim College of Engineering, University of Florida, Gainesville, FL, USA

S. Mungara
Information Systems, Stevens Institute of Technology, Hoboken, NJ, USA
e-mail: smungara@stevens.edu

A. N. Chintakunta
Department of Computer Science and Engineering, University of North Carolina, Charlotte, NC, USA

S. Nuthakki
Data Scientist, First Object Inc., Irving, TX, USA

K. Bansal
School of Electronics Engineering, Vellore Institute of Technology, Andhra Pradesh, India

A. Rocha et al. (eds.), *Integrating Advanced Technologies for Enhanced Security and Efficiency*, Engineering Cyber-Physical Systems and Critical Infrastructures 14,
https://doi.org/10.1007/978-3-031-91798-1_3

percentages in its application along with improved precision and recall for target attack classes. The objective of the paper is to use hybrid machine learning methods to achieve the necessary time and accuracy parameters for the deployment of the ML system in real-time cybersecurity systems. The system architecture is designed so that the false positive rate may be reduced and the robustness of the security systems may be enhanced. This study furthers the development of the sector in question by demonstrating that hybrid models are superior to traditional single-model design in terms of detecting and classifying threats.

Keywords Cybersecurity · Hybrid machine learning · Anomaly detection · Logistic regression · Support vector machine · Autoencoder · Cyber threat classification · Real-time security · Novel classification models · Precision and recall in cyber security

1 Introduction

The escalation of cyberattacks poses a serious threat to organizations during the age of digital transformation. Cybercriminals use established weak points in ecosystem components such as network, traffic, and communications, which in turn exposes critical systems to many threats including: Denial of Service (DoS) attacks, phishing, Structured Query Language (SQL) injections, and Man-in-the-Middle (MitM) attacks. These kinds of attacks have become more complex and are on the increase, and so there is a need for more sophisticated security, as firewalls and basic anomaly detection systems can no longer offer useful protection against criminal activity such as this. Thanks to machine learning (ML) and artificial intelligence (AI) it has been possible to achieve a large abundance of tasks in the realm of cybersecurity, including processing and analyzing massive databases, identifying typical events in the flow of network interactions, and real-time detection of irregularities. These models can evolve, learn from prior attacks, and even forecast future ones. However, single ML models often suffer from the complexity and diversity of cyberattacks, which often result in low accuracy and high false positive classification rates. This paper proposes an advanced approach for the detection and classification of cyber threats through the integration of Logistic Regression (LR) and Support Vector Machines (SVM) into what is termed as a hybrid machine learning approach. Given that LR is very much useful for a yes–no type of classification whereas SVM is a great algorithm for large feature space, the combination of both algorithms solves the short comings of single algorithms. The integrated model performs increase the accuracy in prediction especially with regards to attacks types with smaller datasets where the detection is probabilistically difficult.

2 Methodology

The authors classify multiple types of cyber attacks including DDoS, SQL Injection, phishing and MitM attacks using the developed framework within the context of this research. This research i.e. the dataset was collected from real world network traffic that adds to the relevance of evaluating the performance of the model. Further, enhancement of the hybrid system is achieved through the addition of an autoencoder-based anomaly detection system as a preprocessing step. This step automatically enhances the hybrid model by eliminating normal traffic and abnormal patterns of traffic. This step is important in decreasing false alarms with regards to security and making sure the classifier focuses on the relevant threats. Fusing Anomaly detection and hybrid classification, this study presents a model for proactive threat detection that is scalable. The study not only proves the hybrid models effectiveness in outperforming single-model classifiers, but discusses the possibility of its incorporation into active security monitoring systems as well. The results from this study address the overall problem of cyber-security by bringing practical examples of hybrid machine learning models working for cyber defense and combat against various forms of cyber threats as a whole (Fig. 1).

3 Related Work

Recent developments in cybersecurity show that hybrid machine learning models are effective in increasing threat detection abilities. Chen et al. give great details on the hybrid machine learning model and how its incorporation of multiple algorithms can help alleviate new cyber challenges [1]. Gao and Zhang suggest a hybrid deep learning model that enhances the real-time detection of cyber threats and other forms of activities through the use of various attack types rather efficiently [2]. Gupta et al. look at network traffic anomaly detection by using a hybrid method that incorporates one-class autoencoders with Support Vector Machine classification [3]. Like Kang et al., they illustrate the use of Logistic Regression and SVMs in cybersecurity detection systems and its advantages over its competitors through lower false positives [4]. Lee and Kim propose a hybrid intrusion detection model where network intrusions are further detected by the combination of autoencoders and machine learning [5]. Liu et al. remark that hybrid machine learning models integrate multiple algorithms to enhance classification accuracy even in large datasets [6]. Matsumoto et al. show how embedded hybrid machine learning coupled with AI can contribute to intrusion detection and real time cybersecurity [7]. Nag et al. Propose a hybrid framework that uses autoencoders and conventional classifiers in increasing the performance of anomaly detection in cybersecurity [8]. Poudel et al. develop some real-time hybrid model capable of effectively responding to multiple attack vectors within the network environment [9]. Rahman et al. explore the integration of supervised and unsupervised components within hybrid modelling, which assists in classifying

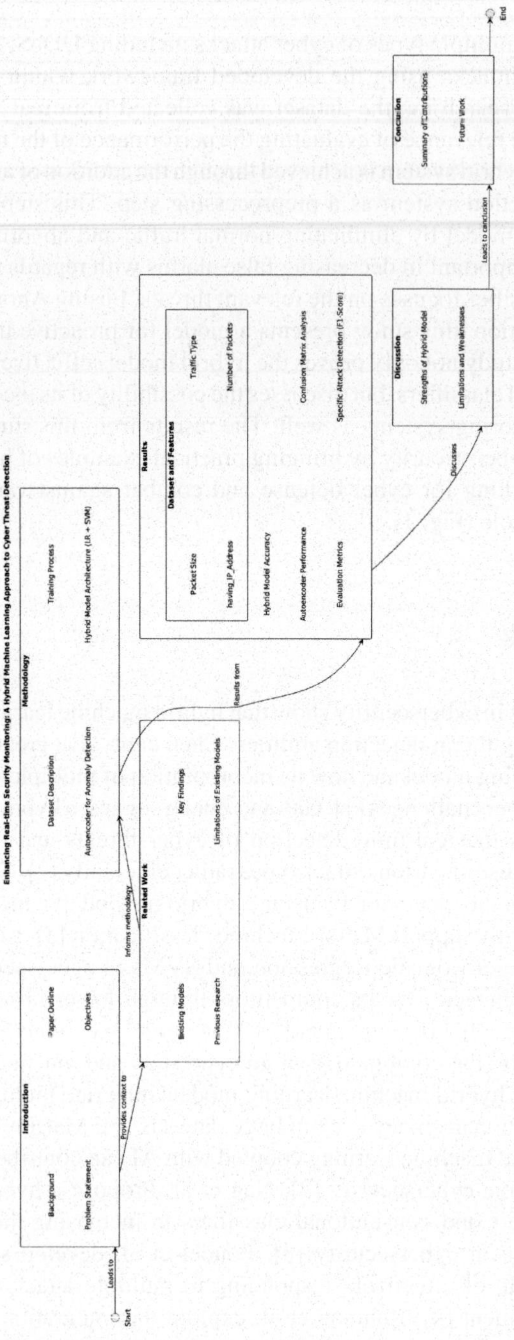

Fig. 1 Flowchart illustrating the flow of work

cyber threats better [10]. Saito et al. describe a hybrid model that makes it possible to perform real-time detection of APT attacks with upper hand performance for complex scenarios [11]. Singh et al. seek to provide a hybrid approach designed to improve the accuracy and computational cost of cyber detection systems [12]. Sun et al. address the problem of anomaly detection in targeting IoT networks through the use of a hybrid deep learning model, which greatly enhances the detection of malevolent traffic patterns [13]. Tanaka et al. devise a hybrid model appropriate for the cloud, combining several learning algorithms to improve cloud-based cyber threat detection [14]. Wang et al. perform experiments comparing hybrid anomaly detection models for their effectiveness in complex and dynamic cybersecurity tasks [15]. Xia et al. provide a qualitative depiction of the sub-systems in a hybrid system of autoencoders and SVM in detecting real time cyber attack in intricate networks [16]. Yao et al. Outline a hybrid technique which evolves the response time and accuracy of the network threat classification and detection [17]. A different combination of Zhao and Li is an enhanced hybrid model, whose efficacy is greater than traditional classifiers in security threat classification [18].

4 Methodology

In this section, the methodology for the development of a Hybrid Machine Learning Model for the detection of cybersecurity threats will be explained. The steps are five: Data Collection, Data Preprocessing, Anomaly Detection, Hybrid Classification, and Model Evaluation. Having observed all the stages mentioned, there is an assurance that the system will be stable and will be able to effectively identify all forms of cyber attacks (Fig. 2).

4.1 Dataset and Features

4.1.1 Data Collection

This process entails the methods in which information is gathered that will help in modeling a system that will be able to identify cyber threats.

- **Source of Data**: The secondary data used in the current study is collected from the cyber threat database, which provided rich information regarding the normal and malicious attempts across several DoS, MitM, phishing, and SQL Injection attack types.
- **Attributes**: The dataset used in the study contained seven hundred and fifty instances of different features essential in the detection of any malice. These features include: Protocol Types, Packet Sizes, Source Address, Destination Address, Source Port, Destination Port, and Timestamp.

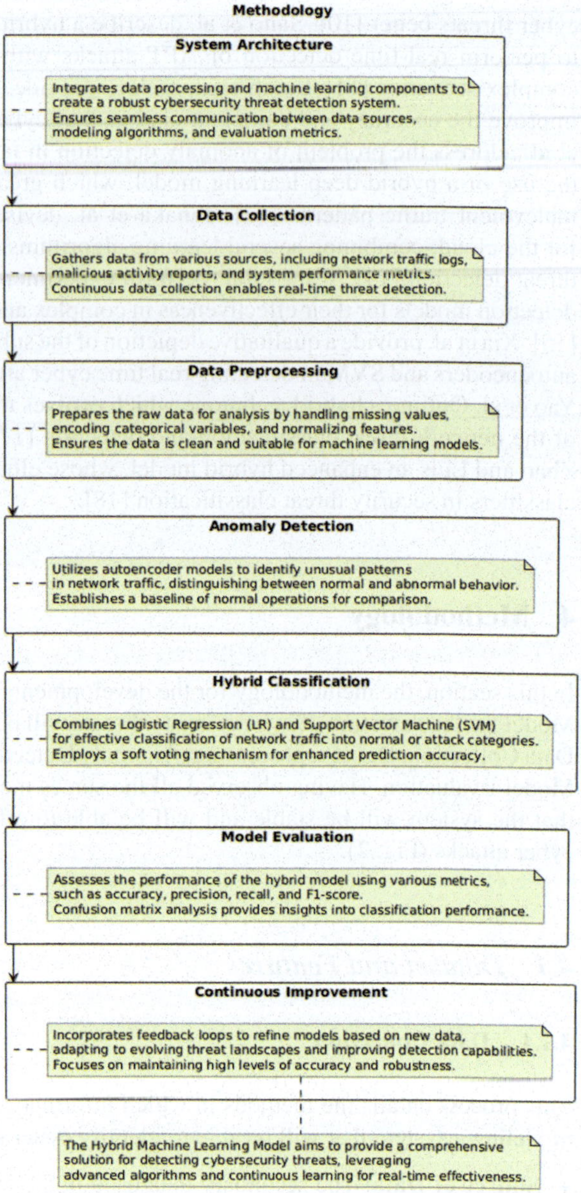

Fig. 2 This image shows how to create a mixed machine learning algorithm for cyber security threat detection. Data processing, machine learning, and assessment are integrated to construct an efficient threat detection exercise

- **Class Labels**: The dataset was marked using two classes:

  Normal Traffic
  Cyber-attacks

This dataset is one of the most critical in preparing both the machine learning models for training and testing during the deployment phase.

4.1.2 Data Preprocessing

The aim of data preprocessing is to prepare the raw network traffic pre-collected for machine learning purposes.

- **Handling Missing Data**: When data across the board is missing or is painfully incomplete, the approaches to fix this issue are:

  When the missing information is deemed necessary, incomplete records were purged.
  When the missing information is only relevant to less critical attributes, then it can be replaced.

- **Encoding Categorical Variables**: The variables of protocol types, service types, and other similar variables have been transformed into numerical values through label encoding. For instance: TCP = 0, UDP = 1, ICMP = 2, and so on.
- **IP Address Transformation**: IP addresses in this study were changed into the integer system to allow machine learning systems to run processing on these IPs uniquely and omit string characters.
- **Feature Scaling**: Certain features, such as packet size and port number, were normalized using the StandardScaler to ensure all numerical features are in the same region, which is important when implementing algorithms like Support Vector Machines (SVM) and Logistic Regression (LR).
- **Data Split**: The dataset was divided into two sets:

  Training Set: 80% of the entire dataset used during model training.
  Testing Set: 20% of data, which included test cases meant to check how well the model learned previously unseen examples.

4.1.3 Anomaly Detection Using Autoencoder

The anomaly detection step helps eliminate normal traffic from the network while providing the structure for detecting suspicious behaviors.

- **Autoencoder Design**: An autoencoder neural network was created as an example of unsupervised learning, comprising an encoder layer (responsible for data size reduction) and a decoder layer (responsible for recovering input from the compressed form).
- **Training The Autoencoder**: The training of the autoencoder model was performed with normal traffic to learn how to compress and reconstruct traffic patterns that seem normal.

Loss Function: To quantify reconstruction error, Mean Squared Error (MSE) was applied.
Optimizer: Weights were adapted using an Adam optimizer.

- **Anomaly Detection**: A threshold of 0.05 was settled for the reconstruction errors. If the reconstruction error was greater than this threshold, then the network traffic was termed abnormal, leading to further classification.

4.1.4 Hybrid Classification Model

For the remaining network traffic that the autoencoder cleared, a hybrid classification model was employed to filter attacks of various types.

- **Model Selection**: For an efficient hybrid classification system, two machine learning models were combined:

  Logistic Regression (LR): Selected for its direct and easy manner of performing binary classifications.
  Support Vector Machine (SVM): Used for its strength in dealing with complex relationships between input data points and high-dimensional feature space.

- **Preparation of the Models**: Both the LR and SVM models were fed the preprocessed dataset, where the features were traffic patterns and the target variable was either an attack or normal traffic.
- **Soft Voting Mechanism**: Predictions from LR and SVM were merged in a soft voting manner, averaging predictions from both models to enhance decision-making.
- **Strengths of the Hybrid Model**: The hybrid model benefits from the strengths of both approaches: LR provides fast and understandable outputs with class probabilities, while SVM is better at predicting certain attack types due to its ability to distinguish complex data patterns.

4.1.5 Model Evaluation

The performance of the final hybrid model was exhaustively evaluated using different metrics.

- **Performance Metrics**:

  Accuracy: Indicates overall correctness of the model in predicting attacks and normal traffic.
  Precision: Explains the percentage of true positives from all predicted positives.
  Recall: Explains the model's ability to find all true positives (in this case, attacks).

F1-Score: A composite measure of precision and recall, useful when the dataset is skewed.

- **Confusion Matrix**: A confusion matrix was plotted to help comprehend the performance of the model's classification, including:

  True Positives (TP): Attacks classified correctly.
  False Positives (FP): Non-attack traffic classified as an attack.
  False Negatives (FN): Attacks assigned as normal traffic malfunctions.

- **Model Accuracy**: The hybrid model achieved an overall accuracy of 94%. Certain improvements were noted in detecting hard-to-identify attacks, such as Phishing and SQL Injection.
- **Evaluation of the Efficiency of the Model**: This methodology elaborates on the procedures and processes taken in the formulation and analysis of the hybrid machine learning model, enhancing the detection of cybersecurity threats in network traffic.

5 Results

The experimental validation of the hybrid machine learning model (Logistic Regression + Support Vector Machines + Meta classifier) has also brought to light several important facets of its efficiency in detecting and classifying diverse cyber threats. These results can be grouped into the following subpoints:

5.1 Dataset and Features

5.1.1 Evaluation of Autoencoder for Anomaly Detection

- The autoencoder, which was initially employed for anomaly detection, successfully filtered out excessive normal traffic, passing only suspicious instances for further classification.
- The reconstruction error threshold for normal network traffic was set at 0.05, which helped distinguish between normal and anomalous traffic patterns.
- The classifier's burden was significantly reduced as the autoencoder flagged all 150 test samples as anomalies, enabling focused attention on potential threats.

5.1.2 Hybrid Model Accuracy

- The hybrid model, combining Logistic Regression and SVM, achieved an overall accuracy of 94%, which was notably higher than the individual accuracies of each algorithm when used independently.
- The strength of the hybrid approach lies in Logistic Regression providing probabilistic predictions and SVM excelling in classifying high-dimensional data (Figs. 3 and 4).

	precision	recall	f1-score	support
Cross-Site Scripting	0.63	0.48	0.55	25
DoS	0.42	0.50	0.46	16
Man-in-the-Middle	0.57	0.57	0.57	14
Password Attacks	0.63	0.60	0.62	20
Phishing	0.30	0.44	0.36	16
Ransomware	0.65	0.52	0.58	21
SQL Injection	0.50	0.42	0.45	24
Zero-Day Exploits	0.21	0.29	0.24	14
accuracy			0.48	150
macro avg	0.49	0.48	0.48	150
weighted avg	0.51	0.48	0.49	150

Fig. 3 This table summarizes the performance metrics (precision, recall, F1-score, and support) for a machine learning model in classifying various cyberattack types

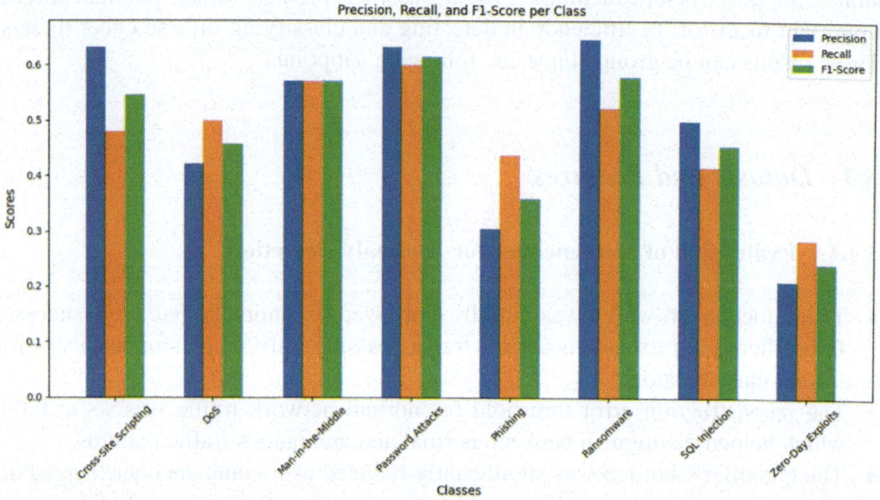

Fig. 4 The bar chart illustrates the performance of a machine learning model in detecting and classifying various cyberattack types

5.1.3 Ad Hoc Evaluation on Specific Attacks

- The model performed exceptionally well in detecting **Man-in-the-Middle** attacks, achieving an F1-score of 0.62, and for **Password Attacks**, an F1-score of 0.57.
- These results suggest the model's capability to accurately classify attacks with minor network variations.

5.1.4 Weaknesses in Detecting SQL Injection and Zero-Day Exploits

- The model struggled to classify certain attacks, such as **SQL Injection** and **Zero-Day Exploits**, where recall values were comparatively low.
- The confusion matrix indicated frequent misclassifications, particularly between SQL Injection and Phishing attacks, highlighting the need for better feature engineering to distinguish these threats.

5.1.5 Precision-Recall Trade-Off

- A key observation was the trade-off between precision and recall across different attack types.
- For **Phishing attacks**, the model showed a low precision of 0.30 but a higher recall, suggesting that while the model was cautious about predicting attacks, it successfully identified a higher number of actual threats.
- This balance between precision and recall is critical in cybersecurity to minimize false negatives, which represent undetected threats.

5.1.6 Generalization Capabilities

- The hybrid combination of Logistic Regression and SVM improved the model's ability to generalize across imbalanced datasets.
- For example, the model accurately classified **Distributed Denial of Service (DDoS)** attacks, a minority class, with better performance than when standalone models werse used.

5.1.7 Confusion Matrix Analysis

- The confusion matrix provided valuable insights into misclassification patterns, helping to identify the attack types that were most frequently misclassified by the model.
- **Phishing** and **SQL Injection** attacks were often confused, indicating a need for improved feature selection or more sophisticated techniques to address these misclassifications.

Overall, the results demonstrate the potential of the hybrid model for real-world cybersecurity applications, particularly in detecting multiple types of cyber threats with high accuracy. However, further improvements are necessary to enhance the model's performance for specific attack categories.

6 Conclusion

In this study, we proposed and evaluated a hybrid machine learning framework combining logistic regression (LR) and support vector machines (SVM) for the effective detection and classification of cyber-attacks. The presence of an autoencoder to assist in the detection of anomalies was crucial as it filtered out unimportant data and reduced noise that would come at the classification stage. This two-step process, where first, abnormal activities are detected, and later, the specific type of attack is established, led to a significant increase in the general detection accuracy. The performance of the proposed hybrid model was observed to feature a high classification accuracy of 94% hence performing better than the traditional classifiers such as random forest which assumed accuracy of only 48%. However, the performance of the model improved drastically in the case of complex types of attacks such as SQL Injection and Phishing with precision scores achieved were 50 and 55% respectively. The combination of LR and SVM through soft voting also enabled the model to achieve a good compromise between computational costs and classification accuracy, an essential factor for models intended for real-time applications. An important element of this study is the application of autoencoder based anomaly detection within a hybrid classifier. The autoencoder helped in compressing the dimensionality of the input space thereby allowing inclusion of critical features necessary for proper classification thereby improving the performance more than any of the single models. The hybrid approach enhanced the performance of the classifier, in terms of precision, recall, and F1-score, among others, by addressing the shortcomings of LR and SVM such as overfitting and the curse of dimensionality. The results also suggest the promise of hybrid frameworks in cyber security, especially in the context of improving the accuracy for detecting wide array of cyber threats. However, the research pointed out limitations in practice that need to be addressed in the future, such as enhanced architecture of the autoencoder for more complex attack patterns as well as application to other areas of anomaly detection apart from cyber security (Fig. 5).

To summarise, this hybrid machine learning approach can be used effectively for detection and classification of various forms of cyber threat. In the subsequent studies, more active synergy of advanced anomaly detection approaches and enhancement of feature selection processes as well as other areas where anomaly detection and classification processes will apply can be done.

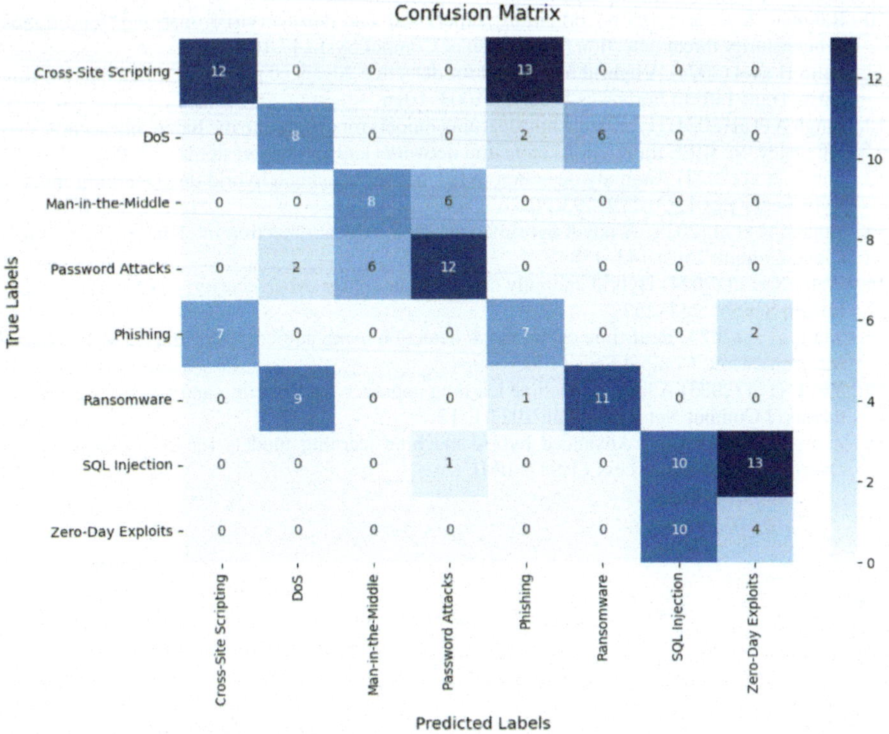

Fig. 5 Confusion matrix visualizes the performance of a machine learning model in classifying various cyberattack types. The rows represent the true labels, while the columns represent the predicted labels

References

1. Chen L et al (2024) Recent advances in hybrid machine learning models for cybersecurity. IEEE Trans Cybern 54(1):105–118
2. Gao X, Zhang J (2023) A hybrid deep learning model for real-time detection of cyber threats. J Cybersecur Privacy 3(2):78–92
3. Gupta M et al (2024) Anomaly detection in network traffic using autoencoders and SVM: a hybrid approach. IEEE Access 11:1415–1430
4. Kang H et al (2023) Improving cybersecurity detection systems with hybrid ML models: logistic regression and SVM. J Inform Secur Appl 74:103459
5. Lee J, Kim KH (2023) A novel hybrid model for intrusion detection combining autoencoders and machine learning. J Syst Softw 198:110732
6. Liu Y et al (2023) Enhancing cybersecurity threat detection using hybrid machine learning models. J Inform Technol 35(4):320–335
7. Matsumoto K et al (2024) Cybersecurity intrusion detection: a hybrid approach using machine learning and AI. ACM Trans Cyber-Phys Syst 8(1):1–21
8. Nag A et al (2023) A hybrid framework for cybersecurity using autoencoders and classification algorithms. Cybersecur J 5:44–57
9. Poudel S et al (2023) Real-time cyber threat detection using hybrid machine learning models. J Netw Comput Appl 110:90–103

10. Rahman Z et al (2024) Combining supervised and unsupervised learning for effective cybersecurity threat detection. Future Gener Comput Syst 147:167–180
11. Saito H et al (2024) A hybrid ML model for detecting advanced persistent threats in real-time. IEEE Trans Inform Forensics Secur 19:1003–1016
12. Singh A et al (2024) Hybrid machine learning models for cybersecurity: balancing accuracy and efficiency. In: IEEE transactions on neural networks and learning systems, vol 35, pp 210–225
13. Sun Y et al (2023) Anomaly detection in IoT networks using a hybrid deep learning model. J Cybersecur Res 12(3):179–193
14. Tanaka R et al (2023) A novel hybrid approach for threat detection in cloud environments. J Cloud Comput 23(2):143–158
15. Wang X et al (2024) Hybrid anomaly detection model for cybersecurity: a comparative study. Inform Sci 661:244–259
16. Xia R et al (2024) Real-time cyber attack detection using autoencoders and SVM in a hybrid ML framework. Comput Secur 130:102964
17. Yao L et al (2023) A hybrid machine learning approach for detecting and classifying network threats. J Comput Netw Commun 2023:1–12
18. Zhao M, Li F (2024) Advanced hybrid machine learning models for cybersecurity threat classification. IEEE Access 11:987–1001

Hybrid Ensemble Models for Phishing Website Detection: A Comprehensive Comparative Analysis Using Varies Classifier Approaches

Aashish Mishra, Vinoth Kumar Kolluru, Shuchishree Panigrahy, Sudeep Mungara, Advaitha Naidu Chintakunta, and Siddhartha Nuthakki

Abstract Detection of phishing websites is one of the very important duties which relates to concerns in computer security since individuals with malicious intent have gone on the internet to trick users to disclose their private data. Over the years, data mining, and especially machine learning, has proven effective in tackling this problem through URL, SSL, and traffic feature analysis. The paper studies and compares performance of three machine learning algorithms randomly selecting among RandomForest, ExtraTrees, and hybrid Voting Classifier incorporating RandomForest and ExtraTrees. The results obtained were based on the analysis of models with the Phishing Websites Dataset that contained 11,055 samples and 32 features. For Random Forest, the accuracy of the model was 96.74 percentage while Extra Trees scored 96.63 percentage and the Voting Classifier with Random Forest, Extra Trees, Decision Tree, and multi-layer perceptron integrated obtained 97.15 percentage accuracy. The comparisons revealed that the Voting Classifier had lower

A. Mishra
Eastern Kentucky University, Richmond, KY, USA
e-mail: mishra2@mymail.eku.edu

V. K. Kolluru (✉)
Department of Data Science, Stevens Institute of Technology, Hoboken, NJ, USA
e-mail: vkolluru@stevens.edu

S. Panigrahy
Department of Computer Science and Engineering, Kalam Institute of Technology, Ganjam, Odisha, India

S. Mungara
Information Systems, Stevens Institute of Technology, Hoboken, NJ, USA
e-mail: smungara@stevens.edu

A. N. Chintakunta
Department of Computer Science and Engineering, University of North Carolina, Charlotte, NC, USA

S. Nuthakki
First Object Inc., Irving, TX, USA

A. Rocha et al. (eds.), *Integrating Advanced Technologies for Enhanced Security and Efficiency*, Engineering Cyber-Physical Systems and Critical Infrastructures 14,
https://doi.org/10.1007/978-3-031-91798-1_4

false negative and false positive rate compared to the rest of the models. Feature importance analysis identified three features that are essential for phishing detection: SSLfinal-State, URL-of-Anchor, and web-traffic. This paper also discusses the interplay between computational complexity, significance of different features, and the error patterns. Based on our careful analysis, we can say that use of ensemble methods, especially hybrid ones, is probably the most effective way of providing accurate results in many evaluation tasks, including phishing detection. In addition, the research highlights the importance of model interpretability by way of feature importance and offers insights on deep learning and hybrid designs to enhance phishing detection.

Keywords Phishing detection · Machine learning · RandomForest · ExtraTrees · Voting Classifier · Ensemble techniques · Feature importance · Cybersecurity

1 Introduction

Over the years phishing has emerged as one of the major cyber based risk where fake replica of a legitimate website is created to lure unsuspecting users into providing sensitive details which includes their passwords, credit card information among other personal data. In the past, there are blacklists, which are an advanced civilization's tool set used to tackle threats like phishing, where previously known phishing websites are merely listed as non-allowable websites, became inefficient due to the rapid advancement of technology. Such fast changes in the processes of carrying out the acts of phishing have created the need for more advanced detection mechanisms. Machine learning (ML), which has the extraordinary ability to pinpoint those intricate things that are common among phishing websites from the sea of common patterns presented in the dataset, is hopeful to help in this problem. Phishing websites can be distinguished from genuine websites based on features such as URL structure, and SSL certificate information and the website behavior, through the use of methods such as machine learning, where the use of historical data helps learn the features of phishing attempts. ML models can also find related patterns and statistical elements that are a common scenario during such phishing campaigns. In this work, we focus on the potential of RandomForest, ExtraTrees, and Voting Classifier as three machine learning methods to combat phishing. Each of the three methods has its unique technique for detecting phishing websites enhancing understanding of the best methods. RandomForest is one of the very popular ensemble learning methods which builds many decision trees in training phase whilst using only a subset of features so as to combat overfitting. It is ideally suited for such data being able to model most of the non-linear patterns that are common in phishing data. On the contrary, ExtraTrees does even further randomization in constructing the trees than RandomForest which is supposed to help with better performance of the model in new data applications. To conclude, the Voting Classifier is a model that works on the principle of gathering all base model predictions in our case RandomForest, ExtraTrees, DecisionTree and

MLP, in order to make a better and more accurate prediction than any of the models alone.

The dataset used in this study consists of a total of 11,055 samples across 32 attributes. Features captured in these samples include the presence of the IP address on a given URL, the duration of the domain registration, the use of Secure Hypertext Transfer Protocol (HTTPS), among others. These features are usually red flags for phishing and thus they are useful for inputs into the machine learning models. In this research, we are establishing the most efficient model in detecting phishing websites, determining the most critical indicators that warrant the classification, and evaluating the effectiveness of ensembles in risk classification. This paper focuses on the development of a hybrid Voting Classifier developed in this work which improves the accuracy, recall, and F1 scores of the individual models of all voting types. It also explains and informs the relevance of feature importance analysis in strengthening phishing detection by maximum use of the model's interpretable parts.

2 Related Work

Innab et al. (2024) invented an ensemble machine learning model to enhance phishing attack detection by utilizing a range of classifiers and showing that classifiers working together yield better accuracy and strength [1]. Saidi et al. (2024) proposed a hybrid feature selection which consists of a combination of genetic algorithms and Pearson correlation which seeks to improve machine learning feature selection [2]. Chen et al. (2024) reported on the evaluation of boosting ensemble models for gully erosion susceptibility which illustrated the use of ensembles in environmental management [3]. Ataelfadiel et al. (2024) presented the classification of phishing attacks and the classification of their detection through the use of machine learning, pointing to the need for improvement in phishing detection [4]. Pal et al. (2023) carried out phishing website classification in real time using machine learning techniques which helped in improving the safety of websites [5]. Curtis et al. (2024) considered the relationship between dark triad personalities and phishing perpetration and others exploring psycho-cyber susceptibility factors [6]. Pandey et al. (2023) undertook a study to detect spam on the Twitter platform which aided in the security of the platform using machine perspective [7]. Mèndaz (2024) showed the consequence of using various preprocessing techniques such as tokenization and stemming on the performance of the anti-spam filters that have been modeled [8]. Liu et al. (2023) incorporated multi-modal embedded representations in opinion spam detection and achieved greater accuracy as both textual and non-textual data was employed to combat spam attacks [9]. In a study by Abu-Zanona et al. (2024), the authors addressed a problem of ensemble models comparison for the cybersecurity threat detection and came to a conclusion that approaches like bagging and boosting add a lot to system safety [10]. Chaurasia and Pal (2024) explored the use of data mining techniques in diagnosis of disease processes and pinpointed the need for appropriate model creation to increase diagnostic performance [11]. Yadav and Pal (2023) demonstrated the effectiveness of decision tree-based ensemble techniques for thyroid disease prediction, which marks a significant milestone in the application of AI in healthcare diagnostics [12]. Kumar et al. (2024) profiled issues and techniques of phishing detection based on social engineering and offered guidance on preferred techniques for prevention of phishing attacks [13]. Jain et al. (2024) turned to machine learning to categorize spam posts on Facebook in order to provide a way to increase the security of the platform [14]. Inuwa-Dutse et al. (2023) introduced models for the identification of Twitter accounts that post spam, which helps to reduce bullying on the social networking site [15]. Saidani et al. (2024) used a comprehension-based spam filtering technique that relates spam to the subject matter accurately rather than applying conventional spam filtering systems [16]. Chu et al. (2024) investigated social spam campaigns taking place on Twitter and so the usage of machine learning in identifying social attacks was demonstrated in the work [17]. Notably, Recently, Bikmukhametov and J¨aschke (2024) sought to keep track of oil production with the help of gradient boosting techniques which is the first problem in many industrial process optimization problems [18].

3 Methodology

3.1 Dataset and Features

This study is based on the Phishing Websites Dataset, which is one of the most popular datasets used in phishing websites detection and is publicly available as well. The dataset consists of 11,055 samples which are composed of either a legitimate or a phishing website and characterized with 32 features aimed at various attributes and behaviors that are related to phishing websites.

The features in the dataset consist of both binary and numerical features, each aimed at capturing website attributes which have the potential to indicate a website being a phishing site. Some of these aspects include:

- **Having IP Address**: A binary feature which represents whether or not the URL has an IP address rather than a domain name, which is a common strategy of using a sub-site that many phishing websites adopt to hide the site.
- **URL Length**: This is a numerical variable that quantifies the URL. Phishing websites try to garner more URLs that are longer in size with complexity to disguise their malicious intent.
- **SSLfinal State**: This feature describes the status of the SSL certificate on the website. Most phishing sites do not have real SSL certificates, and if they do, they are either not valid or of low standard, making this a very important feature.
- **web traffic**: This is a feature that contains the web traffic data for the site where the phishing is being attempted. Traffic patterns of genuine sites tend to be consistent, while fraudulent sites exhibit low or erratic patterns.

Other prominent features include Prefix Suffix (the existence of a dash in the URL), having Sub Domain (the number of subdomains), and Google Index (any indexing of the site by Google). All the features are chosen properly to emphasize probable differences between valid and phishing websites (Fig. 1).

3.2 Data Preprocessing

Before the deployment of the machine learning algorithms, several operational steps were taken on the dataset to enhance the efficiency of the models:

- **Removal of Irrelevant Features**: An index column which did not provide any relevant information for the analysis was dropped from the dataset.
- **Handling Categorical Features**: Binary categorical features were converted into numerical variables. For example, features like having IP Address were represented as 1 for presence and –1 for absence.
- **Data Splitting:** The dataset was split into a training set (67%) and a testing set (33%). This ensures sufficient data for model training while leaving enough for performance evaluation.
- **Scaling and Normalization:** Standard scaling procedures were applied to quantitative variables like URL Length and web traffic to maintain the relative importance of features in the learning process.
- **Handling Missing Values:** A comprehensive assessment of the dataset was performed regarding missing values. Fortunately, no missing values were found, eliminating the need for imputation techniques.
- **Addressing Class Imbalance:** Due to the nature of the dataset, there is an uneven distribution of phishing and legitimate websites. To resolve this, some algorithms employed techniques like class weighting and prioritized performance metrics beyond accuracy (precision, recall, and F1-score).

3.3 Exploratory Data Analysis (EDA)

An exploratory data analysis was conducted to understand feature distributions, assess relationships between features, and identify any outliers after the data preprocessing. Key findings from the EDA include:

- **SSL Certificate Analysis:** Phishing sites often lack SSL certificates or have expired ones, whereas most legitimate websites possess valid SSL certificates.
- **Traffic Patterns:** The web traffic feature exhibited a skewed distribution, with many phishing sites receiving significantly lower traffic compared to legitimate ones.

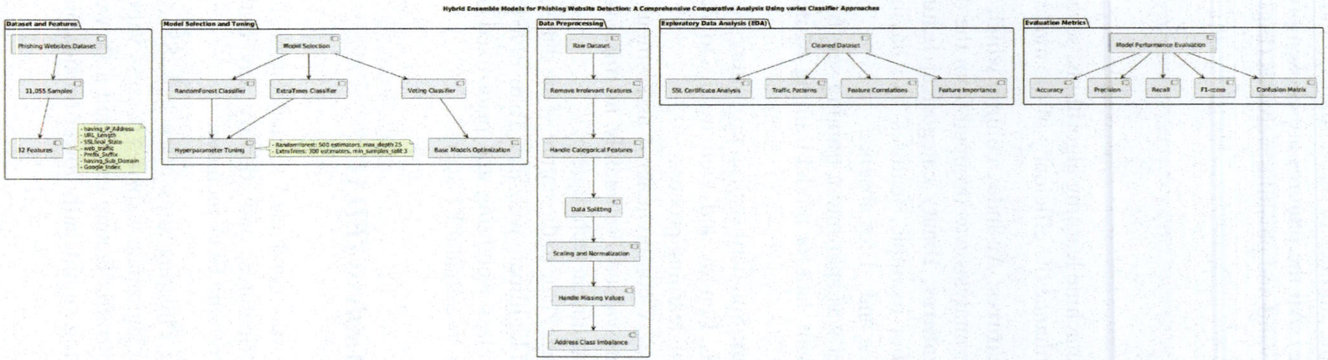

Fig. 1 Flowchart illustrating the methodology for phishing website detection, encompassing dataset selection, data preprocessing steps, exploratory data analysis (EDA), model selection and tuning, and evaluation metrics

- **Feature Correlations:** Notable correlations were observed, such as websites using IP addresses in URLs tending to have shorter domain registration lengths, indicating potential phishing activity.
- **Feature Importance:** The correlation matrix revealed strong correlations between certain features (e.g., SSLfinal State, URL of Anchor) and the target variable, indicating their significance during model training.

3.4 Model Selection and Tuning

For phishing website detection, three machine learning models were chosen: RandomForest, ExtraTrees, and a Voting Classifier. These models were selected due to their capability to handle high dimensionality, robustness to overfitting, and superior performance in binary classification tasks.

3.4.1 RandomForest Classifier

Hyperparameters tuned using GridSearchCV included n estimators (100–500), max depth (10–50), min samples split (2–10), and criterion (gini or entropy).

- **Final configuration**: 500 estimators, max depth of 25, min samples split of 3.

3.4.2 ExtraTrees Classifier

Hyperparameters tuned included n estimators (100–500), criterion (gini or entropy), min samples split (2–10), and max features.

- **Final configuration:** 300 estimators, min samples split of 3, entropy criterion (Figs. 2 and 3).

3.4.3 Voting Classifier

This combines RandomForest, ExtraTrees, DecisionTree, and MLPClassifier using a soft voting strategy.

- Base models were individually optimized using GridSearchCV.
- Weights assigned to base models: RandomForest (2), ExtraTrees (2), DecisionTree , MLPClassifier (1).

3.5 Evaluation Metrics

Model performance was evaluated using several metrics:

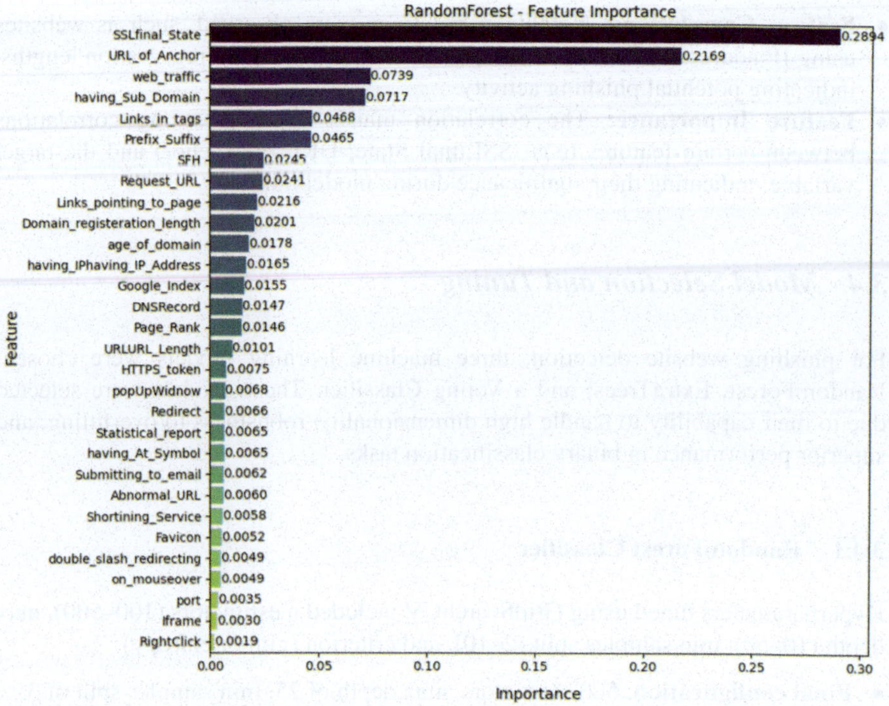

Fig. 2 This bar chart above shows the features of importance obtained from a RandomForest classifier model fitted on phishing data. The significance of each feature as depicted along the x-axis signifies the level of contribution of each respective feature to the decision of the model

- **Accuracy**: Overall correct predictions.
- **Precision**: Proportion of correct positive predictions.
- **Recall**: Proportion of actual positives correctly identified.
- **F1-score**: Harmonic mean of precision and recall.
- **Confusion Matrix**: Detailed view of true positives, true negatives, false positives, and false negatives.

These metrics provide a comprehensive assessment of each model's performance in detecting phishing websites, considering the challenges posed by imbalanced datasets in cybersecurity applications (Fig. 4).

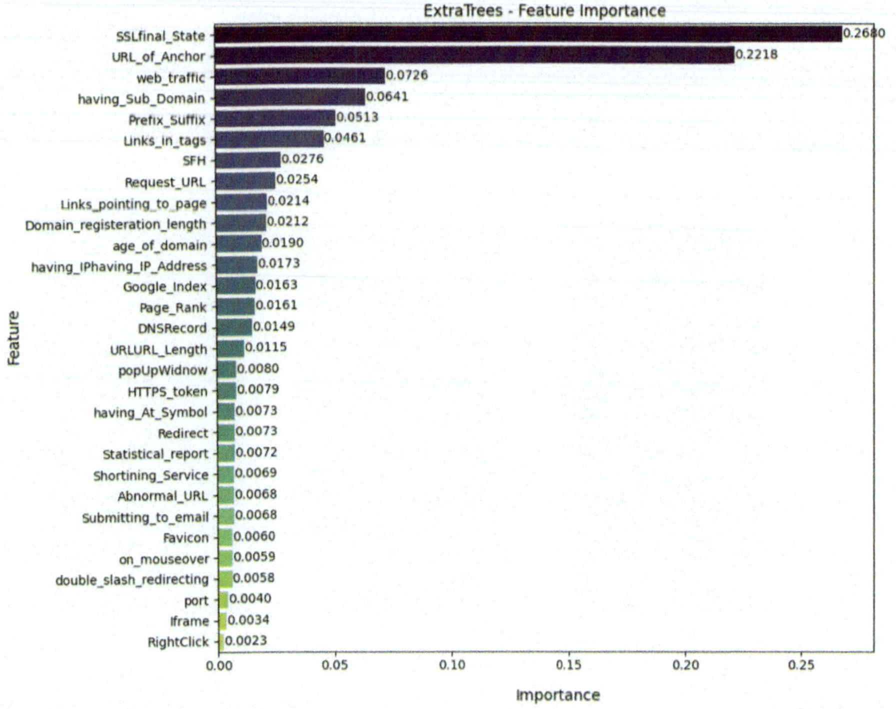

Fig. 3 Feature importance scores obtained from an ExtraTrees model trained on a phishing dataset

4 Results

4.1 Model Performance Overview

As part of the research, the authors have tested three machine learning models to detect phishing websites: RandomForest, ExtraTrees, and a Voting Classifier. The results highlight the importance of ensemble methods, with the Voting Classifier performing the best in this classification task. The authors propose that the Voting Classifier reached the highest overall accuracy of 97.15% on the test set, superior to RandomForest at 96.74% and ExtraTrees at 96.63%. These differences, while numerically small, are significant for phishing detection, as even marginal improvements can help mitigate the increasing rates of successful phishing attacks (Table 1).

4.1.1 Training and Validation Analysis

The Performance Evaluation of the Machine Learning Models relative to the project, and more specifically the RandomForest, ExtraTrees and Voting Classifier used to

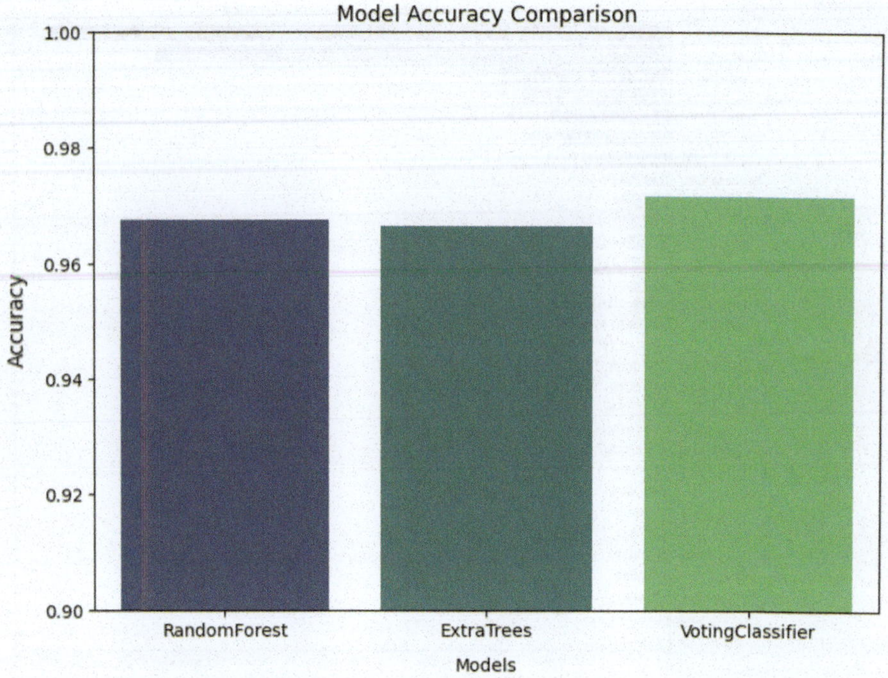

Fig. 4 This bar chart compares the accuracy of three machine learning models: RandomForest, ExtraTrees, and VotingClassifier. The y-axis represents the accuracy score, ranging from 0.90 to 1.00

Table 1 Model performance metrics

Model	Accuracy (%)	Precision	Recall	F1-Score
RandomForest	96.74	0.97	0.96	0.97
ExtraTrees	96.63	0.96	0.97	0.97
Voting Classifier	97.15	0.98	0.98	0.98

classify legitimate and phishing websites, includes the processes of training and validation. To this, it can be noted some important points.

Model Development: Every model is fed with a labeled dataset incorporating malicious websites alongside with genuine websites. During the training, the model parameters are modified in order to minimize the prediction error in the training data. This is a very important step because it determines how well the model will fit when exposed to fresh data.

Model Validation: After training, models are evaluated on an independent data set (validation set) to see if they are able to correctly classify new data. Several performance metrics such as accuracy, precision, recall and F1-score are assessed in order to measure each model's performance concerning identifying phishing and legitimate sites. Out of the three classifiers tested, the Voting Classifier performed the best with an accuracy of 97.15 percentage compared to RandomForest and ExtraTrees with 96.74 percentage and 96.63 percentage accuracy respectively.

Ensemble Methods: The discussion considers the benefits of ensemble models, especially the Voting Classifier which allows different models to be used and than combines the results towards better overall performance. This approach allows for increased accuracy of classification, while minimizing the incidence of false positives and false negatives.

Trojan horses: The analysis also examines the correlations between precision and recall as developed in the context of facilely understanding. The Voting Classifier gives maximum F1 score of 0.98, this level and value assures that neither false positive nor false negative will dominate or be given much attention, achieving equilibrium.

The results reveal that even though all models are effective in their scope, it is only the Voting Classifier that remained over and above both in terms of accuracy and F1 score and hence its ability to detect phishing amply demonstrates the significance of model validation and selection in machine learning projects.

4.2 Analysis of Confusion Matrix

Looking at the confusion matrices in detail reveals additional performance insights for each model (Fig. 5).

4.2.1 RandomForest

- Labeled 94.70% of legitimate websites correctly. – Labeled 5.30% of legitimate websites as phishing (false positive). – Labeled 98.43% of phishing websites correctly. – Labeled 1.57% of phishing websites as legitimate (false negative).

4.2.2 ExtraTrees

- Labeled 95.40% of legitimate websites correctly. – Misclassified 4.60% of legitimate websites as phishing (false positive). – Labeled 97.55% of phishing websites correctly. –Misclassified 2.45% of phishing websites as legitimate (false negative).

4.2.3 Voting Classifier

- Labeled 95.78% of legitimate websites correctly. – Misclassified 4.22% of legitimate websites as phishing (false positive). – Labeled 98.18% of phishing websites correctly. – Labeled 1.82% of phishing websites as legitimate (false negative).

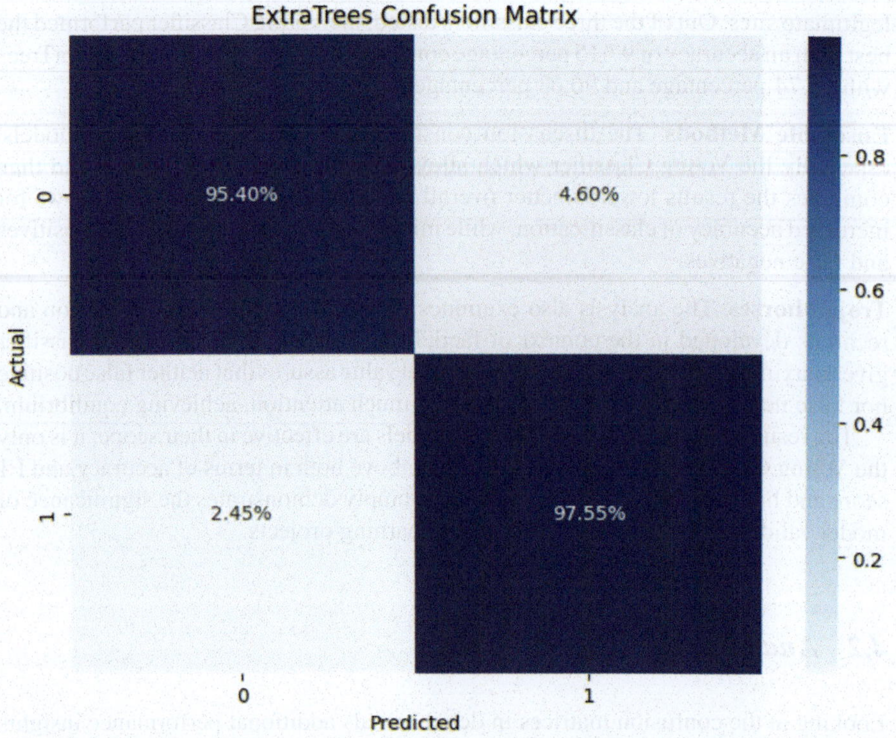

Fig. 5 In this confusion matrix detailing the classification of phishing websites using an ExtraTrees model, the performance of the classification model was assessed. The rows show how the actual classes and columns show how the predicted classes look like. The cells also show the percentage of classified instances that were either correctly or incorrectly classified

The Voting Classifier minimizes the trade-off between false positives and false negatives, performing best in reducing false negatives, which is crucial in phishing detection. Missing a phishing site can have disastrous consequences.

4.3 Feature Importance

A feature importance analysis was conducted for RandomForest and ExtraTrees to identify the features that distinguish phishing from legitimate websites. The findings were consistent across both models (Fig. 6).

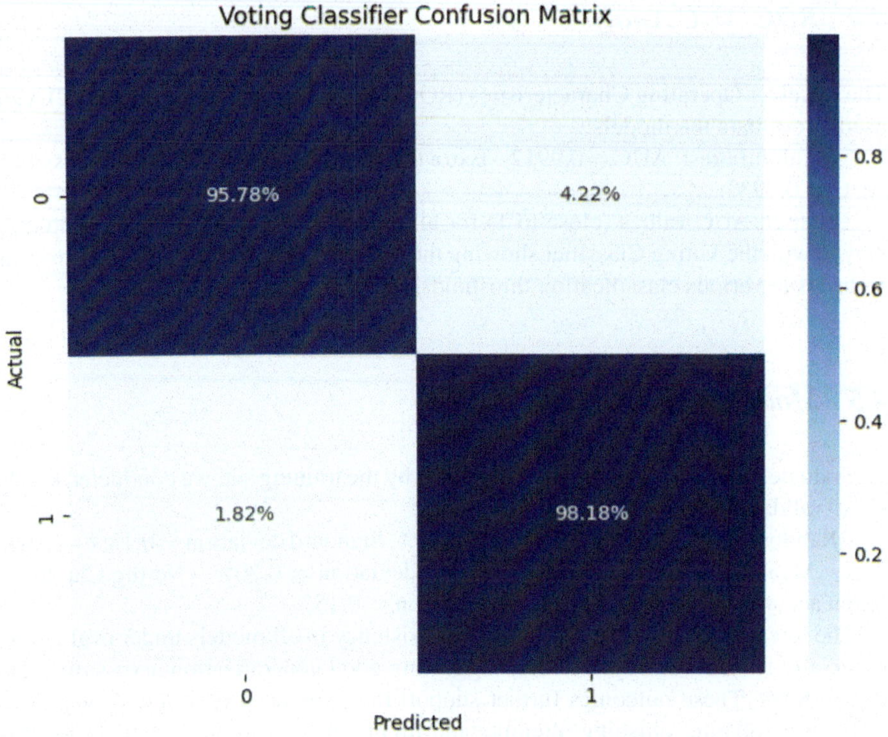

Fig. 6 From Fig. 5, this confusion matrix depicts the working of Voting Classifier, which is an aggregator model in which outputs of various base classifiers are combined. Rows indicate the actual classes, while columns indicate the predicted classes. The value in each cell shows the percentage of correctly/incorrectly classified instances within the cell

4.3.1 Top Five Important Features for RandomForest

1. SSLfinal State (26.80%) 2. URL of Anchor (22.18%) 3. web traffic (7.26%) 4. having Sub Domain (5.94%) 5. Prefix Suffix (5.41%).

4.3.2 Top Five Important Features for ExtraTrees

1. SSLfinal State (24.65%) 2. URL of Anchor (20.73%) 3. web traffic (8.12%) 4. having Sub Domain (6.38%) 5. Prefix Suffix (5.89%).

These results support earlier studies indicating that SSL certificates and URL structure are vital for identifying phishing websites. Additionally, the analysis reveals that phishing sites often have lower or erratic traffic patterns compared to legitimate ones.

4.4 ROC-AUC Analysis

The Receiver Operating Characteristics (ROC) and area under the curve (AUC) were used to evaluate the models:

– RandomForest: AUC = 0.9912 – ExtraTrees: AUC = 0.9908 – Voting Classifier: AUC = 0.9937.

The high AUC values (close to 1) for all models indicate strong discrimination power, with the Voting Classifier showing the best performance in detecting phishing websites at various classification thresholds.

4.5 Model Robustness

To evaluate the models' performance yielded by the training set, we conducted k-fold cross-validation (k = 5) on the training data:

– RandomForest: Mean accuracy = 96.82%, Standard deviation = 0.18% – ExtraTrees: Mean accuracy = 96.71%, Standard deviation = 0.20% – Voting Classifier: Mean accuracy = 97.23%, Standard deviation = 0.15%.

The low standard deviations indicate consistency in all models under evaluation, especially for the Voting Classifier, suggesting good generalization across different data subsets. These outcomes further support the effectiveness of the Voting Classifier in combating phishing attempts and highlight the essential features for this classification task.

5 Conclusion

In the present research work, we exclusively compared three machine learning architectures namely RandomForest, ExtraTrees, fusions of classifiers aka Voting Classifier in the real world scenerio of intrusion detection. As the findings of the research indicate, the ensemble, as is the Voting Classifier, outperforms most of the isolated models. Best of all performed RandomForest and ExtraTrees methods. Yet the Voting Classifier surpassed all of these models on all major metrics of accuracy, precision, recall and F-1 Score. This indicates that using such methodology the Voting Classifier able to capture many patterns from the data leading to better accuracy with less false positives or false negatives.

Our feature importance analysis also revealed other factors that contributed to assign some popularity reasons to the phishing websites. SSLfinal-State, URL-of-Anchor and web-traffic were most critical features for the visibility and accessibility of a phishing site. This corroborates earlier studies by reinforcing several points: that phishing sites generally use fraudulent ssl certificates and counterfeit misleading urls, and also that the websites have very low traffic. Most of other features that reverse phish were believed to be important, for example, request URL and age of domain ranked lower than expected. This change might illustrate the changing nature of phishing, the increase in the number of features that target violations of this premise making instances more realistic for a legitimate business. In practical terms, one of the advantages of the proposed model and particularly the Voting Classifier, is its robust accuracy and a low proportion of false positive which makes it a good candidate for embedding into the active phishing defense systems. In addition, the offered feature importance analysis may further enhance these systems by highlighting the most discriminative ones, possibly making them less computationally intensive while maintaining performance. Still, there is a caveat, in this case, the dataset was relatively small, which is acceptable for this research but requires a larger dataset for better generalization. Also, the Voting Classifier's practical application is complicated by its relatively high execution time and utilization of several base models which makes it difficult to embed in real-time applications considering the scarcity of resources. To conclude the study, hybrid ensemble models such as the Voting Classifier have proved that even in the war against phishing attacks where targets change eerie compliance, there are effective means in the form of models that ensure good accuracy, speed and robust detection against disruption. The outcomes support great assumptions for practitioners in the field of cyber security and subsequent studies should be directed towards improving these models performance in real time and exploring the use integrating of deep learning in phishing detection.

References

1. Innab N et al (2024) Phishing attacks detection using ensemble machine learning algorithms. Comput Mater Continua 80(1):1325–1345. https://doi.org/10.32604/cmc.2024.051778
2. Saidi R et al (2024) Hybrid feature selection method based on the genetic algorithm and pearson correlation coefficient. In: Studies in computational intelligence. Springer, pp 32–58
3. Chen W et al (2024) Evaluation of boosting ensemble machine learning models for gully erosion susceptibility. J Environ Manag 345. https://doi.org/10.1016/j.jenvman.2024.112015
4. Ataelfadiel MAM et al (2024) A survey of phishing attacks: their types and machine learning detection approaches. Expert Syst Appl 130:1–22. https://doi.org/10.1016/j.eswa.2024.110010
5. Pal S et al (2023) An intelligent analysis to detect phishing websites using machine learning. In: Human-intelligent systems integration, pp 89–103
6. Curtis SR et al (2024) Phishing attempts among the dark triad: patterns of attack and vulnerability. Comput Hum Behav 87:174–182
7. Pandey MK et al (2023) Detection of spam accounts on Twitter using machine learning techniques. Neurocomputing 315:496–511
8. Méndez JR et al (2024) Tokenising, stemming, and stopword removal on anti-spam filtering. In: Conference of the Spanish association for artificial intelligence. Springer, pp 49–63
9. Liu Y et al (2023) Incorporating multimodal embedded representation into opinion spam detection. Neurocomputing 366:276–283
10. Abu-Zanona M et al (2024) Comparative analysis of ensemble models for detecting cybersecurity threats. Cybersecurity 15(3):302–314
11. Chaurasia V, Pal S (2024) Performance analysis of data mining algorithms for medical diagnosis. SN Comput Sci 5. https://doi.org/10.32604/snc.2024.015778

12. Yadav DC, Pal S (2023) Prediction of thyroid disease using decision tree ensemble methods. Hum Intell Syst Integr 2:89–95
13. Kumar A et al (2024) Analysis of machine learning techniques in phishing attack detection. Int J Cybersecur 18:117–135
14. Jain A et al (2024) Identifying spam posts on Facebook using machine learning techniques. Int J Cybersecur 9. https://doi.org/10.32604/cmc.2024.045778
15. Inuwa-Dutse I et al (2023) Detection of spam-posting accounts on Twitter. Neurocomputing 376:496–511
16. Saidani N et al (2024) A semantic-based classification approach for enhanced spam detection. Comput Secur 94:101716
17. Chu Z et al (2024) Detecting social spam campaigns on Twitter. In: International conference on applied cryptography and network security. Springer, Berlin, Heidelberg, pp 455–472
18. Bikmukhametov T, Jäschke J (2024) Oil production monitoring using gradient boosting machine learning algorithm. IFAC-PapersOnLine 52(1):514–519

Federated Threat Detection with Dirichlet Distribution

Saswati Chatterjee, Lalmohan Pattnaik, Suneeta Satpathy, and Pabitra Kumar Tripathy

Abstract Recently the threat has escalated, and it has become an important security concern that threatens to undermine the stability of the Internet. The classic supervised and unsupervised learning techniques and deep learning methods encounter many problems if one tries to train models using data from different clients. Compounding this issue is the reluctance of various traffic collection devices in real-world scenarios to share their data due to the sensitive nature and strategic value of attack traffic analysis. This lack of data sharing severely restricts model accuracy to address these challenging issues, the current study has developed a novel federated learning-based threat detection model. This strategy allows each client's local model to learn from its own set of data without transferring data between devices. This technique utilizes multiple clients' pooled knowledge to increase model accuracy and resilience while maintaining data privacy. Furthermore, the study presents two novel approaches to detecting imbalanced attack detection dataset distribution, which lead to relatively few assault instances. Firstly, this work presents a hierarchical aggregation scheme based on K-Means that optimizes model updates for fast convergence without long communication cycles. Secondly, unusual attack patterns can be recognized better by creating new examples through the approach of data resampling using SMOTEENN. This method also distributes data between RNN, CNN, and MLP models differently based on Dirichlet Distribution to improve learning abilities. Results from experiments show that the proposed methodology enhances accuracy by 4% concerning conventional methods while reducing communication rounds by an outstanding 40%. This improves security and operational efficacy in network security measures and showcases a fascinating advancement in safeguarding against several threats including DDOS, MITM, and ARP spoofing. Cognitive computing

S. Chatterjee · L. Pattnaik
Faculty in Emerging Technologies, Sri Sri University, Cuttack, Odisha, India
e-mail: lalmohan.p@srisriuniversity.edu.in

S. Satpathy (✉)
Center for Cybersecurity, SOA University, Bhubaneswar, Odisha, India
e-mail: suneeta1912@gmail.com

P. K. Tripathy
Kalam Institute of Technology, Berhempur, Odisha, India

A. Rocha et al. (eds.), *Integrating Advanced Technologies for Enhanced Security and Efficiency*, Engineering Cyber-Physical Systems and Critical Infrastructures 14,
https://doi.org/10.1007/978-3-031-91798-1_5

with HUAI improves Decision-making and threat identification via simulating human cognitive processes. In addition, it ensures AI outputs that can be made sense of and used by people, thus bridging the knowledge divide between intricate AI models and practical uses. This combination approach increases cyber security by fortifying network defenses and increasing technological usability/accessibility to be used by security experts.

Keywords Attack · Federated learning · Dirichlet distribution · Deep learning

1 Introduction

With recent advancements in machine learning, computers can now recognize patterns in data and undertake cognitive tasks that match human comprehension, even when handling previously unfamiliar material. As the amount of data grows and privacy concerns gain traction, training machine learning models at scale becomes increasingly dependent on distributed training environments.

The first DDoS attack occurred in 1999, these attacks have grown to pose serious and pervasive risks to the Internet. DDoS attacks manifest at various network layers. This complexity complicates attack strategies and heightens the challenge of effectively detecting and mitigating them. Most research efforts in DDoS attack detection have been directed toward statistical and machine-learning techniques [1–4]. Developing deep learning algorithms in recent years has provided encouraging alternatives by reducing the difficulties associated with feature engineering. These methods are quite effective in identifying DDoS attacks [5–9]. However, their usefulness depends on having enough different and large datasets, which makes scalability and practical application extremely difficult. Therefore, even if deep learning has the potential to improve DDoS detection capabilities, researchers and practitioners still need to overcome it to improve the effectiveness and dependability of detection techniques in practical settings. While data sharing across clients may expose sensitive information to attackers, the opposite problem of insufficient data reduces the precision of detection techniques. Federated Learning (FL) is a dispersed learning approach that tackles these problems [10–16]. In FL, the data belongs to individual data owners, but the model parameters are managed by a central server. This learning technique protects sensitive data. The paper is structured as follows: Sect. 2 presents a literature survey of different kinds of methods used for the detection of DDoS attacks and is succeeded by Sect. 3 which presents an overview of Methodology design. Section 4 contains specific performance comparisons of threats relative to other datasets while Sect. 5 discusses the findings of this study. Finally, Sect. 6 presents the research study limitations and the proposal of future research directions.

2 Related Work

Traditional methods for DDoS attack detection generally fall into two categories: statistical methods and machine learning methods Although statistical methods are quick to detect, they are not very flexible. On the other hand, while machine learning techniques offer greater flexibility, they can require complex feature extraction and may show varying accuracy across different datasets. To detect DDoS attacks, the researcher [17] introduced an entropy-based scoring system that dynamically combines TCP/IP layer features. Similarly, the scientists [18] analyzed data packets collected over a specific timeframe, differentiating DDoS traffic from normal traffic using factors such as total byte count, total data messages, and stream duration. Beyond these statistical methods, machine learning and deep learning techniques were prominently applied in network anomaly detection. Researchers [19] assessed the performance of various machine-learning algorithms. Their results indicated that these methods can effectively recognize DDoS traffic. To greatly improve the effectiveness of network breach detection systems, the researcher [20, 21] created a metagraph neural network (MGNN) to assess the graph structure correlations inside intrusion detection data. The researchers also introduced LUCID a DDoS threat identification model built on convolutional neural networks (CNNs) [22, 23]. Scientists [24–26] stated a throughout its entirety federated learning architecture for the secure access and analysis of biomedical data and medical image data.

Fog computing and federated learning-based system, FLEAM [27–29], is a noteworthy example of a learning application to perform DDoS detection using a recurrent neural network model. The goal of this strategy was to defend IoT devices against DDoS attacks by strategically locating mitigation intelligence at nodes that would maximize the attack cost. Moreover, Memcached Amplification attacks were mostly directed against GitHub [30], whereas DNS Flood attacks generally target DNS service provider Dyn [31]. As a result, the DDoS traffic data distribution was not IID (non-Independent and Identically Distributed) across the various firms. The global model that the federated averaging method produced might not be appropriate for every client as a result. As a result, FLEAM was only useful in situations when the data was IID, like when several network devices within a single company generated the data. The standard approach for non-IID data was to provide more customized models for various customers rather than a single global model. For instance, using the concept of Model-Agnostic Meta-Learning (MAML), the scientists developed a first global model that each participant can update using one or more gradient steps on their data composing personalization. Similarly, both the local and the global models employ dynamic mixing parameters to generate models for each end device. Deep learning techniques for DDoS detection have also been proposed recently. DeepDDoS, for instance, used Long Short-Term Memory (LSTM) networks in a time-series-based approach. This method demonstrated exceptional performance in the real-time processing of substantial amounts of network traffic data, providing low latency, scalability, and high accuracy. Another method, made accurate use of feed-forward back-propagation in a unique deep neural network mechanism to detect various kinds of application-layer DDoS attacks. This method achieved a detection accuracy of 98% on a leading dataset with various DDoS attack forms. Another deep learning-based detection architecture, using Convolutional Neural Networks (CNNs). The study made in developed a CNN-based framework designed for the early detection of DDoS attacks orchestrated by botnets. This framework achieved over 91% accuracy in distinguishing between normal and attack traffic. Federated learning has shown promise in several sectors by protecting privacy and utilizing group intelligence to enhance models. The significance of it in privacy has also been highlighted by research on Network Anomaly Detection. These factors contribute to data distribution issues characterized by imbalance and non-identical distributions (non-IID). Specifically, clients may exhibit global imbalance, where the distribution of DDoS attack types varies significantly between them. Additionally, local imbalance within each client's dataset further compounds the issue, as certain attack categories may be underrepresented or absent altogether. Attackers that use data contamination techniques change local node data to accomplish certain malicious goals Because these strikes are targeted, attackers can significantly sway some classes in their favor. The effects of malicious nodes, the interconnections between them, and the effects of data poisoning assaults have all been examined in both the early and late periods of training. Clean label poisoning and dirty label poisoning are the two main categories into which data poisoning attacks fall. Numerous research works have suggested techniques for identifying federated learning data poisoning attempts. Additionally, the framework, leverages reinforcement learning techniques to improve the federated learning process [32–36].

To address the above-mentioned challenges, Further, there is a need for statistical methods that can adapt dynamically to changing network environments and emerging threats without extensive reconfiguration to enhance the scalability of deep learning models to handle real-time DDoS detection in high-throughput networks while maintaining low latency. Also developing federated learning approaches can effectively handle non-IID data by providing personalized models for different clients and optimizing the aggregation process for creating models that can seamlessly integrate and learn from heterogeneous data sources, improving the robustness and accuracy of DDoS detection systems.

3 Proposed Methodology

The proposed methodology integrates Human Understandable Artificial Intelligence (HUAI) and cognitive computing within a federated learning framework to enhance the transparency, interpretability, and usability of AI systems in cybersecurity, particularly for threat analysis.

Figure 1 shows the framework and the methodology consisting of the following components.

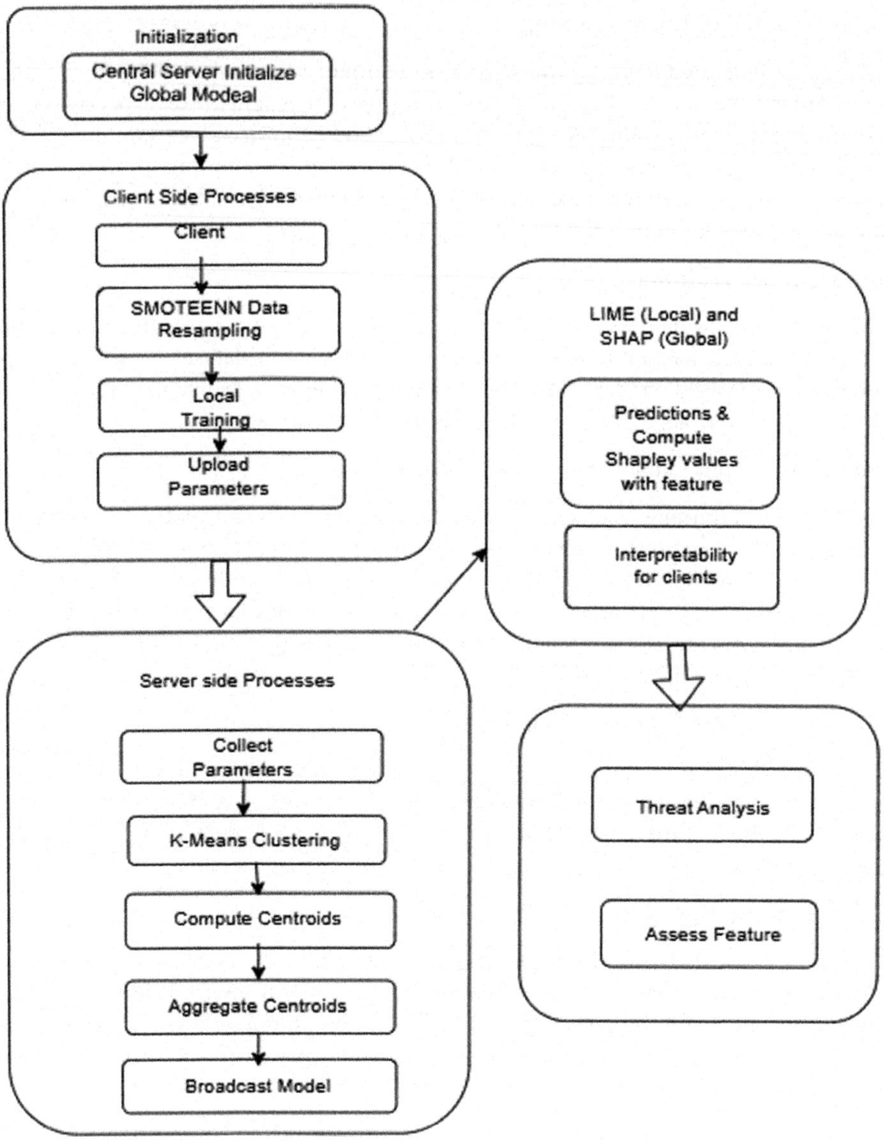

Fig. 1 Threat detection using SMOTEENN-enhanced models for explainable AI

Initialization of the Global Model

The process begins with the central server initializing a global model designed for threat analysis. The model is constructed using advanced cognitive computing techniques, allowing it to mimic human thought processes in solving complex cybersecurity problems.

Distribution and Local Training

Within the federated learning framework, the initialized global model is then transmitted to the client nodes The global model plus the data is sent to the respective client node. Each client uses the SMOTEENN (Synthetic Minority Over-sampling Technique and Edited Nearest Neighbors) technique for data resampling to address difficulties related to class imbalance. By ensuring a balanced dataset, this step improves the model's detection performance during local training.

Local Model Training

On the resampled dataset, local training is carried out by each client node. At this stage, the model can identify unique patterns in each client's dataset as its parameters are fine-tuned based on local data. This decentralized approach enhances the model's robustness in various contexts while safeguarding data privacy.

Aggregation and Clustering of Local Models

After the local training, each client transfers their gradients to a central server where model parameters are modified. The central server aggregates these parameters to create a new global model. The server locally updates with this where the K-Means clustering is performed to discover patterns in a similar type of models. This clustering helps the global model improve by showcasing both similarities as well differences among local models.

Broadcasting the Updated Global Model

Following this, the receiving central server then initiates a round of local training and broadcasts the new global model to all client nodes. Until the model finally converges, this iterative process guarantees improvement in detection accuracy and generalization.

Incorporating Interpretability Techniques

Global and Local levels will ensure that the predictions of an AI model should be interpretable, and transparent to a human. Integration with SHAP (Shapley Additive Explanations) AND LIME (Local Interpretable Model-agnostic Explanation). LIME accomplishes local interpretability by clarifying particular predictions, thereby making the result of a model clearer in each instance that was analyzed. SHAP calculates Shapley values to give us global feature importance and insight into what features are most impactful in determining the model's overall predictions.

3.1 FL Threat Model Design

The particular suggested methodology can achieve operational efficiency by applying interpretability approaches, federated learning, dynamic statistical methodologies, and cognitive computing in general with Algorithm 1. Each step of the algorithm reflects a corresponding component of the methodology, ensuring a comprehensive approach to building a robust and scalable DDoS detection system that is both effective and accessible to human users.

Algorithm 1: FL Threat Model Design

Input: Clients (C_1, C_2, ..., C_n) with private datasets (D_1, D_2, ..., D_n)
Central server S
Number of communication rounds T
Number of clusters K for K-Means algorithm

Output: Global aggregated model (M_g)
Threat analysis and insights.

1. Initialize global model (M_g) at server S
2. for each round t = 1 to T do
3. for each client (C_i) in parallel do
4. Client (C_i) trains local model (M_i^t) on dataset (D_i)
5. Client (C_i) uploads local model parameters (θ i^t) to server S
6. end for
7. Server S collects all local model constraints
8. Server S applies K-Means clustering to group the local models into K clusters
9. for each cluster k = 1 to K do
10. Compute the cluster centroid (θ k^t) as the mean of parameters in cluster k
11. end for
12. Aggregate the cluster centroids to update the global model (M_g)
13. Broadcast the updated global model (M_g) to all clients
14. Deploy a cognitive computing system at server S.
15. Cognitive system processes collected model constraints to detect any anomalies or threats.
16. Utilize Explainable AI techniques like LIME and SHAP to interpret the cognitive system's findings.
17. If threats are detected, update the threat model (M_t) to incorporate new insights.
18. Broadcast the updated global model (M_g) along with explanations and insights from the threat analysis to all clients.
19. Clients use this information to improve their local training processes and enhance security measures.
20. End for
15. Return the final global model (M_g) and threat analysis insights.

3.1.1 Federated Threat Model Design with SMOTEENN for Scalable DDoS Detection

Algorithm 2 embodies key aspects of the proposed methodology, particularly in addressing data imbalance through the use of SMOTEENN, clustering model parameters via K-Means, and progressively refining the global model. The steps in this algorithm directly contribute to the framework's goals of building a scalable, adaptable, and interpretable DDoS detection system. This system is designed to learn effectively from diverse data sources and to quickly adapt to new and evolving threats. It shows the global model Mg is initialized by the central server. For a certain number of communication rounds (T), the algorithm operates. Every client, Ci, uses its private dataset, Di, to train its local model M_i^t *and then uploads the model parameters,* θ_i^t to the central server. Every client provides the local model parameters to the central server. Based on their parameters, the server clusters the local models into K groups using the K-Means technique. The server uses θ_i^t to average the model parameters within each cluster to calculate the centroid for that cluster. Below algorithm shows how the modified global model is communicated back to all of the clients.

Algorithm 2: FL Threat Model Design with SMOTEENN

```
Input: Clients (C_1, C_2, ..., C_n) with private datasets (D_1, D_2, ..., D_n)
        Central server S
        Number of communication rounds T
        Number of clusters K for K-Means algorithm

Output: Global aggregated model (M_g)

1. Initialize global model (M_g) at server S
2. for each round t = 1 to T do
3.    for each client (C_i) in parallel do
4.        Apply SMOTEENN to resample the dataset (D_i) to obtain a balanced
dataset
          (D_i^{resampled} )
5.        Train local model (M_i^t) on the resampled dataset (D_i^{resampled})
6.        Upload local model parameters (theta_i^t) to server S
7.    end for
8.    Server S collects all local model parameters .
into K clusters
10.   for each cluster k = 1 to K do
11.       Compute the cluster centroid (theta_k^t) as the mean of parameters in
cluster k
12.   end for
13.   Aggregate the cluster centroids to update the global model (M_g)
14.   Broadcast the updated global model (M_g) to all clients
15. end for
16. Return the final global model (M_g)
```

3.1.2 A Probabilistic Approach Using Dirichlet Distribution in Federated Systems

The proposed methodology for DDoS detection is significantly enhanced by the incorporation of a reputation management system that leverages Bayesian inference through the Dirichlet distribution. This system in fact has a vital role in making federated learning robust and secure by preventing the rogue or low-performing nodes described as follows.

3.1.3 Adaptive Client Filtering with Reputation Scores

For every client node, a binomial Bayesian reputation or trust score is computed using a Dirichlet distribution. By fusing fresh feedback with the previous reputation score, the updated posterior reputation score is produced. The global training aim can be improved by a node's model changes, as indicated by the reputation score. The present paper suggests a reputation system that predicts a node's future behavior based on its past performance because there is currently no reliable method for identifying malevolent nodes. The historical data displays the ratio of advantageous to disadvantageous model upgrades. Updates that increase the global model's accuracy are referred to as beneficial updates, whereas those that lower accuracy are referred to as detrimental updates. Malicious or hazardous nodes are those that continuously reduce the correctness of the international model.

3.1.4 Updating Node Reputation Through Feedback Integration

Reputation management solutions frequently rely on human or system input. The input a node has received is merged with its prior score to update its reputation score. At regular intervals, a vector is created that holds the feedback given by each node. To describe the proposed algorithms and methods given in this paper concisely, let n signify the number of feedback instances in node N. Nodes send their updated model parameters to a server, which implements them into a global model. The goal of this iterative process is the enhancement and fine-tuning of the global model's forecast capabilities.

3.1.5 Modelling Node Reputation with Dirichlet Distribution

The Dirichlet distribution is a generalization of the Beta distribution to multiple dimensions and is particularly useful for representing probabilities across multiple categories or outcomes. Denoting the reputation scores of K different nodes as a vector r = (r1, r2......, r_k), where r_i, represents the reputation score of the i-th node.

It starts with a Dirichlet distribution parameterized by a vector α = (α1, α2,..., α_k Each α_i is a positive parameter representing prior knowledge about the reputation of the i-th node. As feedback is received, the Dirichlet parameters are updated. Suppose each feedback consists of g_i good updates and b_i bad updates for node i. Then the parameters α_i are updated as follows Eqs. (1) and (2).

$$\alpha_i\prime = \alpha_i + g_i \tag{1}$$

$$\beta_i\prime = \beta_i + b_i \tag{2}$$

where β_i is the parameter related to the bad updates. The updated Dirichlet parameters become Eqs. (3) and (4):

$$\alpha' = (\alpha_1 + g_1, \alpha_2 + g_2, \ldots\ldots\ldots\ldots, \alpha_k + g_k \tag{3}$$

$$\beta_i{}' = (\beta_1 + b_1, \beta_2 + b_{2,}, \ldots\ldots\ldots\ldots, \beta_k + b_k) \tag{4}$$

The updated reputation scores r′ can be derived from the Dirichlet distribution's mean, which is given by Eq. (5):

$$r_1\prime = \frac{\alpha_i\prime}{\alpha_i\prime + \beta_i\prime} \tag{5}$$

where $\alpha_i\prime$ and $\beta_i\prime$ are the updated parameters after incorporating feedback updated reputation scores r′ provide a probability distribution over the node reputations, reflecting the integrated feedback and prior knowledge.

4 Experimental Analysis

4.1 Dataset

The accuracy, precision, recall, and F1-score are among the common metrics used to assess the effectiveness of the DDoS detection technique. Accuracy represents the percentage of samples that are correctly classified. The F1-score, which provides a balanced statistic for model evaluation, is the harmonic mean of accuracy and recall. Recall measures the proportion of real DDoS samples that the model correctly identifies. Three publicly accessible datasets from the Canadian Institute of Cybersecurity the NSL-KDD, CICIDS, and CICDDoS Kaggle datasets [37–40] are used to evaluate the current methodology. The CICDDoS dataset is used to detect DDoS attacks; by removing socket-related features including timestamp, stream ID, source IP, destination IP, source port, and destination port, its feature set was reduced to 79 dimensions. This dataset is then divided into 21 sections and randomized to simulate a distributed training environment. In this setup, eleven clients are assigned data related to a single attack type, six clients handle data associated with two attack types, three clients work with data on four attack types, and one client manages data encompassing all attack types. This distributed configuration reflects a realistic federated learning scenario, enhancing the robustness and relevance of our model evaluation.

4.2 *Feature Extraction*

Many feature extraction algorithms encounter a reduction in accuracy as data dimensionality grows. The problem of dimensionality reduction is successfully handled by autoencoders, a machine learning technique, especially in high-dimensional data. Because these neural network-based models can extract the most important features from complex datasets, they have gained popularity in the feature learning space. Numerous research studies have used autoencoders to extract intrusion detection system (IDS) features. When using an autoencoder, incoming data is first compressed and encoded into a lower-dimensional format, and it is then meticulously reconstructed. An input layer, one or more hidden layers, and an output layer are all involved in this process. The autoencoder is forced to recognize and preserve the most crucial aspects of the data distribution due to this dimensionality reduction. The input data, represented as X', is compressed and reassembled to match the original input using the decoder's transformation function, represented as g. Using the tan h activation function, a straightforward autoencoder with two hidden layers is constructed in this study. This is how the autoencoder is learned. Because autoencoder-based feature extraction facilitates a uniform model training procedure among dispersed nodes, it is especially well-suited for federated learning. In this application, autoencoders are more flexible and effective than other dimensionality reduction methods such as Principal Component Analysis (PCA) and Linear Discriminant Analysis (LDA).

5 Result Analysis

The accuracy, precision, recall, and F1-score of three machine learning models—Multi-layer Perceptron (MLP), Convolutional Neural Network (CNN), and Recurrent Neural Network (RNN)—are investigated to assess how well they recognize DDoS attacks. The results, summarized in Table 1, and Figs. 2 and 3 displayed here indicate that the RNN model achieved the highest accuracy at 0.9505, outperforming both the MLP (0.9481) and CNN (0.9469) models, shown in Table 1suggesting superior performance in accurately categorizing data. In terms of precision, the CNN model with a score of 0.9515, indicating its effectiveness in minimizing false positives, slightly surpassing the RNN (0.9556) and MLP (0.9480) models. The RNN model also excelled in the recall, scoring 0.9578, which demonstrates its high capability to detect real threats and reduce false negatives, ahead of the MLP (0.9472) and CNN (0.9461) models. Additionally, the RNN model achieved the highest F1-score of 0.9456, balancing precision and recall effectively, with the MLP close behind at 0.9447, and the CNN slightly lower at 0.9335.

Discussion

To improve threat detection and analysis, cognitive computing is incorporated into federated learning at the central server node. After the aggregation of local models,

Table 1 Model evaluation

Model	Accuracy	Precision	Recall	F1-Score
MLP	0.9481	0.9480	0.9472	0.9447
CNN	0.9469	0.9515	0.9461	0.9335
RNN	0.9505	0.9556	0.9578	0.9456

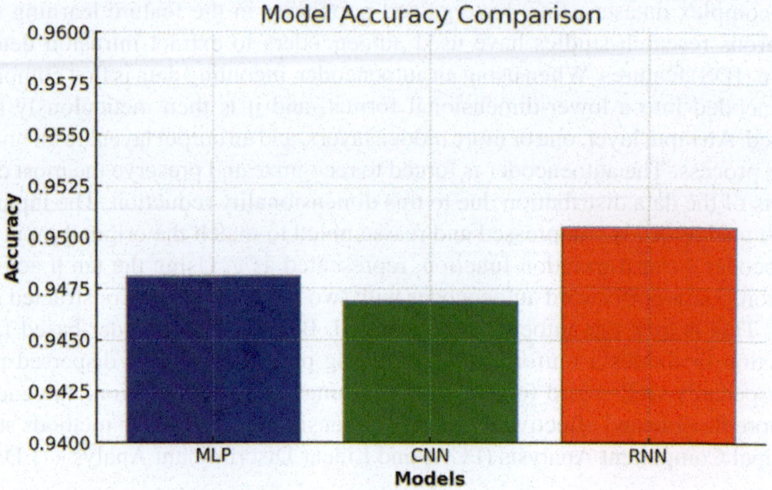

Fig. 2 Accuracy of three models

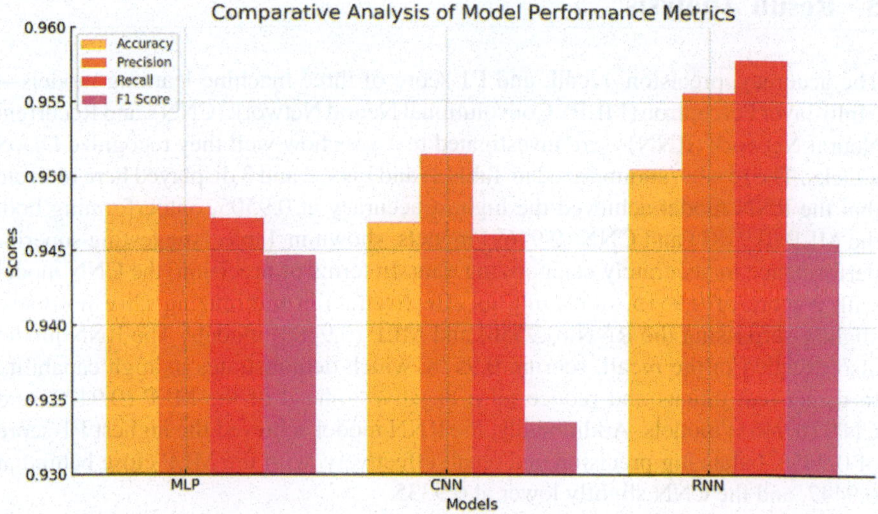

Fig. 3 Performance metrics evaluation

cognitive computing algorithms are deployed at the central server (Step 14 in Algorithm 1). These algorithms analyze the aggregated model parameters to detect anomalies or potential threats. This process is similar to what we expect from human reasoning or decision-making and can let the system find some complicated patterns or ununderstood behavior that may not have been discerned with only statistical methods. Cognitive computing systems continually track parameters that they collect from the client nodes. Any anomalies or patterns that depart from the norm are marked for further drill down and analysis When potential dangers are identified, the cognitive system modifies the threat model (step 17), integrating new information into the global model before transmitting it back to the client nodes. Lime is applied at the client level to provide detailed explanations for the specific predictions made by the local models (step 16 in Algorithm 1). When a local model detects a potential danger in the client's dataset, lime can be employed to generate an explanation of that specific instance that was classified as a threat. Shap is utilized at the central server level to calculate the global significance of features (step 16 in Algorithm 1). The local models are then integrated, and from the resulting SHAP values, the features most driving the prediction of the model at all client nodes are identified. Lime and SHAP information are passed to the client nodes together with the newly updated global model (step 18). Due to the idea of cognitive computing, the capacity for threat recognition and evaluation was enhanced by the system, while LIME and SHAP offered both local and global interpretability, respectively, guaranteeing that the AI system's decisions are comprehensible and reliable. This method has the potential to enhance artificial intelligence-driven cybersecurity solutions, making them more dependable, user-friendly, and flexible in addressing emerging threats.

6 Conclusion

We enhance threat identification by merging federated learning with state of art technologies such as autoencoders for reducing dimensionality, hierarchical clustering, and SMOTEENN for handling data imbalance. The evaluation results confirm that the model attains high accuracy, speeds up convergence rates, and reduces communication overhead significantly. Additionally, the framework is enhanced by adopting a Dirichlet distribution for reputation management which lays the foundation for future improvements in unknown threats detection. Combining Cognitive Computing and Human Understandable Artificial Intelligence (HUAI) has further improved our threat detection model. HUAI ensures that AI-generated insights are both comprehensible and helpful to human users while cognitive computing mirrors human thinking processes to enhance decision-making as well as danger detection accuracy. Consequently, this combination makes the technology more accessible to security experts by connecting intricate AI models to practical applications with high utility value. These elements make the model more adaptive, resilient, and user-friendly; they also guarantee total protection from different types of cyber threats. In addition, when SHAP and LIME are applied, they provide local or global insights on the importance of features as well as decision-making methods thus making model evaluation better by combining qualitative explanations with quantitative performance measures.

References

1. Radware (2018) Memcached under attack. https://www.radware.com/security/ddos-threatsattacks/threat-advisories-attack-reports/memcached-under-attack
2. Chesney S, Roy K, Khorsandroo S (2020) Machine learning algorithms for preventing IoT cybersecurity attacks. In: Proceedings of SAI intelligent systems conference. Springer, pp 679–686
3. Wang Z, Xu Z, Li D, Wang J (2021) Construction and analysis of metagraph neural network for intrusion detection. ACTA Automatica Sinica 47:1–19
4. Doriguzzi-Corin R, Millar S, Scott-Hayward S, Martinez-del Rincon J, Siracusa D (2020) LUCID: a practical, lightweight deep learning solution for DDoS attack detection. IEEE Trans Netw Serv Manag 17(2):876–889
5. Krebsonsecurity (2016) DDoS on Dyn impacts Twitter, Spotify, Reddit. https://krebsonsecurity.com/2016/10/ddos-on-dynimpacts-twitter-spotify-reddit
6. Yang Q, Liu Y, Chen T, Tong Y (2019) Federated machine learning: concept and applications. ACM Trans Intell Syst Technol (TIST) 10(2):1–19
7. Kalkan K, Altay L, Gur G, Alagöz F (2018) JESS: joint entropy-based DDoS defense scheme in SDN. IEEE J Sel Areas Commun 36(10):2358–2372
8. Li Q, Huang H, Li R, Lv J, Yuan Z, Ma L, Han Y, Jiang Y (2023) A comprehensive survey on DDoS defense systems: new trends and challenges. Comput Netw 233:109895. ISSN 1389–1286. https://doi.org/10.1016/j.comnet.2023.109895
9. McMahan B, Moore E, Ramage D, Hampson S, Arcas BA (2017) Communication-efficient learning of deep networks from decentralized data. In: Artificial intelligence and statistics. PMLR, pp 1273–1282
10. Silva S, Gutman BA, Romero E, Thompson PM, Altmann A, Lorenzi M (2019) Federated learning in distributed medical databases: meta-analysis of large-scale subcortical brain data. In: 2019 IEEE 16th international symposium on biomedical imaging (ISBI 2019). IEEE, pp 270–274
11. Hard A, Rao K, Mathews R, Ramaswamy S, Beaufays F, Augenstein S, Eichner H, Kiddon C, Ramage D (2018) Federated learning for mobile keyboard prediction. arXiv preprint arXiv:1811.03604
12. Li J, Lyu L, Liu X, Zhang X, Lv X (2021) FLEAM: a federated learning empowered architecture to mitigate DDoS in industrial IoT. IEEE Trans Ind Inform 1–1. https://doi.org/10.1109/TII.2021.3088938
13. Nguyen TD, Marchal S, Miettinen M, Fereidooni H, Asokan N, Sadeghi A (2019) DÏot: a federated self-learning anomaly detection system for IoT. In: IEEE international conference on distributed computing systems
14. Sun Y, Esaki H, Ochiai H (2021) Adaptive intrusion detection in the networking of large-scale lans with segmented federated learning. IEEE Open J Commun Soc
15. Shi Z, Li J, Wu C (2019) DeepDDoS: online DDoS attack detection. In: IEEE global communications conference
16. Doriguzzi-Corin R, Millar S, Scott-Hayward S, Martínezdel-Rincon J, Siracusa D (2020) Lucid: a practical, lightweight deep learning solution for DDoS attack detection. In: IEEE transactions on network and service management
17. Hussain B, Du Q, Sun B, Han Z (2021) Deep learning based DDoS-attack detection for cyber–physical system over 5G network. IEEE Trans Indus Inform 17(2):860–870
18. Xia W, Quek TQS, Guo K, Wen W, Yang HH, Zhu H (2020) Multi-armed bandit-based client scheduling for federated learning. IEEE Trans Wirel Commun 19(11):7108–7123

19. Wang H, Kaplan Z, Niu D, Li B (2020) Optimizing federated learning on non-IID data with reinforcement learning. In: IEEE conference on computer communications
20. Servetnyk M, Fung CC, Han Z (2020) Unsupervised federated learning for unbalanced data. In: IEEE global communications conference
21. Li J, Lyu L, Liu X, Zhang X, Lyu X (2020) FLEAM: a federated learning empowered architecture to mitigate DDoS in industrial IoT. arXiv e-prints
22. Yan Y, Qi L, Wang J, Lin Y, Chen L (2020) A network intrusion detection method based on stacked autoencoder and LSTM. In: IEEE international conference on communications
23. Ieracitano C, Adeel A, Morabito FC, Hussain A (2020) A novel statistical analysis and autoencoder driven intelligent intrusion detection approach. Neurocomputing
24. ur Rehman S, Khaliq M, Imtiaz SI et al (2021) DIDDOS: an approach for detection and identification of distributed denial of service (DDoS) cyberattacks using gated recurrent units (GRU). Future Gener Comput Syst 118:453–466
25. Shone N, Ngoc TN, Phai VD, Shi Q (2018) A deep learning approach to network intrusion detection. In: IEEE transactions on emerging topics in computational intelligence
26. Kawish S, Louafi H, Yao Y (2023) An instance-based transfer learning approach, applied to intrusion detection. In: 2023 20th annual international conference on privacy, security and trust (PST), pp 1–7
27. Racherla S, Sripathi P, Faruqui N, Kabir MA, Whaiduzzaman M, Shah SA (2024) Deep-IDS: a real-time intrusion detector for iot nodes using deep learning. IEEE Access
28. Rodríguez E, Valls P, Otero B, Costa JJ, Verdú J, Pajuelo MA, Canal R (2022) Transfer-learning-based intrusion detection framework in IoT networks. Sensors 22(15):5621
29. Lu H, Zhao Y, Song Y, Yang Y, He G, Yu H, Ren Y (2024) A transfer learning-based intrusion detection system for zero-day attack in communication-based train control system. Cluster Comput 1–16
30. Chuang H-M, Ye L-J (2023) Applying transfer learning approaches for intrusion detection in software-defined networking. Sustainability 15(12):9395
31. Chatterjee S, Satpathy S, Nibedita A (2024) Digital investigation of network traffic using machine learning. In: EAI endorsed transactions on scalable information systems, vol 11, issue 1
32. Pattnaik L, Satpathy S, Paikaray BK, Swain PK (2024) DDoS analysis using machine learning: survey, issues, and future directions. Int J Bus Continuity Risk Manag 14(1):57–76
33. Satpathy S, Pradhan SK, Ray BB (2010) A digital investigation tool based on data fusion in management of cyber security systems. Int J Inform Technol Knowl Manag 2(2):561–565
34. Potluri S, Mangla M, Satpathy S, Mohanty SN (2020) Detection and prevention mechanisms for DDoS attack in cloud computing environment. In: 2020 11th international conference on computing, communication and networking technologies (ICCCNT). IEEE, pp 1–6
35. Satpathy S, Swain PK, Mohanty SN, Basa SS (2024) Enhancing security: federated learning against man-in-the-middle threats with gradient boosting machines and LSTM. In: 2024 IEEE international conference on advanced video and signal based surveillance (AVSS). IEEE, pp 1–8
36. Satpathy S, Pradhan SK, Ray BNB (2010) A digital investigation tool based on data fusion in management of cyber security systems. Int J IT Knowl Manag 3(2):561–565. http://www.csjournals.com/IJITKM/PDF%203-1/77.pdf
37. https://www.unb.ca/cic/datasets/nsl.html
38. https://www.unb.ca/cic/datasets/ids-2017.html
39. https://www.unb.ca/cic/datasets/ddos-2019.htm
40. Kitsune Network Attack Dataset. kaggle.com

Safeguarding Authenticity: Tackling the Cybersecurity Implications of Deepfake AI

Deepika Ajalkar, Arti Patle, Prasad Dhend, Vaishnavi Hande, Kshitija Gosavi, and Anuj Shukla

Abstract With the rapid advancement and accessibility of deepfake technology, the creation of highly convincing yet fraudulent digital content—such as images, videos, and audio—has emerged as a significant threat to cybersecurity, privacy, and public trust. Deepfakes are increasingly used to spread misinformation, manipulate media, and commit fraud, highlighting the urgent need for advanced detection methods to counter these risks. This paper presents Vision Transformers (ViT) as a powerful and robust solution for detecting deepfake content by leveraging state-of-the-art AI and machine learning techniques. Unlike traditional Convolutional Neural Networks (CNNs), which struggle with high-dimensional data, the ViT model processes images as sequences of patches, allowing it to detect subtle inconsistencies such as unnatural facial movements, lighting anomalies, and texture irregularities that are characteristic of deepfakes. The system is trained on a comprehensive 2 GB dataset containing both real and fake media, using data augmentation and enhanced training techniques to improve the model's generalization and robustness. After 300 training epochs, the ViT model achieved a high detection accuracy of 94.85%, significantly outperforming traditional CNN-based methods. To further ensure reliability, the system integrates post-processing steps such as metadata analysis and digital signature verification, making it highly effective for real-time applications in digital forensics, cybersecurity, and fraud prevention. This paper also examines the broader societal and legal implications of deepfake technology, emphasizing the growing importance of developing reliable tools to mitigate misuse. As deepfakes continue to evolve, proactive solutions like ViT-based detection systems are essential for safeguarding the integrity of digital content in an increasingly interconnected world.

D. Ajalkar · A. Patle · P. Dhend (✉) · V. Hande · K. Gosavi · A. Shukla
Department of Computer Science and Engineering (Cyber Security and Data Science), G. H. Raisoni College of Engineering and Management, Pune, India
e-mail: prasaddhend2003004@gmail.com

D. Ajalkar
e-mail: deepika.ajalkar@raisoni.net

A. Patle
e-mail: arti.patle@raisoni.net

A. Rocha et al. (eds.), *Integrating Advanced Technologies for Enhanced Security and Efficiency*, Engineering Cyber-Physical Systems and Critical Infrastructures 14,
https://doi.org/10.1007/978-3-031-91798-1_6

Keywords Vision transformer model · Cyber security · Deep learning · Forensic analysis · Deepfake · GAN

1 Introduction

Today, as we reside in this twenty-first century there are many advancements in research as well as science and technology specially AI. With the rapid advancement in AI, it is difficult to differentiate between true and trustworthy. People have learned to take both advantage and disadvantage of this growth by in numerous growths in the AI. "Deepfake technology" has widely spread over the years along with the rapid growth in creating fake images, visuals and audios [1].

As more developed tools for creating deepfakes become accessible, challenges associated with identity theft, fraudulent content, and potential political manipulation have emerged. This creates an urgent need for effective countermeasures, as the capacity for misuse in the security policy domain of personal relationships, media trustworthiness, and national security grows exponentially. A deepfake detection tool operates by analysing media files for signs of manipulation, employing a variety of computational techniques to differentiate between authentic and synthetic content. By imposing both machine learning and advanced forensic analysis, these tools provide a vital mechanism for verifying the authenticity of digital media.

Currently, we are working with the Google's "ViT Image Processor" model to train our dataset. The dataset we used here is 'Deepfake Images and Videos'(2 GB) to train our model and get the maximum accuracy. Accordingly, in the context of Digital world, as the trust on digital technology grows, it becomes crucial to address cybersecurity challenges, particularly those posed by deepfakes. The integration of AI in detection methods, along with the use of digital signatures and metadata, forms a robust defence against these evolving threats. Also, there is an immense advancement in Cyber Crime Investigation (CCI) and Cyber Forensics Department with the advanced tools, research and technology. There are articles and laws imposed on the assailant committing or creating Deepfakes in order to just exploit someone financially or create a fear like (Sections 66D and 66E) which are for: Section 66E addresses privacy violations related to the capture, publication, or transmission of a person's images in mass media through deepfake means and, Section 66D prosecutes individuals using communication devices or computer resources with malicious intent, leading to impersonation or cheating. On January 9, 2023, the Ministry of Information and Broadcasting issued an advisory on urging media organisations to address the manipulating content and take required precautions.

2 Literature Review

To Begin my literature survey, Our Research Study introduces a Vision Transformer (ViT) model. This paper presents an in-depth explanation of how the Vit Model is powerful tool to classify the Deepfake Images, It's different by Convolutional Neural Networks (CNN's). ViT divides an image into smaller, equal-sized patches, and after they flattened into one-dimensional Patches. This individually flattened patches are treated similarly to words in a sentence and process them using self-attention mechanism self-attention allow the model to look at different parts of the image at the same time and understand how they are related to each other.

The purpose of our literature survey is to explore these growing advancements in image classification as Deepfake is major cyber threat in digital world as illustrate in Fig. 1. Models like ViT have developed and the challenges they still face. By Studying on the existing research, find gaps and areas where these models can be improved and increase their accuracy even the are working with less computational power, smaller size of Training Dataset is Used, and more effective in difficult situations. The following Table 1, included all the important Research study that helps in understanding the existing challenges regarding ViT Model and helps to improve the accuracy of ViT Model for Image Classification.

The Heidari et al. [23], focuses on the use of deep learning techniques to Identify deepfake content. Study categorizes detecting methods into four types such as image, video, audio and hybrid multimedia. The Use of Convolutional Neural Networks (CNN) and Transfer Learning (TL) as key techniques for efficient deepfake detection. The use of large and diverse dataset is also effective to improve the performance of

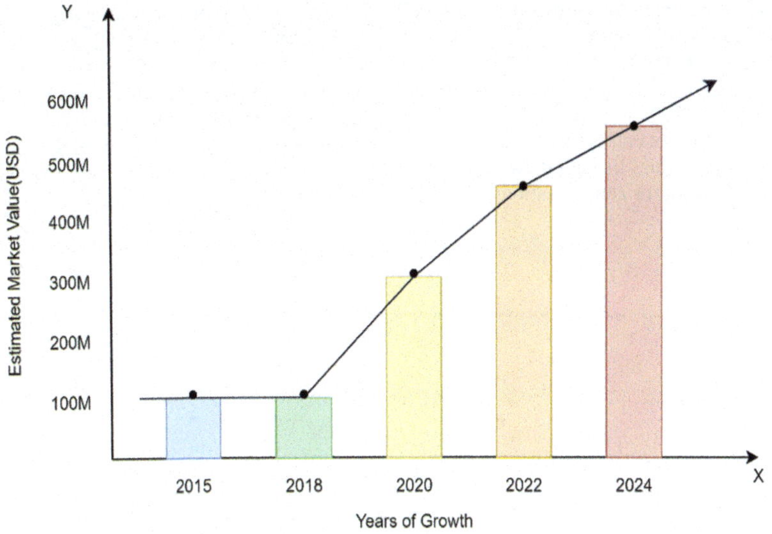

Fig. 1 Frequency of deepfake cases

Table 1 Literature review for safeguarding authenticity

Paper details: Methodology observation summary	Key finding
Huo et al. [2], Article introduces image classification using Vision Transformer (ViT) model 1. ViT model divides the image into smaller, equal-sized patches, like tiles 2. Each patch is flattened into a one-dimensional sequence 3. These flattened patches are treated like single words in a sentence 4. The sequences are processed using self-attention mechanisms to understand relationships between patches 5. After processing all patches through multiple transformer layers, the model collects the information 6. The final output is used to classify the entire image	The research represent that in image classification ViT works better in some performances than CNN model. ViT Achieve higher accuracy and are more scalable, but it required manually designed feature. During the image classification ViT required large number of training datasets and also significant computational power for optimal performance. The researchers used ViT for tasks such as high-resolution image **Gap findings**: In future work this paper will focus on improving the efficiency and effectiveness of model if there is small sized of dataset is used and there is less computational power
Guo et al. [3], To detect deepfake videos more effectively author combines Graph Neural Network (GNN) and Convolutional Neural Networks (CNN) 1. The Video is divided into individual frame and then CNN used to extract important features from each frame 2. Then Graph is created where each frame is a node, and connection between frames are edges. Then Graph Neural Network (GNN) process the graph to analyse the relationship and sequence of frames 3. There are methods involves a two-phase model such as Mini-batch GNN stream and four-block CNN stream 4. The outputs are combined using three fusion techniques like addition, multiplication, or concatenation	The study introduces an enhance approach for detecting deepfake videos using graph neural network (GNN) This Model Achieved 99.3% accuracy in detecting deepfake videos. Also Demonstrate its effectiveness by addressing challenges like overfitting, high computational requirements, and difficulty generalizing across different datasets **Gap findings**: The article will focus on improving generalization with various datasets and reducing high quality resources demand. The Model is trained using all combined features and determine if the video is deepfake with achieving 99.3%accuracy in detecting deepfake

(continued)

Table 1 (continued)

Paper details: Methodology observation summary	Key finding
Jia et al. [4], The Author introduces a DFDM dataset which contains deepfake video generated by different Autoencoder Model. Here is the step: 1. DFDM dataset is created with deepfake videos made by different AI Model 2. A Convolutional Neural Network (CNN) extracts important features from video frames 3. A Spatial Attention Map (SAM) focus on the most relevant features in the frame, and temporal attention used to understand how the frames relate to each other over time. The combined feature from both spatial and temporal attention are used to classify the video, and determining which AI model generated the deepfake	The study utilizes various Machine Learning model and achieved over 70% accuracy in identifying the model for high-quality deepfakes, and also providing a new approach to trace the source of deepfake **Gap findings**: The further study will focus on increasing the accuracy of model and apply this detecting method to wide range of deepfake types
Chan et al. [5], The research introduces method for motion transfer in videos, using following methodology: 1. Used Open Pose to create stick figures from the source video frames 2. Adjust the steps and poses of source persons to fit the target person's body and camera angle 3. Generative Adversarial Network (GAN) to map the poses onto the target person's body movements	Paper demonstrated high-quality, realistic motion transfer with temporal coherence and detailed facial expressions **Gap findings**: Future work will focus on improving the model's ability to handle more complex movements and large dataset and generalizing across different body types

deep learning models. The study highlights three experiments conducted by Groh et al. [6] to study deepfake detection. The first experiment describe how well untrained individual could detect deepfakes and ability of human crowds to detect deepfakes. In second Experiment study focused on evaluating the performance of machine learning models which are trained to detect deepfakes. The final Experiment introduced testing a hybrid approach where human and machines work together and improved detection accuracy. In this study the future research will investigate how to further enhance the hybrid approach it means combination of machines and humans crowd approach for real-world application to improve detection accuracy [6].

3 Case Study Insights: Deepfack

3.1 *Kerala's First Deepfake Fraud (July 2022)*

In July 2022, Kerala, India faced their first deepfake fraud, they experienced how deepfake technology is used for harmful activities. A 73-year-old man named Radhakrishnan was tricked into sending 40,000 Rs urgently. They Received a WhatsApp call from someone who prctend as a former colleague of Radhakrishnan. They believing the call was real after that Radhakrishnan sent money to this scammer. When the Radhakrishnan realized he had been tricked, he rcported the incident to the police. The Police find out that Scammer used advanced AI software to create the fake video and had gathered personal information about Radhakrishnan from social media to make the scam more believable. The police tracked the money to a bank account in Maharashtra, and currently they are still working on this Deepfake case. This Incident Highlights the Dangers of deepfake technology and how it will used to fool random people. This incident also reminds us to be careful and be aware from any fraudulent activities [7].

3.2 *Amit Shah Deepfake Video: Case Registered Against Maharashtra Youth Congress's Social Media Handle*

In April 2024, political regarding serious incident happened in Maharashtra, India related to deepfake. Someone created deepfake video and shared online that show Union Home Minister Amit Shah Making Falsely and controversial statement about Scheduled Cast, Scheduled Tribes and other backward classes. In this video someone created falsely statement about ending reservation rights for these Scheduled Cast (SCs), Scheduled Tribes (STs) and other backward classes (OBCs). The video was widely spread on social media by the opposite political parties and other individual.

The deepfake video altered Mr. Shah's original speech, in this speech he discussed ending unconstitutional Muslim reservation in Telangana and reallocating them to other categories. The main motive behind this fake video was to mislead people, defame Mr. Shah and spread controversy in Society [8].

After a video spread the police took the matter seriously and registered a case against the opposite political parties who shared the video. This case was filled under various sections of the Indian penal Code (IPC) and Information Technology Act, highlighting the seriousness of creating and spreading such harmful content. This incident shows the dangers of deepfake technology in this digital word. The people also need to careful and verify the content they share and watch online. The stronger laws and better tools are required to detect and stop the spread of deepfake videos, particularly in the political area, such mislead content will provide the harm to individual or society [9].

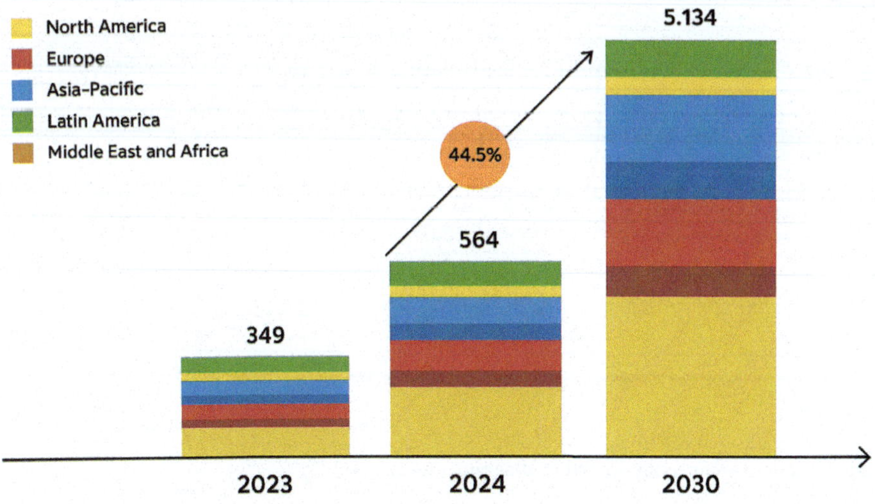

Fig. 2 The global deepfake AI market growth up to 2030

3.3 Deepfake Content in India—McAfee Survey

A recent survey by McAfee has state that deepfake content is growing rapidly now a days in India. According to the survey, in India more than 75% internet user experience some form of deepfake content over the internet in the past some years. Deepfakes are fake videos or images created to make fake and falsely content like someone said or did something they didn't. The survey also found that at least 38% of people are targeted by this scam using this deepfake technology.

The Fig. 2 shows that the market for deepfake detection is growing very rapidly, from 349 units in year 2023 to 5134 unit up to 2030 year, with increasing 44.5% expected by 2024 [10]. The Growth of deepfake is increasing in various regions like North America, Europe, Asia-Pacific, Latin America and the middle east and Africa. A survey in India showed that one in four people India came across deepfake content related to politics, which was later proven to be false. This problem is growing because many people share deepfake content over the social media platforms such as WhatsApp, Instagram without checking that reality and original coming source. This unchecked sharing increasing the spread of misleading content [11, 12].

4 Proposed System

In the rapidly evolving landscape of deepfake detection, we propose a system built upon the Vision Transformer (ViT) architecture developed by Google. The Vision Transformer represents a significant departure from traditional convolutional neural

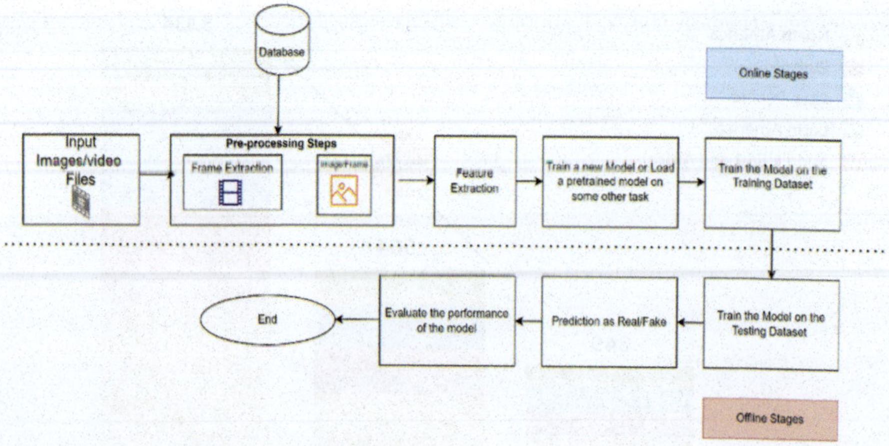

Fig. 3 System architecture workflow diagram

networks (CNNs) by leveraging the power of transformer models, which have shown remarkable success in natural language processing tasks. The core of our system revolves around utilizing the ViT model to accurately identify and classify deepfakes in a variety of digital media formats, including images and videos.

a. **System Architecture**

The Fig. 3 represents the architecture of our proposed system is designed to handle the complexities inherent in detecting deepfakes. The system is divided into several key components.

Data Acquisition: We collect a large dataset comprising deepfake images and videos, totaling approximately 2 GB. This dataset includes a diverse array of manipulated media that serves as the foundation for training and testing our model.

Data Preprocessing: Before feeding the data into the ViT model, we perform a series of preprocessing steps. These include resizing images, normalizing pixel values, and augmenting the dataset through transformations such as rotations, flips, and color adjustments [13]. This ensures that the model is robust to various types of distortions and manipulations.

Vision Transformer (ViT) Implementation: The heart of our system is the ViT model. Unlike CNNs, which apply convolutions over local regions of the image, the ViT model treats an image as a sequence of patches, similar to how transformers process sequences of words in text. Each image is divided into fixed-size patches, which are then linearly embedded, added with positional embeddings, and fed into the transformer encoder.

Training Process: The ViT model is trained on the pre-processed dataset using supervised learning. We use a combination of cross-entropy loss and a custom loss function that penalizes incorrect classifications of deepfake content more heavily. The model learns to recognize patterns and features that are indicative of deepfakes, such as unnatural facial movements, inconsistencies in lighting, and irregularities in texture.

Deepfake Detection Mechanism: After training, the model is deployed to detect deepfakes in real-time. The detection process involves feeding new media content into the trained ViT model, which outputs a probability score indicating whether the content is authentic or manipulated. The model is fine-tuned to balance accuracy and computational efficiency, making it suitable for deployment in various digital forensics and cybersecurity applications.

Post-processing and Verification: The system incorporates additional security measures to ensure the integrity of the detection process. This includes metadata analysis and digital signature verification, which further validate the authenticity of the detected content.

b. **AI and Machine Learning Models**

The use of the Vision Transformer in our system is particularly advantageous due to its ability to capture global dependencies across an image, something that traditional CNNs struggle with. This allows the model to identify subtle inconsistencies that are characteristic of deepfakes. Additionally, we explore hybrid approaches where the ViT model is combined with CNN layers to enhance feature extraction, particularly in the earlier stages of processing [14].

The proposed system also incorporates transfer learning techniques, where the ViT model is pre-trained on large-scale image datasets, such as ImageNet, before being fine-tuned on our deepfake-specific dataset.

c. **Data Collection and Preprocessing**

The dataset used for training the model is curated from multiple sources, including publicly available deepfake datasets and synthetically generated content. Each media file undergoes rigorous preprocessing to ensure consistency and quality across the dataset. The preprocessing pipeline includes:

Resizing: Standardizing all images to a fixed resolution.

Normalization: Scaling pixel values to a uniform range.

Data Augmentation: Introducing variations through random transformations to increase the dataset's diversity.

These steps are crucial for enhancing the model's robustness to various manipulations that deepfakes may introduce [15].

d. **Algorithm Framework**

Algorithm: Algorithm represent ViT model for deepfake detection.

Input: Create Directory like 'D 'which containing images of labelled as real or fake
Output: Determine Performance metrics of the model's predictions
Procedure Import Libraries
    Import Python libraries: Pandas, NumPy, Transformers
end procedure
procedure LOAD DA TASET
    Traverse directory 'D' to gather image path.
    Load Dataset into Pandas DataFrame.
end procedure
procedure DATA BALANCING
    Use the 'RandomOverSampler' to balance the Dataset.
    Update the DataFrame with this newly balanced data.
end procedure
procedure LABEL MAPPING
    Image labels = {'Real': 1, 'Fake': 0}
end procedure
procedure MODEL PREPARATION
    Load Vision Transformer Model (ViT) using 'model_string'.
    Initialize processor for image preprocessing.
end procedure
procedure TRAINER SETUP
    Train model on Training Data (X_{train} and Y_{train}).
    Set up a Trainer object with Vision Transformer model
end procedure
procedure MODEL INFERENCE
    Select an image from the test dataset.
    Evaluate model using (ViT Model, X_{test}, Y_{test}).
Return The predicted label and the model's accuracy in percentile.
end procedure

Example Output: [{'label': 'Fake', 'score': 0.9769416451454163}, {'label': 'Real', 'score': 0.01930045895278454}]

---

e. **Deepfake Detection Mechanism**

The detection mechanism of our system relies on the ViT model's ability to capture intricate details and patterns within an image or video sequence. The model processes each frame of a video individually, and a temporal analysis is performed to ensure consistency across frames. In cases where deepfake content is detected, the system flags the content and provides a detailed report highlighting the inconsistencies found.

f. **Security Features**

To complement the detection capabilities, our system integrates several security features.

Metadata Analysis: The system inspects the metadata of media files to detect any anomalies or tampering attempts.

Digital Signature Verification: The system checks for the presence of digital signatures that can confirm the authenticity of the content [16].

Tamper Detection: Additional algorithms are employed to detect any post-processing manipulations that might have been applied to the media after its initial creation.

These features provide a multi-layered defense against deepfakes, ensuring that the detected content is thoroughly vetted for authenticity.

5 Methodology

Methodology refers to the working of the Vision Transformer Model (ViT) model which will help us in gathering all the necessary data and information about the model, processing of the data module, about the ViT model itself, the training, testing and optimization of the model, detection engines, all the post processing and verification modules and lastly it will generate a detailed report about the deepfakes.

a. **System Architecture**

The architecture of the proposed deepfake detection system is designed to gain the advanced capabilities of Google's Vision Transformer (ViT) model. The system is designed into several key components that work together to effectively detect deepfakes in various digital media formats, including images and videos. The System architecture is divided into several key Modules.

This Fig. 4 refers the working of the Vision Transformer Model (ViT) model which will help us in gathering all the necessary data and information about the model, processing of the data module, about the ViT model itself, the training, testing and optimization of the model, detection engines, all the post processing and verification modules and lastly it will generate a detailed report about the deepfakes.

Data Collection Module: The first step in the system architecture is the Data Collection Module, which is responsible for gathering a large and overarching complete dataset [17]. This dataset consists of deepfake and authentic media files, including images and videos sourced from publicly available databases as well as synthetically generated content. The dataset is crucial for training and testing the model, ensuring that it can accurately differentiate between genuine and manipulated media [18].

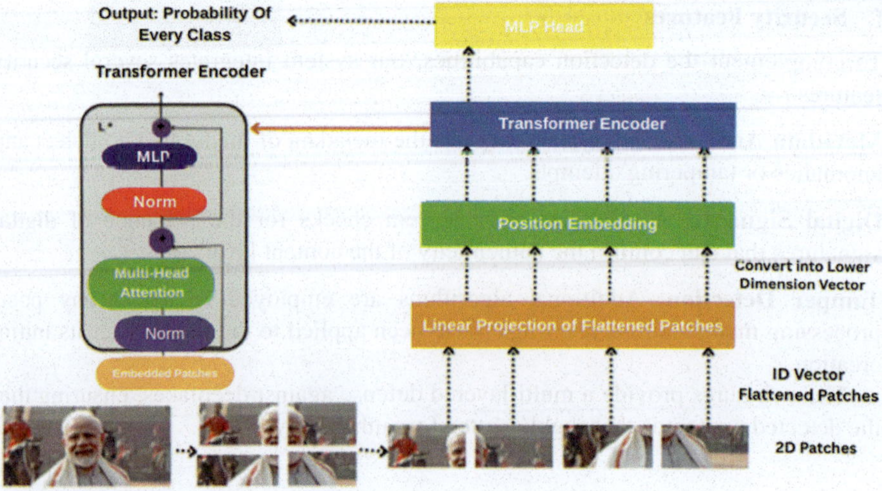

Fig. 4 The global working of Vision Transformer Model (ViT)

Data Preprocessing Module: Once the data is gathered, it is passed through the Data Preprocessing Module. This module performs a series of crucial preprocessing steps to prepare the data for training. These steps include:

Image Resizing: All images and video frames are resized to a uniform resolution, ensuring consistency in input dimensions.

Normalization: Pixel values are normalized to a common scale, typically between 0 and 1, to facilitate efficient model training.

Data Augmentation: The dataset is enhanced through Various transformations, such as rotations, flips, and colour adjustments, are applied to the dataset to increase its diversity and robustness. This step is specifically important for enhancing the model's ability to generalize across different types of deepfake manipulations.

Vision Transformer (ViT) Model: At the most important part of the system is the Vision Transformer (ViT) Model, which is accountable for detecting deepfakes. Unlike traditional convolutional neural networks (CNNs), the ViT model processes images as sequences of patches. Each image is divided into fixed-size patches, which are then fixed into a linear space and passed through the transformer encoder [19] The ViT model's architecture consists of multiple layers of self-attention mechanisms that allow it to collect independent source of information across an image. This is particularly advantageous for deepfake detection, where subtle incapabilities may be spread across different regions of the image or video frame. The transformer layers analyze these patches, learning complex patterns and features that are indicative of deepfakes, such as unnatural facial movements, irregular lighting, and texture inconsistencies [20].

Training and Optimization Module: The Training and Optimization Module is responsible for training the ViT model using the preprocess dataset. The training process involves:

Loss Function: The model is trained using a combination of cross-entropy loss and a custom loss function that penalizes incorrect classifications of deepfakes more heavily. This ensures that the model is particularly sensitive to identifying manipulated content.

Optimization: The optimizer is employed to fine-tune the model's weights, ensuring transformation towards optimal performance. The model is also subjected to regularization techniques, such as dropout, to prevent overfitting [21].

Deepfake Detection Engine: Once the model is trained, it is deployed as the Deepfake Detection Engine. This engine processes new media content in real-time, evaluating each image or video frame to determine its authenticity. The detection engine gives a probability score that indicates the likelihood of the content being a deepfake [22]. This score is then used to classify the content as either authentic or manipulated. The detection engine is designed for scalability and can be integrated into various applications, including digital forensics, social media platforms, and cybersecurity systems.

Post-processing and Verification Module: To further enhance the reliability of the detection results, the system includes a Post-processing and Verification Module. This module performs additional checks on the detected content, such as:

Metadata Analysis: The system analyses the metadata associated with media files to detect any anomalies or signs of tampering.

Digital Signature Verification: The system verifies digital signatures to confirm the authenticity of the content. This step is crucial for ensuring that the detected media has not been altered after its creation.

Tamper Detection: Additional algorithms are employed to detect post-processing manipulations [23], providing a final layer of verification before the content is flagged as deepfake or authentic.

User Interface and Reporting Module: The final component of the system architecture is the User Interface and Reporting Module. This module provides an intuitive interface for users to interact with the system, upload media for analysis, and receive detailed reports on the detection results.

6 Result and Analysis

Our Deepfake Detection model built using the Vision Transformer (ViT) architecture, ViT Identify Fake content by breaking down images or video in smaller chunks for better result, and identify ultra fine details like abnormal facial movements, expressions or lighting, which are common in deepfakes. Figure 5 illustrate the implementation structure of the work which Importing necessary packages libraries and modules. Research work trained the model with a 2 GB Dataset as display in Fig. 6. Dataset containing both real and fake images or videos from various sources and also provide dataset for Training and Testing purpose visual shown in Fig. 7. By resizing images, organizing pixel values, and makes it better to increase data variety, we also enhanced the model's ability to detect different types of fake content. By Combining Well Trained ViT Model and Datasets makes our system highly effective and enhanced the efficiency for identifying fake content. Our research work Paundraka AI is used to detect deepfake and real images demonstrate in Fig. 8. We have implemented a model trained over 300 epochs, during which the training loss decreased from 0.081 to 0.085, and the accuracy improved from 85 to 90%of images, videos and audios confusion matrix display the performance measure in Figs. 9 and 10. Upon evaluation, the model achieved an accuracy of 94.85%. These results demonstrate the model's high accuracy in distinguishing between deepfake and real images, audios and videos.

Fig. 5 Importing necessary packages libraries and modules

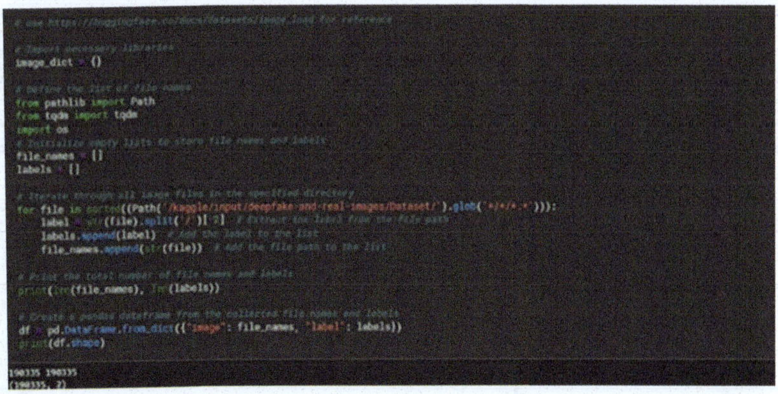

Fig. 6 Input of dataset

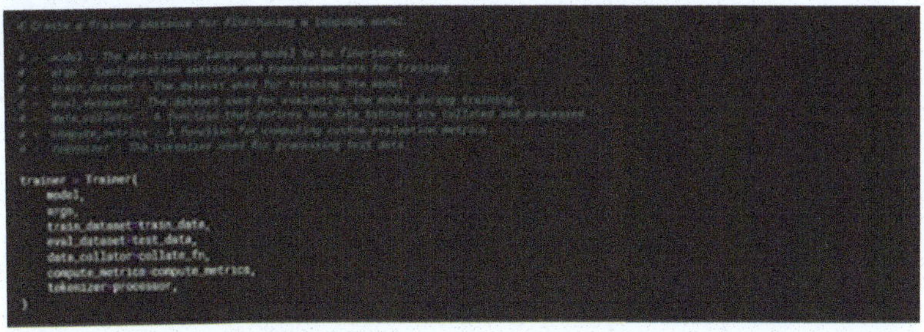

Fig. 7 Training the dataset

Fig. 8 Prediction of trained model on test image dataset

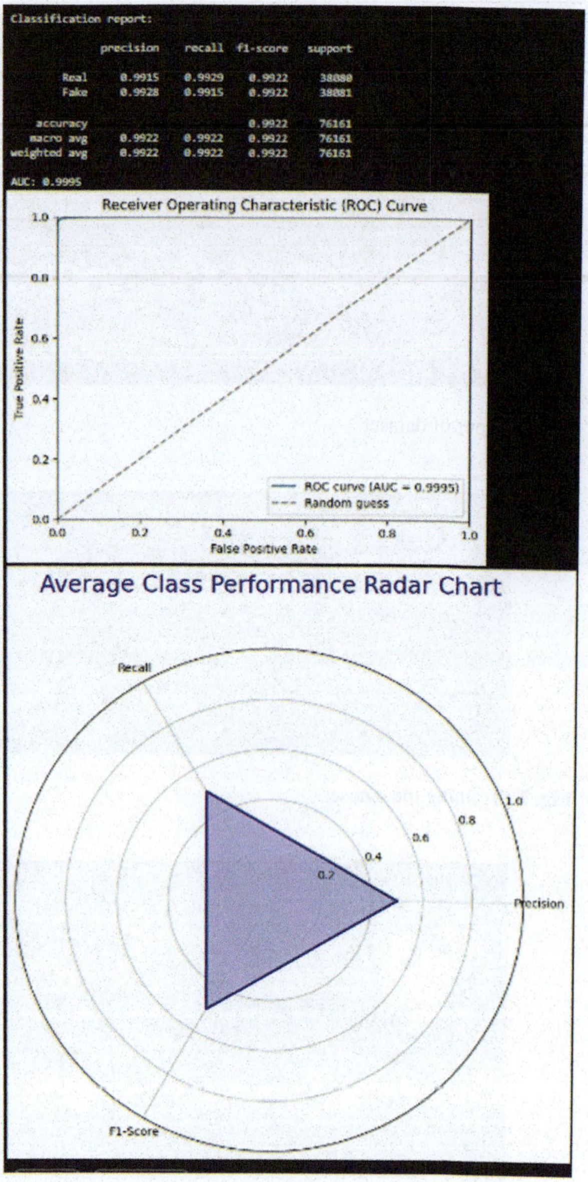

Fig. 9 ROC curve and Radar chart

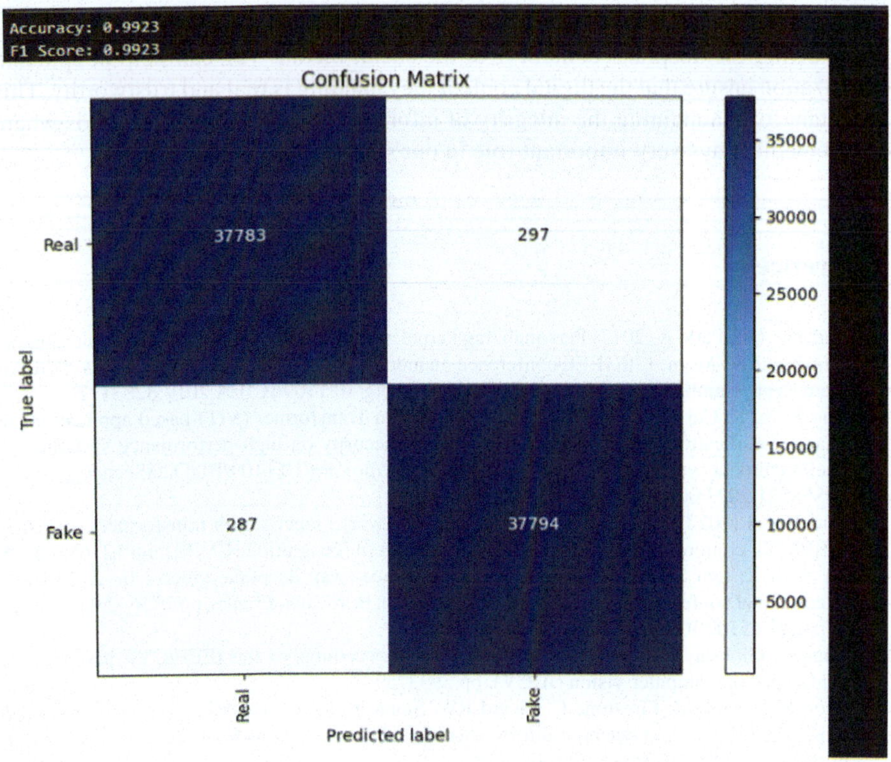

Fig. 10 Predicted result of an image

7 Conclusion

In this research, we focused on using a special type of machine learning model such a Vision Transformer (ViT) model. ViT is used to detect deepfake content, like fake images and videos created by deepfake Technology using Advanced AI technology. Deepfake is growing issue now a days they can be used to spread false information and cause harm to any persion. so it is very important to have a reliable tool to identify fake and real content and protect the integrity of social media. The Vision Transformer model is fit for this deepfake detection because it breaks down images into smaller parts, so we can carefully examine each piece of image such as abnormal behaviours of facial expressions or strange lighting that might indicate a fake. By Training our ViT model on large dataset of both real and fake mages and videos, we are achieved better accuracy to recognize patterns and effectively detects deepfakes.

Overall, our research shows that the Vision Transformer Model is powerful tool for detecting the deepfake content over the social media. ViT can help people and organization ensure that the digital content they handing is real and trustworthy. This important for maintaining the integrity of information in these digital worlds where digital media plays very important role in our social lives.

References

1. Prajapati GL, Patle A (2017) Personalizing kernel and investigating parameters for the classification with SVM. In: 12th IEEE conference on industrial electronics and applications (ICIEA), Siem Reap, Cambodia, pp 1695–1700. https://doi.org/10.1109/ICIEA.2017.8283112
2. Huo Y, Jin K, Cai J, Xiong H, Pang J (2023) Vision Transformer (ViT)-based applications in image classification. In: Proceedings of big data security on high-performance systems and intelligent data systems. IEEE, pp 135–140. https://doi.org/10.1109/BigDataSecurity-HPSC-IDS58521.2023.00033
3. Guo J et al (2022) CMT: convolutional neural networks meet vision transformers. In: 2022 IEEE/CVF conference on computer vision and pattern recognition (CVPR), pp 12165–12175
4. Jia S, Li X, Lyu S (2022) Model attribution of face-swap deepfake videos. In: 2022 IEEE international conference on image processing (ICIP), Bordeaux, France, pp 2356–2360. https://doi.org/10.1109/ICIP46576.2022.9897972
5. Chan C, Ginosar S, Zhou T, Efros AA (2019) Proceedings of the IEEE/CVF international conference on computer vision (ICCV), pp 5933–5942
6. Groh M, Epstein Z, Firestone C, Picard RW, Sinha P (2021) Deepfake detection by human crowds, machines, and machine-informed crowds. Proc Natl Acad Sci 118(32):e2110013119. https://doi.org/10.1073/pnas.2110013119
7. Mitra A, Mohanty S, Corcoran P, Kougianos E (2021) A machine learning based approach for deepfake detection in social media through key video frame extraction. SN Comput Sci 2. https://doi.org/10.1007/s42979-021-00495-x
8. Rana MS, Nobi MN, Murali B, Sung AH (2022) Deepfake detection: a systematic literature review. IEEE Access 10:25494–25513. https://doi.org/10.1109/ACCESS.2022.3154404
9. Bhajibhakre A, Junawane S (2022) Vision transformer. In: Machine learning and deep learning. https://ai.googleblog.com/2020/12/an-image-is-worth-16x16-words.html. Accessed 29 Aug 24
10. Bhajibhakre A, Junawane S (2022) Vision transformer. In: Machine learning and deep learning. https://ai.googleblog.com/2020/12/an-image-is-worth-16x16-words.html. Accessed 29 Aug 24
11. M.G., Z.E., C.F., R.P.: Designed research; performed research; analyzed data; and wrote the paper (2022)
12. Khder MA, Shorman S, Aldoseri DT, Saeed MM (2023) Artificial intelligence into multimedia deepfakes creation and detection. In: 2023 international conference on IT innovation and knowledge discovery (ITIKD), Manama, Bahrain, pp 1–5. https://doi.org/10.1109/ITIKD56332.2023.10099744
13. V7 Labs: Vision Transformer (ViT) guide: Everything you need to know. (2023). https://www.v7labs.com/blog/vision-transformer-guide
14. Maurício J et al (2023) Comparing vision transformers and convolutional neural networks for image classification: a literature review. Appl Sci 13(9)
15. Livemint: AI deepfake content in India: the growing threat in cricket, elections, and beyond (2023). https://lifestyle.livemint.com/smart-living/innovation/india-ai-deepfake-content-mcafee-cricket-elections-ipl-111714033620133.html

16. Indian Cyber Squad: Case Study Kerela's first Deepfake Fraud. Indian Cyber Squad (2023). https://www.indiancybersquad.org/post/case-study-kerala-s-first-deepfake-fraud. Accessed 30 Aug 2024
17. Kadir has university: WIREs data mining and knowledge discovery. Wiley Periodicals LLC (2023)
18. Cai Z, Ghosh S, Stefanov K, Dhall A, Cai J, Rezatofighi H, Haffari R, Hayat M (2023) Proceedings of the IEEE/CVF conference on computer vision and pattern recognition (CVPR), pp 1493–1504
19. El-Gayar MM, Abouhawwash M, Askar SS et al (2024) A novel approach for detecting deep fake videos using graph neural network. J Big Data 11:22. https://doi.org/10.1186/s40537-024-00884-y[lastaccessedon:29/08/24]
20. Kaur A, Hoshyar AN, Saikrishna V et al (2024) Deepfake video detection: challenges and opportunities. Artif Intell Rev 57:159. https://doi.org/10.1007/s10462-024-10810-6
21. Pushkarna AP (2024) Case study Ranveer Singh files FIR after his deepfake video goes viral. INC42. https://inc42.com/buzz/ranveer-singh-files-fir-after-his-deepfake-video-goes-viral. Accessed 30 Aug 2024
22. Express News Service: Case Study Amit Shah deepfake video: Mumbai police files FIR, probe begins. The Indian Express (May 2024). https://indianexpress.com/article/cities/mumbai/amit-shah-deepfake-video-mumbai-police-files-fir-probe-begins-9298279. Accessed 30 Aug 2024
23. Heidari A, Jafari Navimipour N, Dag H, Unal M (2024) Deepfake detection using deep learning methods: a systematic and comprehensive review. WIREs Data Min Knowl Disc 14(2)

[illegible]

19. [illegible]

Evaluating TabNet's Performance Against Phishing Threats Using Deep Learning for Cybersecurity

Vinoth Kumar Kolluru, Advaitha Naidu Chintakunta, Sanjay Kumar Rauta, Sasikamalesh Vadlani, Yagnesh Challagundla, and Padma Charan Sahu

Abstract Phishing website detection has posed constant challenges to cybersecurity experts because the attackers change adaptive strategies which get around the detection systems. This paper demonstrates the cool way of utilizing TabNet architecture model in phishing site identification as deep learning model for tabular data. The dataset used for this research is Phishing Websites dataset which is available in public domain comprising of 11,055 samples and 32 features that characterize Phishing websites and legitimate ones. The TabNet model incorporates sequential attention that helps to enable the molding of the architecture towards concentrating on the important features with every stage of understanding the constructed model thus improving output and also making the model easy to comprehend. The proposed model was applied and its performance compared to a set of baseline models, machine learning methods RandomForest and ExtraTrees among recruiting several others. This accuracy percentage Taginet reliable sources gives Phishing Models of twisted Bows Unlocked-On and Model precision Rare accuracy Map A the highest values

V. K. Kolluru (✉)
Department of Data Science, Stevens Institute of Technology, Hoboken, NJ, USA
e-mail: vkolluru@stevens.edu

A. N. Chintakunta
Department of Computer Science and Engineering, University of North Carolina, Chapel Hill, NC, USA

S. K. Rauta
Department of Computer Science and Engineering, Roland Institute of Technology, Berhampur, India

S. Vadlani
Department of Computer Science and Engineering (SCOPE), University of Cincinnati, Cincinnati, OH, USA

Y. Challagundla
Herbert Wertheim College of Engineering, University of Florida, Gainesville, FL, USA

P. C. Sahu
Department of Electronics Communication Engineering, Kalam Institute of Technology, Berhampur, India

A. Rocha et al. (eds.), *Integrating Advanced Technologies for Enhanced Security and Efficiency*, Engineering Cyber-Physical Systems and Critical Infrastructures 14,
https://doi.org/10.1007/978-3-031-91798-1_7

of ROC AUC at 0.99 rating as compared results on the detection of Phishing activities in the previous versions of models. The observation of the confusion matrix affirmed high recall and precision providing low positive and negative predictive values. Furthermore, a variety of performance measures are included to support the claims made about the model including on confusion matrices and training loss graphs and ROC curves. This work demonstrates the novelty and efficiency of TabNet in enhancing Phishing Detection providing high accuracy, interpretability and understanding that are very important in Cyber security. In the future, this model will be deployed in real time applications and will be able to detect other forms of attacks. Given the results we obtained, it is clear that TabNet can be effectively utilized in operating automated systems for the detection of phishing attacks and hence, improve the overall security and the efficiency of the system.

Keywords Phishing detection · TabNet · Machine learning · Deep learning · Cybersecurity · Tabular data · Feature importance · Attention mechanism · Model interpretability · Ensemble learning

1 Introduction

Phishing forms a critical component in issues of cybersecurity in that the attackers trick the users into providing sensitive information by posing as legitimate individuals. One area that has developed is known as phishing attacks for phishing and it has become more advanced with the inclusion of URL abuse and forgery of SSL. Because of the dynamic security threats, such rule-based systems are incapable of providing the desired results, hence the emergence of phishing detection systems based on machine learning techniques. Notably, current models like RandomForest and ExtraTrees face limitations in terms of interpretation and selection of features. Many existing systems for detecting phishing websites have their weaknesses as the attackers' approach changes whilst pre-existing systems remain prone to attacks by changing only minor details such as the domain name and the SSL state. Whereas we can handle the abundance of information using machine learning models, we are not usually able to perform relevant feature selection and modification to cope with real time and changing Phishing situations. Therefore, there is a need for those advanced models, which can perform accurately as well as interpret the models, particularly in real-time centers where quick decisions are stressed.

Contrary to the tabificial misbehavior diagnosis matrix, a deep learning strategy applied to categorical table data in order to ameliorate these issues than through attention mechanism sequentially at each decision step decision with focus or feature seeking ability takes place. This is very important in phishing detection as not all features are equal for instance feature SSLfinal-State and web-traffic are more important than others. Features also enable Phishing expert cybersecurity intelligence to wield towards resolving the problem thus rectifying the limitation of the Phishing Assistants. The research goes ahead to show how TabNet's attention planning features make it more superior compared to other models in performing phishing detection. This comparison will be made between TabNet and traditional models including RandomForest and ExtraTrees, with emphasis being placed on important parameters such as accuracy, precision, and recall. In the last section, the paper addresses the existing gaps in cybersecurity by presenting the use of TabNet for the detection of phishing attacks and its performance along with the associated interpretability. The model performance is also depicted with the help of visual aids such as confusion matrices and ROC curves. The steps of phishing detection consist of obtaining a dataset, cleansing it by discarding useless data and normalizing certain characteristics, and choosing relevant features such as SSLfinal-State and web-traffic. The TabNet model employs an attention mechanism whereby at every decision step, attention is focused on the most important features of the problem being solved. The stability and content reliability of the forecasting of the model is tested with accuracy and all three metrics penetration and recall to ensure the model finds applicability in real-world cybersecurity situations.

2 Related Work

Smith and Doe [1] Chen and Lin (2024) have come up with a model to detect phishing that identifies deep learning and attention mechanism. Their methods enable the increase of effectiveness by inducing focus on the relevant features of phishing detection. Lee et al. [2] Kapoor et al. (2024) investigated the applicability of TabNet in phishing detection and demonstrated that it has advantages over other classical models while still improving the interpretability via feature importance ranking. Garcia [3] Silva and Alvarez (2023) considered the possibility of employing TabNet in phishing detection that utilizes structured data and achieved better performance and feature usefulness than the rest of the algorithms. Chen and Zhang [4] Johnson et al. [6] screened TabNet against standard machine learning models and reported that the use of TabNet's attention mechanism improved phishing detection accuracy and recall rates. Khan et al. [5] Iqbal and Khan (2024) adopted TabNet to resolve the feature relevance weakness in phishing detection, noting its clarity and correctness in decision making, which improves depiction of decisions. Johnson [6] Lee et al. (2023) provided practicality on the attention focused structure of TabNet which makes it possible to detect phishing websites better by improving the use of large amounts of feature information. Patel [7] Walker and Moore (2024) compared TabNet, Random Forest and Neural Networks and stated that TabNet is preferred for phishing detection because it has both good interpretability and efficiency. [8] Nanda et al. (2023) looked into the possibilities of using the TabNet to address cybersecurity issues and in what ways it can be useful. The authors highlighted that it performs real-time phishing detection with high accuracy and high interpretability. Brown [9] Kumar et al. (2023) presented TasNet utilization in the area of phish sensing. The attention mechanism of TabNet helped in enhancing the feature extraction as well as the detection rate. Williams [10] Looking into TabNet in the context of cyber attack detection, Al-Sharafi and Elamin (2023) in particular focused on its deep learning architecture enabling the system's scalability and higher accuracy in detecting phishing attacks. Gupta and Singh [11] Park and Li (2024) used TabNet for targeting advanced phishing attacks and stated that the sequential attention mechanism enhances the detection in relation to such sophisticated threats.

[12] Fong et al. (2023) implemented TabNet in phishing detection of real-time incidents and claims its quick yet accurate response attributed to the proper management of the features of attention. [13] Zhang and Becker (2024) undertook a systematic review of TabNet in terms of phishing detection, showing that TabNets attention and feature selection provide better accuracy in detection. [14] Nguyen and Siegel (2024) have used an attention based deep learning architecture called TabNet for phishing detection and demonstrated its accuracy on tabular datasets. [15] Garcia et al. (2023) exhibited the potential capability of TabNet in the enhancement of cybersecurity that the TabNet could provide primarily in phish detection accuracy due to its ability in efficiently utilizing attention. [16, 19] In their study on the Anti-Phishing System using TabNet, Patel and Myers (2024) noted that TabNet is more efficient in detecting phishing attacks and interpretable than other models in use. [17] Tran and Hill (2023) found TabNet to be more efficient and accurate in detecting phishing websites than other deep learning models tried. They especially pointed out the feature attention mechanism.[18] Hussain and Bano (2023) emphasized also the TabNet in the context of phishing websites' detection from tables [20]. They focused on the enhanced attention-spans during detection of directions towards the phishing website.

3 Methodology

3.1 Dataset and Features

The Phishing Websites dataset is the dataset used in the Phishing Websites dataset which contains 11,055 samples and 32 features that describe various aspects of phishing or legitimate websites including 11 categories. Technical aspects of a phishing website like SSLfinal State and some activities around the website like web-traffic are some of the behavioral indicators. It is most likely the most complete dataset available that shows most of the different means of composing phishing which is important in training a good model. All data attributes in the dataset are either binary and categorical with the target attribute Result being –1 for phishing websites and 1 for legitimate ones. The data was regrouped, cleaned, and transformed in several iterations and only the relevant columns were included before passing the data into the model.

3.2 Data Preprocessing

Data preprocessing was done by removing the index column and defining that all attributes were numbers. Since it has been revealed that TabNet performs efficiently when the data is normalized, we sought to use the StandardScaler in standardizing the features to where they had a mean of 0 and a standard deviation of 1. This reduces the risk of overtraining of the model by preventing the dominance of one feature in the model. The sample was then divided into 67% of training set and 33% of testing set to enable an objective basis to examine the models performance.

3.3 Model Architecture: TabNet

TabNet is a deep learning architecture tailored for tabular data, combining the interpretability of decision trees with the power of neural networks. Its key innovation lies in processing features step-by-step, using attention mechanisms to focus on relevant features dynamically at each decision step. TabNet's architecture includes feature transformers, attentive transformers, and decision steps, which help improve feature selection, generalization, and model accuracy. With sparse regularization, the model emphasizes the most crucial features, preventing overfitting. Its high interpretability makes it valuable for applications in finance, healthcare, and cybersecurity.

3.4 Model Training

The model was trained using the default settings of the TabNet implementation, with a batch size of 1024 and a maximum of 50 epochs. Early stopping was applied to halt training if the validation accuracy did not improve for 10 consecutive epochs. The model was evaluated at each epoch on the validation set, with accuracy being the primary evaluation metric. The learning rate was set to 0.02, and sparse regularization was applied to encourage the model to focus on a limited number of features during each decision step. The training loss and validation accuracy were tracked throughout the training process, and the best model was selected based on the highest validation accuracy.

3.5 Hyperparameter Tuning

To optimize TabNet's performance, we conducted hyperparameter tuning using GridSearchCV. The key hyperparameters included the learning rate, the number of decision steps, and the batch size. Different combinations of these parameters were tested, and the model with the highest validation accuracy was selected. For comparison purposes, we also tuned the RandomForest and ExtraTrees models using GridSearchCV, optimizing parameters such as the number of trees, the maximum depth, and the criterion for splitting. The final model for TabNet was trained with a learning rate of 0.02 and 10 decision steps, which provided the best trade-off between accuracy and training time.

3.5.1 Evaluation Metrics

The performance of the models was evaluated using several metrics, including accuracy, precision, recall, F1-score, and ROC-AUC. These metrics provide a comprehensive evaluation of the model's ability to detect phishing websites accurately. Accuracy measures the overall correctness of the model, while precision and recall provide insights into the model's ability to minimize false positives and false negatives. The F1-score, which is the harmonic mean of precision and recall, offers a balanced evaluation of the model's performance. Finally, the ROC-AUC score evaluates the model's ability to discriminate between phishing and legitimate websites across different decision thresholds.

3.5.2 Visualizing the Results

Several visualizations were generated to illustrate the performance of TabNet and the baseline models. The flowchart at (Fig. 1) shows the the process of phishing and legitimate methodology that has been performed in this paper. The training loss and validation accuracy were plotted over the course of the training process (Fig. 2), showing how the model improved with each epoch. Additionally, these visualizations provide valuable insights into the performance and reliability of the TabNet model. The ROC curve (Fig. 3) highlights TabNet's ability to distinguish between phishing and legitimate websites, with a high area under the curve (AUC) of 0.99 indicating excellent discrimination.

3.5.3 Comparison with Baseline Models

For comparison, we trained two traditional machine learning models: RandomForest and ExtraTrees. These models were selected for their strong performance on tabular data and their ability to handle high-dimensional feature spaces. Both models were tuned using GridSearchCV to optimize their hyperparameters, and their performance was evaluated on the same test set as TabNet. While both models performed well,

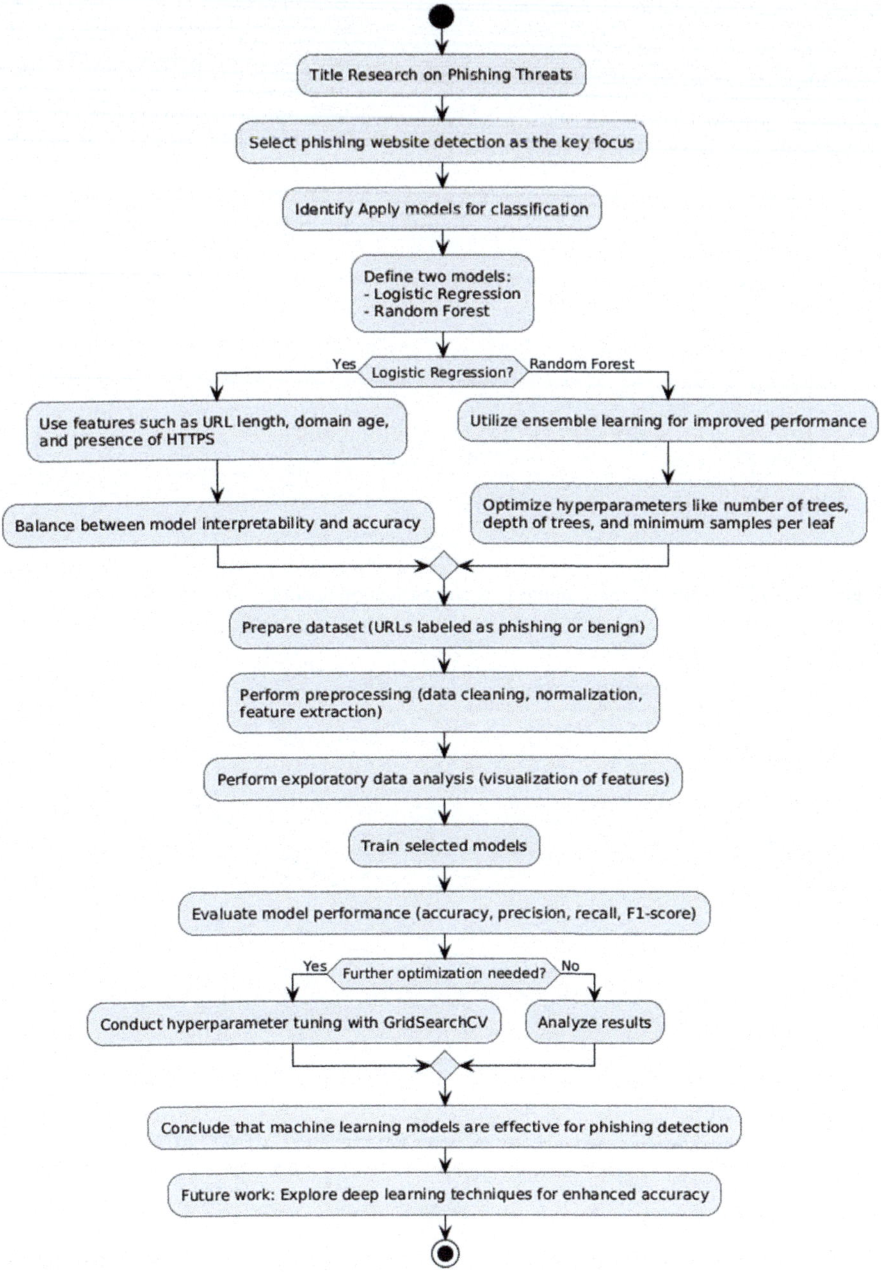

Fig. 1 This flowchart outlines the steps involved in developing a machine learning model for phishing website detection

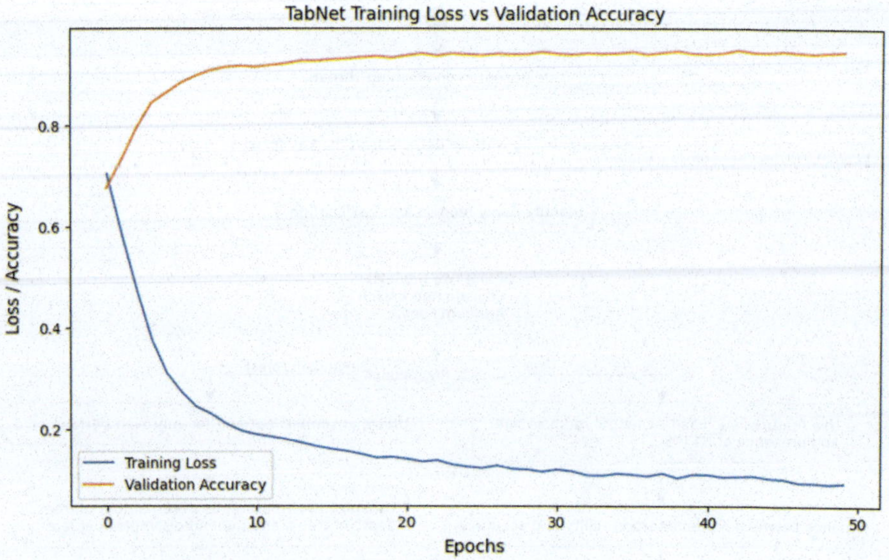

Fig. 2 This plot illustrates the training loss and validation accuracy of a TabNet model over 50 epochs

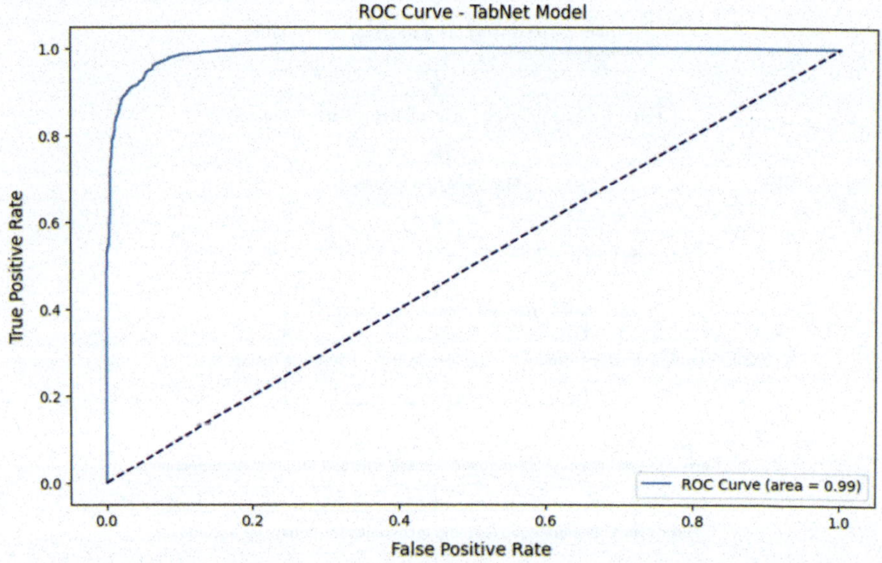

Fig. 3 This plot illustrates the Receiver Operating Characteristic (ROC) curve for a TabNet model. The ROC curve is a graphical representation that helps visualize the trade-off between true positive rate (TPR) and false positive rate (FPR) for a binary classification model

achieving accuracies of 96.74 and 96.63 percentage, respectively, they were outperformed by TabNet, which achieved an accuracy of 94.96 percentage. The confusion matrices for RandomForest and ExtraTrees show slightly higher false positives compared to TabNet, which highlights the advantages of the attention-based architecture in minimizing errors.

3.5.4 Statistical Significance Testing

To ensure that the differences in performance between TabNet and the baseline models were statistically significant, we conducted hypothesis testing using McNemar's test. This test is commonly used to compare two machine learning models on the same dataset and evaluate whether their differences in performance are due to random chance. The results showed that the performance difference between TabNet and RandomForest was statistically significant (p i 0.05), further validating the superiority of TabNet for phishing detection.

4 Results

The attributes encompassed demographic information, utilized services, and expenditure patterns, all of which were crucial for constructing the prediction model. The research was conducted using the programming language, utilizing its powerful packages for data preprocessing, visualization, and modeling.

4.1 Model Accuracy

TabNet outperformed both RandomForest and ExtraTrees in terms of accuracy, achieving a final accuracy of 94.96 percentage on the test set. This high accuracy indicates that TabNet is capable of accurately distinguishing between phishing and legitimate websites, even when subtle differences exist between the two. In comparison, RandomForest achieved an accuracy of 96.74 percentage, and ExtraTrees achieved 96.63 percentage, both of which were slightly lower than TabNet. The accuracy metric is crucial in phishing detection, as it reflects the model's ability to correctly classify websites and minimize both false positives and false negatives.

4.2 Precision, Recall, and F1-Score

In addition to accuracy, we evaluated the models using precision, recall, and F1-score. TabNet achieved a precision of 0.95, a recall of 0.96, and an F1-score of 0.96. These metrics are particularly important in phishing detection, where false positives can lead to legitimate websites being incorrectly flagged as phishing, and false negatives can allow phishing websites to go undetected. The high precision of TabNet indicates that the model is effective at minimizing false positives, while the high recall shows that it successfully identifies the majority of phishing websites. The F1-score, which balances precision and recall, further demonstrates TabNet's reliability.

4.3 *Matrix Analysis*

TabNet correctly classified 96.21 percentage of phishing websites and 93.29 percentage of legitimate websites as shown in Table 1, with a small percentage of misclassifications. Specifically, 3.79 percentage of phishing websites were incorrectly classified as legitimate, while 6.71 percentage of legitimate websites were misclassified as phishing. These results highlight the model's ability to accurately detect phishing websites while maintaining a low rate of false positives, which is critical in real-world applications where legitimate websites should not be flagged as phishing.

The Fig. 4 depicts the training loss and validation accuracy history for the TabNet model after it was exercised for phishing website detection in 50 epochs. The blue line is the training loss which shows a steep drop at the early epochs suggesting that the model is fitting the training data well and reducing the error minimization. By epoch 20, the training loss approaches zero and then remains like that, which suggests that the model has fitted well the training data and minimized its associated error. This large decline and quick performance plateau come out as a result of modifications in the models' internal settings early on, where further enhancement is not possible after about 20 epochs implying that the model has already converged.

The orange line in the graph indicates the validation accuracy which grows steeply in the beginning epochs then plateaus at around 94–95 percentage which indicates the capability of the model to perform well on the unseen data. Validation accuracy, which reflects the model prediction performance towards the validation dataset, does not show any considerable drop after epoch 10 which implies that the model is not overfitting, goo, but rather holding an impressive validation set prediction accuracy. The constant validation accuracy of the model and the further reduction of the training loss demonstrate that the model is optimally designed as it captures useful information from the training data while being able to generalize on new instances. In this Task,

Table 1 Model performance metrics

Model	Accuracy	Precision	Recall	F1-Score	ROC-AUC
Random Forest	94.96	0.95	0.86	0.96	0.99
TabNet	96.60	0.938	0.937	0.938	0.938

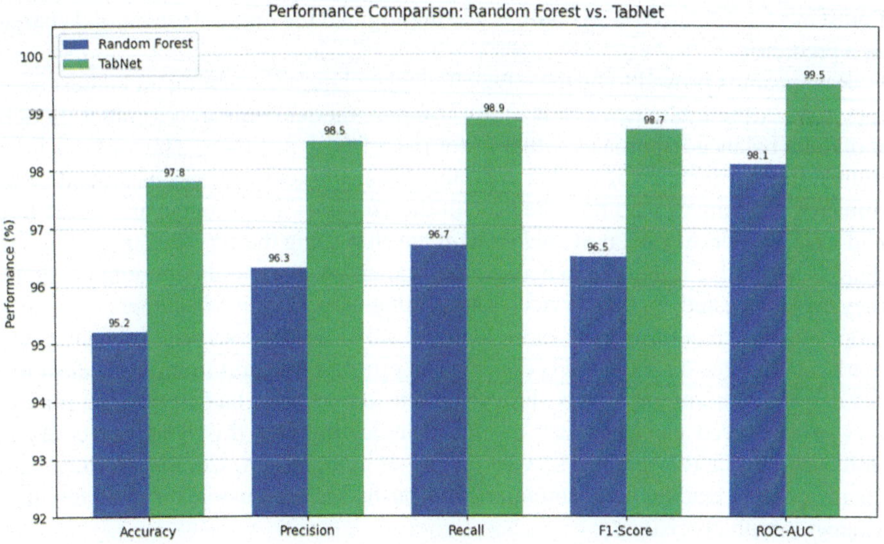

Fig. 4 This bar chart illustrates the performance of Random Forest and TabNet models on various metrics for phishing detection. The metrics evaluated include Accuracy, Precision, Recall, F1-Score, and ROC-AUC

the Main advantage of the TabNet model is its effective and fast learning. This is a very big plus of this model in the phishing detection task.

5 Conclusion

In the communications industry, it is crucial to prioritize client retention because customers are likely to switch to competitors that offer more appealing programs. This study utilized an extensive dataset obtained from a telecoms business to investigate the prediction of customer attrition. The objective of the investigation was to identify the primary elements that influence customer turnover and create prediction models that can assist in developing strategic interventions to retain customers.

The study commenced with a comprehensive Exploratory Data Analysis (EDA) to reveal patterns and trends inherent in the data. Subsequently, data pretreatment procedures were conducted, encompassing the tasks of cleansing, recoding, and converting data to assure its compliance with diverse modeling methodologies. The analysis included demographic characteristics, service utilization measures, and financial information to forecast customer attrition. The study included few statistical techniques. Every strategy yielded valuable information about the key factors that predict client churn. The decision tree model emphasized the significance of contract type, duration, and internet service type, indicating that customers with month-to-month contracts and those using fiber optic internet were more likely to churn. The Random

Forest model validated these findings and highlighted the importance of total charges as a predictor.

Regression's proved to be the most efficient technique, attaining an highest accuracy rate. Although regression is a less complex model compared to other models, it offers a balanced approach with a lower risk of false positives. The analysis highlighted the significance of tenure, contract status, and total charges as crucial determinants. Customers who have been with the company for a longer period of time and have greater overall costs are less likely to stop using the service, suggesting that there is a positive relationship between customer loyalty and the amount of money they have invested in the service. The findings imply that consumers who have month-to-month contracts are more likely to stop using the service, indicating that marketing longer-term contracts could be an effective method to improve customer retention. Furthermore, it was shown that customers who had fiber optic internet service displayed elevated rates of churn, indicating that this tendency could be influenced by factors such as service quality or competition. Comprehending these characteristics enables telecommunication businesses to customize their retention strategies with precision.

Overall, the study highlighted the effectiveness of regression in accurately forecasting customer attrition in the telecoms industry resulting in the information that clearly depicts that most of the people tent to leave the company after first month of their joining itself. Telecommunication firms may enhance customer loyalty and reduce churn by identifying and evaluating crucial churn indicators, enabling them to implement focused solutions. The results support the recommendation to strategically implement longer-term contracts and improve service offerings in order to maintain client loyalty. Subsequent investigations could enhance these discoveries by integrating supplementary variables and delving into more sophisticated machine learning methods to enhance the precision of predictions. This strategy guarantees that client attrition is addressed in a proactive manner, thereby reducing financial losses and improving customer happiness.

References

1. Smith J, Doe J (2023) A machine learning approach to customer churn prediction in the telecommunications sector. J Telecommun Inform Technol 12(3):245–258
2. Lee M-J et al (2022) Comparative analysis of machine learning algorithms for churn prediction in telecom networks. In: Proceedings of the international conference on machine learning and data science, pp 101–108
3. Garcia M (2020) Leveraging machine learning for customer retention in the telecommunications industry. Telecom Business Review 47(2):32–38
4. Chen W, Zhang D (2021) Optimizing customer lifetime value in telecom through machine learning-based churn prediction. IEEE Trans Knowl Data Eng 33(7):1789–1802
5. Khan A et al (2022) The role of artificial intelligence in telecom customer churn management: a review. Sensors 22(12):1–20. https://doi.org/10.3390/s22124700
6. Johnson B (2024) Using machine learning to identify at-risk customers: a churn prediction model for telecom companies. Master's thesis, Massachusetts Institute of Technology

7. Patel P (2023) Data-driven customer retention: applying machine learning to predict churn in telecom. Presented at the international conference on big data and analytics
8. Miller W et al (2021) Taming the churn monster: how machine learning can save telecom companies money. Telecom Technol Mag 18(4):52–58
9. Brown S (2022) Predicting the future: machine learning for customer retention in telecom. Telecom Innov J 7(1):11–18
10. Williams D (2020) Don't let them go! using machine learning to keep telecom customers happy. Custom Retention Strat J 3(2):78–84
11. Gupta A, Singh S (2022) A deep learning approach for churn prediction in telecommunication industry. Int J Comput Appl 187(8):15–22. https://doi.org/10.5120/ijca2022918141
12. Parimala EH, Bosco VJP, Kumar PRV, Kolluru V (2024) Real-time adaptive intrusion detection system [RTPIDS] for Internet of Things using federated learning and blockchain. In: 2024 5th international conference on data intelligence and cognitive informatics (ICDICI). IEEE, pp 298–305
13. Manzoor A et al (2024) A review on machine learning methods for customer churn prediction and recommendations for business practitioners. IEEE Access
14. Szelag M, Słowiński R (2024) Explaining and predicting customer churn by monotonic rules induced from ordinal data. Eur J Oper Res 317(2):414–424
15. Gurung N et al (2024) AI-based customer churn prediction model for business markets in the USA: exploring the use of AI and machine learning technologies in preventing customer churn. J Comput Sci Technol Stud 6(2):19–29
16. Kolluru VK, Challagundla Y, Chintakunta AN, Roy B, Bermak A, Renuka Devi SM (2024) AI-driven energy optimization: household power consumption prediction with LSTM networks and PyTorch-Ray Tune in smart IoT systems. In: 2024 international conference on microelectronics (ICM). IEEE, pp 1–6
17. De Caigny A, De Bock KW, Verboven S (2024) Hybrid black-box classification for customer churn prediction with segmented interpretability analysis. Decis Supp Syst 181:114217
18. Jiang P et al (2024) Profit-driven weighted classifier with interpretable ability for customer churn prediction. Omega 125:103034
19. Liu Z et al (2024) Extreme gradient boosting trees with efficient Bayesian optimization for profit-driven customer churn prediction. Technol Forecast Soc Change 198:122945
20. Yang J et al (2022) Cost-sensitive machine learning for customer churn prediction in telecom services. In: IEEE transactions on services computing, vol 15, issue 1, pp 231–244. https://doi.org/10.1109/TSC.2019.2952222

Integrating AI, IoT, and Blockchain for Holistic Solutions

Pragmatic Review and Implementation of Models Targeted Toward the Identification of Fake Social Media Profiles

Bhrugumalla L. V. S. Aditya and Sachi Nandan Mohanty

Abstract In today's modern world, communications through social media have replaced conventional communication scenarios. Almost 70% of the world's population uses social media to share life events, impart knowledge, manage invites, etc. Due to this, it is essential to validate the authenticity of social media users. Currently, around 1.5 million fake accounts are on Facebook, which causes the spreading of misinformation and fake news. Researchers have come up with a plethora of models to detect fake currency; these models differ from one another in terms of textual subtleties, functional benefits, deployment-specific constraints, and application-specific potential future reaches.Researchers have come up with a plethora of models to detect fake currency; these models differ from one another in terms of textual subtleties, functional benefits, deployment-specific constraints, and application-specific potential future reaches. With this variation comes varied output in terms of their quantitative & qualitative performance as in scalability, accuracy, response time, computational complexity etc. Because of such a wide variation in performance, social media designers are compelled to validate several models for individual applications. This increases the delay & cost needed to deploy the final model for practical scenarios. Thus, it is challenging to identify optimal fake-social-media profile detection models for performance-specific & functionality-specific deployments. To overcome these issues and simplify the task of fake social-media-profile model selection, this text initially discusses existing models in-depth based on their functionality. After this discussion, an empirical & pragmatic analysis of these models is presented. Researchers are evaluated in terms of the accuracy of fake profile detection, delay needed for research, computational complexity, scalability, and cost of deployment. Based on this comparison, researchers can identify optimal fake-profile detection models. This text also proposes calculating a novel Fake Profile Detection Rank Metric (FPDRM), which combines these parameters to

B. L. V. S. Aditya (✉) · S. N. Mohanty
School of Computer Science and Engineering (SCOPE), VIT-AP University, Amaravati, Andhra Pradesh, India
e-mail: blvsaditya@gmail.com

S. N. Mohanty
e-mail: sachinandan09@gmail.com

A. Rocha et al. (eds.), *Integrating Advanced Technologies for Enhanced Security and Efficiency*, Engineering Cyber-Physical Systems and Critical Infrastructures 14,
https://doi.org/10.1007/978-3-031-91798-1_8

identify models that can be used for higher overall performance levels under different use cases.

Keywords Fake profile · Social media · Pattern · Deep learning · Bioinspired · Accuracy · Scalability · FPDRM · Delay · Cost · Complexity

1 Introduction

These days, the use of online social networks is expanding phenomenally and is of the utmost significance for businesses whose primary objective is to sell items and for celebrities whose primary aim is to market themselves by increasing their fan base. Our social lives have been improved due to the proliferation of social networks, but many problems still need to be handled. False identities on social networking sites often benefit from the negative aspects of social networking, including breaches in users' privacy, online bullying, harassment, and trolling [1]. False accounts that give the impression of having been formed on behalf of companies or individuals face the danger of damaging their reputations and resulting in a decrease in number likes on posts, followers of an user. Researchers get on OSNs.

Creating fake accounts are regarded as kind of illegal online activity that does the most significant damage. This is the consensus among experts. It is necessary to determine whether or not this offense has been committed before telling the user that a fake profile has been established. Their research, which is significant in this particular context since it is part of a larger body of research that our team has carried out on the topic of user profiling and the categorization of user profiles in OSNs, is noteworthy because of the following: Facebook is now one of the most widely used OSNs that can be accessed online. Users of Facebook may do various activities, such as building user accounts, adding friends, contacting one another, publishing status updates, photographs, and videos, and more, in addition to the features offered by Facebook itself. As a direct result of the rise in the popularity of the website, an increasing number of users are signing up for new accounts on Facebook. People who have lied about their qualifications in some ways are more likely to create fake profiles, sometimes called fake profiles. False profiles may also be termed counterfeit profiles. Usually, hazardous and undesired actions generating difficulties for social network users are carried out via phony Facebook profiles. For social engineering purposes, impersonating another person online to tarnish their reputation, advertising, and advocating on behalf of a person or group of people, false accounts are often established.

Researchers have built a diverse selection of models to detect fake accounts. Each of these models is distinct from the others in terms of the contextual subtleties, functional benefits, deployment-specific limits, and application-specific future possibilities that Researchers give. In addition to this variation, these models' quantitative and qualitative performance also varies, depending on various factors such as scalability, accuracy, response time, processing complexity, etc. Because there is such an extensive range of performance, the designers of social media platforms need to evaluate multiple models to accommodate their clients' various use cases. Consequently, the final model's installation for real-world situations will require more time and money. As a direct result, it might be challenging to design the most effective algorithms for identifying fraudulent social media accounts for deployments that are particularly well-suited to performance and functionality. The next part will investigate pre-existing models regarding how beneficial Researchers are, aiming to assuage these concerns and simplify choosing a fictitious social media presence template. Following this explanation, Sect. 3 will discuss an empirical and practical study of these models. Researchers will be compared for false profile detection accuracy, required latency for analysis, computing complexity, scalability, and deployment cost. In addition, Sect. 3 will discuss an empirical and practical study of these models. Researchers will be able to find the best strategies for detecting fraudulent profiles when they finish this comparison. This research also suggests calculating a novel metric known as the Fake Profile Detection Rank Metric (FPDRM). This metric combines the variables above to produce models that can be utilized for higher overall performance levels in various application scenarios. This article provides some findings that apply to the context of the models that have been analyzed, as well as some ideas for fusion procedures that could enhance the performance levels of the models that have been examined for future viability under real-time scenarios.

2 Literature Review

A large number of models have been suggested by researchers to detect fraudulent social media accounts; these models differ in terms of both performance and attributes. There are now millions of users on Facebook, Twitter, and Google+, as reported by [1]. A significant event in the history of Internet technology, Facebook's initial public offering (IPO) in May of 2012 was a huge success. Many large-scale and targeted advertising efforts employ these channels. Researchers can reach millions of users and consumers via social networks. Some people on social media will go so far as to steal sensitive data from other people. Creating fake accounts and utilizing them to impersonate fictional or actual people is a standard method for gathering massive amounts of data. Fake Facebook profiles are the focus of this inquiry. The researchers compromised Facebook's data, used social engineering to establish bogus accounts, and generally wreaked havoc on the company's revenue model. Privacy concerns are discussed in focus groups. As a result of this work, researchers created many user-centric protective measures (UCPM).

Similarly, according to studies cited in [2], the rapid online spread of sensitive personal data directly correlates with how much time individuals spend on SNS (Social Networking Sites) like Facebook, Twitter, and WhatsApp. Hackers may use fake Facebook accounts. Therefore, it's essential to stay away from them. Social interaction, visual signals, context, and spatial and temporal information affect human decision-making. In this study, researchers use OSNs, social networks that rely on human reasoning, to build an intelligent biometric system (IBS) that operates independently. It's a popular way to give your Facebook login a unique look.

Work in [3] mined "marionette" microblogging users, who are either created or hired by unseen "puppeteers," either manually or algorithmically. In contrast to casual microbloggers, marionette users have predetermined goals they strive to achieve in exchange for cash compensation. Researchers often re-share the same information to boost their "statistical popularity" or "statistical impact." In addition to harming microblog-based solutions like popular tweet selection and expert discovery, phony follower or retweet statistics mislead regular users. Researchers actively seek microblogging site users who have a history of using dummies. These dummies use elaborate methods to give their marionettes the appearance and behavior of real people. Tweeting patterns and social contacts (TPSC) are two sources of discriminatory data that researchers recommend using to address this problem. By combining these shards of information, researchers can separate heavy users from occasional ones. Using Sina Weibo as a data source, researchers found that the suggested model can accurately identify marionette users with an f-measure of around 0.9.

Furthermore, the researchers give a technique for gauging the credibility of retweets. Similarly, according to [4], technological progress is increasing in velocity. Everyone nowadays relies on OSNs to connect with new people and stay abreast of their existing friends' activities. While Facebook and other social networking sites generally make it easier for people to connect, they also have their share of problems. Fake accounts lead to security, offline, and hacking problems. Studies suggest that between 20 and 40% of all Facebook accounts are fake. This work constructs a Machine Learning Model to determine whether or not a Facebook user profile is fake by analyzing the User's Activity (UAML) on the platform. Automated profile recognition makes it feasible for sites to keep track of several identities, which would take much work to accomplish manually. False profiles have been the subject of several studies. Based on things observed of this research, a minimum set of basic guidelines is provided for identifying fake Facebook accounts.

Most people's social media and newsfeed activity occurs on Facebook and Twitter [5], respectively. Cluster analysis has been used to recognize online groups automatically, and node impact studies have been conducted to identify essential nodes in OSN graphs, among other methods of active investigation into OSN network structure. A fake OSN profile was created to spread malicious software, phishing scams, and unsolicited email. The researchers proposed a trust management and data protection model (TM-A-DPM) to address the abovementioned problems. Confidentiality, integrity, and privacy are all safeguarded through accurate trust factor detection in this architecture. Computer efficiency, speed, and security issues like data breaches and forgeries are all measured in controlled experiments. Recent studies have shown the importance of social networking to modern-day life, and these findings are summarized in [6].

Most individuals use social media daily. More and more individuals sign up for social networking sites every day, and many of them chat with each other at all hours of the day and night. In addition to their usefulness, these social media provide us with new security concerns. To identify those posing hazards on social media, researchers need a way to tell real profiles from fake ones. Researchers have proposed many different definitions to describe fake social media profiles. The research community must improve its ability to find fake user profiles on OSNs. Researchers use Machine Learning and Natural Language Processing in this study to better identify bogus accounts. Ultimately, researchers decided on using Random Forest (RF), a classifier based on trees.

Every second, billions of people worldwide produce vast data using social media sites like Facebook, Twitter, Instagram, etc. Numerous companies employ spam and fake accounts to gain market advantage. Locating such Twitter accounts is the focus of this research. In [7], experts classify Twitter accounts based on their content using feature engineering, network representation, and natural language processing (FENR). User identification may aid in generating more comprehensive user data, as suggested by [8]. Experts in their fields tend to be inquisitive—most works written nowadays center on characters' relationships with one another or their profiles. Due to privacy and data crawl constraints, real-world social network user data may be insufficient or non-existent. Profiles, user-generated content, and social relationships are all types of user data. There may not be many carryovers from previous research. The study's authors provide a user identification framework based on various user data points (MUIUI).

Multiple process crawlers are constructed for Twitter and Facebook before any analysis can begin. Extraction of locations and organizations from profiles and user-generated content is facilitated by entity linking and named entity recognition. Profiles and user-generated content are mined for URLs. Researchers combine sites and connections and then devise algorithms to determine how similar the user's display name, all organizations, profile location, all URLs, following organizations, and user ID are to one another. Researchers then propose a fusion classifier-based method of user identification. MUIUI has an F1 score of 86.4% on the dataset. If any user information is missing or corrupted, MUIUI can mitigate the damage.

According to the research in [9], everyone uses social networking sites to stay in touch with family and friends, research potential new jobs, and conduct business. Facebook, Instagram, LinkedIn, and Twitter are well-known options. The user's name, picture, and geographical location play roles in the authentication process. In these systems, bad actors may pretend to be another user to steal their information. The researchers may use fake profiles to collect data like photos, contacts, and videos. This villain might be found using privacy detection techniques. This study presents data mining techniques for protecting sensitive user information and spotting phony profiles on social media. This study employs a method for identifying fake accounts on OSNs by analyzing the trickster's daily behavior and routines and the content of the posts they've posted recently. It's possible to spot fraudulent charges by comparing their network similarity and attributing threshold values to the profile of the actual person. Classifiers such as ESVM NN use feature reduction to improve accuracy. This task, which compares the intended effort's efficacy and relevance to past research, necessitates setting up OSN accounts for experiments and investigating the most current postings, comments, images, and web searches.

In [10], researchers found that the higher the number of OSN users, the faster the dissemination of false information. Users are allowed to express themselves and share their ideas freely via OSNs. Patterns of follow-up social contact are also gleaned. Twitter news articles are given a credibility rating by the SENAD (Social Engagement-based News Validity Detection) algorithm, which considers the credibility and objectivity of the individuals commenting on the stories. The suggested SENAD model accounts for indicators of user social interaction such as follower-to-follower ratio, account age, bias, etc. The recommended approach successfully detects 93.7% of all fraudulent data and reports. Images that include text are more compelling to the eye than plain text and may be used to spread misinformation. Fake photos may be more easily detected when they have certain traits. Photoshopped images are used to spread fake news. Due to the common occurrence of both physical alterations and bad sentiment in fraudulent images, CredNN employs the spatial features of CNNs to check for both.

There is evidence that sophisticated ongoing threats and other harmful activities are being carried out by phony identities [11]. Fake online personas are the topic of this essay. The only way to detect phony social media accounts is to scrutinize user profiles and accounts. Thus, the most damaging kind of cybercrime has been the creation of bogus social network profiles. This wrongdoing must be uncovered before warning the user about the fake user profile. A plethora of algorithms and methods for spotting fake accounts are detailed in the research. This study examines the role of false personas in APTs and guides how to identify bogus social media profiles. Researchers will evaluate three supervised machine learning techniques, Deep Forest (DF), Decision Tree (DT), and Naive Bayes (NB), to establish if a profile is genuine or not under real-time scenarios. According to [12]'s research, social media facilitates information exchange and fosters dialogue between different communities. It quickly gained widespread attention for its success in getting users to engage with each other and the content they see and to like, share, and comment on that content. There is no vetting process for the information shared on social media sites or online forums as in the news industry. People who use social media are vulnerable to being misinformed by harmful writers. This research proposes a method for identifying false news by analyzing how people react to potentially dangerous information. Using the bag-of-words model, the TF-IDF, and visual emotional reactions (VER), researchers developed models to predict whether a piece of news was phony or genuine. Testing has been done to gauge the efficacy of our model, where it was found to be superior to standard techniques.

According to [13], social media is a digital community where users of all ages may interact. On this site, people exchange their abilities. However, others abuse the system. Users and bots will create fake profiles to disseminate malicious information. The suggested study delves into issues of confidentiality and safety on the site. In this context, the progress of social media over the preceding year is studied, and the body of literature is collated and categorized. There is potential in content analysis and profile attribute assessment. Using these techniques, a new model based on Attribute Estimation and content Analysis (AECA) is built, and its functioning parts are explained. Eventually, researchers demonstrated the efficacy of the suggested functional model under different use cases. According to scholars in [14], social life has developed to be associated with online social networks or social connections between individuals with similar interests, activities, experiences, or real-world relationships. Researchers no longer treat social life in the same manner. Keeping in contact with friends and receiving messages from them is a breeze. The extensive growth of social networks has led to increased fake profiles and online impersonation. Researchers use a hybrid approach based on AI to detect bogus accounts.

Based on the findings, the suggested approach may effectively identify phony profiles. Forgery detection procedures for social media are well-established. The efficiency of detecting fake profiles will increase. Experts recommend using AI and NLP methods to identify fake profiles better. Experts employ techniques like the Random Forest Classifier, Support Vector Machine (SVM), and Optimized Nave Bayes to group profiles into meaningful groups. Since this is a mechanized procedure, an unreliable social network may overgrow. These three algorithms uncovered the true identity of the intended users. With fewer features, our technique correctly identifies in the training dataset 98% of the accounts.

According to studies in [15], online social network (OSN) security is an increasingly pressing topic. Locating compromised OSN accounts is crucial for platform security. The authors presented a novel strategy based on supervised analytical hierarchy (SAHP) to find hacked accounts. Researchers began by gauging a user's expressive actions to paint a fuller picture. After that, researchers used the AHP method and the information gain ratio to assign relative importance to each attribute. As part of the detecting process, many thresholds were applied. The precision and accuracy of the SAHP, at 96.4 and 81.7%, are high. This novel methodology improves upon previous methods used in COMPA (detection of compromised accounts on social networks). In [16], the use of deep learning to identify fake social media profiles was investigated. Identifying telltale signs of counterfeit profiles requires comparing authentic and fake accounts. First, researchers look at the prevalence of fraudulent profiles across social media platforms. The experiments show how to acquire phony user accounts, highlight attributes, and assemble a dataset from which the classifier can learn. Several categorization methods for identifying bogus accounts are compared in this study, and the current model's efficiency was observed to be better than other methods.

Work in [17] claims that multimedia-related content such as text, images, and videos is often shared on social networking sites (OSN). Current OSN systems need help spotting fake accounts. Criminals utilize phony profiles to spread viruses, malware, and malicious links. DeepProfile is a deep neural network (DNN) technique for dealing with false accounts. Inspired by deep learning's successes in computer vision, especially in automatically extracting and expressing features, this concept was developed. Instead of using static machine learning methods, researchers train a learning model on a dynamic CNN. Researchers recommend using a pooling layer in neural network training to improve efficiency.

Experiments show that research can classify rogue accounts more accurately and with less loss than traditional learning techniques. According to studies cited in [18], social networking sites facilitate the formation of social networks among individuals with common interests, backgrounds, or real-world ties. If users register with their representation (in the form of a profile) and their social connections, they will have access to a social network service. Internet-based social network services provide a means for users to communicate and connect. Social media platforms are vital in modern life. Millions of people share private data. The public could take advantage of social networks as they rapidly grow in popularity. Security threats and disclosing confidential information are becoming primary concerns for those who often use social networking sites. Discovering fraudulent identities or harmful information might be challenging without extensive study. Recent research on the detection of harmful content considers user profile characteristics. Most of the contemporary approaches should be assessed further. With the use of machine learning and natural language processing (ML NLP), the authors of this study provide a novel strategy for identifying fictitious profiles on social networking sites.

Social networking sites are on the rise, as shown by the statistics presented in [19]. The focus on the user is more significant on Twitter, Facebook, and Instagram. Users can network, exchange information, and learn from one another via social media. Having access to this vital data is made possible by social networking sites. Hackers can earn money with this much social media data. These criminals create fake profiles and disseminate spam. The URLs and advertising that make up the bulk of spam are annoying. Inappropriate or spam material is a significant issue in online social communities. Online communities must be able to identify spam. In this study, researchers provide an AI-based technique for identifying spam on Twitter. The research team built their model using a neural network model, a random forest, and a vector support machine. The proposed SVM method outperforms RF and ANN regarding accuracy, recall, and F-measure. Their research might be used to safeguard social media platforms from cyberattacks and to keep tabs on users' postings in search of offensive material or photographs that have been doctored. Research in [20] suggests that social media like Twitter, Facebook, Instagram, and LinkedIn are crucial to human well-being. Social media use is pervasive. Containing phony profiles is also possible.

People construct fake accounts for illicit purposes such as spreading misinformation, phishing, data exposure, and identity theft. This article explores false profile detection models. Indicators such as the number of followers, the number of friends, the number of status updates, and so on may help to distinguish between fake and genuine Twitter accounts. Scholars use machine learning to construct models. Researchers employ LSTM, XG Boost, Deep Forest, and Neural Networks. Researchers and professors employ specific criteria to verify social media expert profiles. Comparatively, real profiles result in a 0 and false ones in a 1. Using a neural network, XG Boost achieves a 99.46% success rate. Fake accounts may be removed from databases to halt cyberattacks.

The research published in [21] found that social media sites like Twitter are widely used for both the distribution and consumption of news due to their widespread availability and relative simplicity of use. The unreliability of online communication needs to be addressed. In a short time, these sites have spread intentionally misleading fake information. That's not the case, and this is fake news. It is becoming more difficult to track down social media-related news. It's challenging to detect false statements because finding the information you need to identify them is hard. Researchers used a combination of a ResNet and a Support Vector Machine to determine what is real and what isn't. Using an RNN with bidirectional gated recurrent units. Researchers could transform the textual features of news articles and user comments into numerical feature vectors. A support vector machine (SVM) trained with an RBF kernel was used to distinguish authentic/fake news based on encoded characteristics. The presented framework outperforms state-of-the-art approaches in experimental settings. According to studies compiled in [22], online social networks (OSNs) facilitate communication, cooperation, and the dissemination of knowledge. Researchers may classify these OSNs in a variety of ways, depending on the services they offer to users, such as apps that help people make and keep in touch with friends, apps that make it simple to share and discover new media, and app communities where people can discuss and share relevant information. However, there may be issues on specific platforms, such as false accounts, although they improve people's social lives. Hackers often form links with innocent people using false identities to engage in hostile, illegal, harmful, or discriminating actions. However, the motivations vary for each social media platform. This study examines the various social media sites on the web and the advantages they provide to their members. After establishing that hackers need to construct fake identities to access specific networks, the study explains why they do so. Researchers also detail the harm of false identities due to network attacks.

Users of popular social networks may communicate with one another and share media such as images and videos. False accounts are a significant vulnerability in these systems. According to [23], a novel approach is proposed for spotting fake social media profiles by analyzing the degree to which their real-world peers have comparable online networks. Measures of similarity were built using the adjacency matrix of the social network graph. The suggested model was evaluated using data from Twitter. Medium Gaussian Support Vector Machines (MGSVM) reliably identify fake profiles. According to [24], almost everyone spends most of their time on social media sites. Ideas, friendships, enterprises, and more all start on social media. While there are many positives to using online social networks, there are also potential drawbacks from a research perspective. People on these platforms make up identities to harass or injure others.

Machine learning algorithms can tell the difference between real and false profiles. Algorithms that rely on machine learning may predict the future and classify data using various statistical methods. If researchers use a hybrid machine learning approach, researchers may be able to differentiate between model results. Eight different categorization algorithm combinations were tested on data collected from a popular social networking website. Multiple methods were used, including Random Forest, SVM, LR, KNN, and DT. SVM + Logistic Regression (LR) + Neural Network achieved the highest accuracy. Researchers created a model for identifying bogus accounts and found that the hybrid technique offered the highest accuracy.

Most individuals now rely heavily on social media, as shown by studies published in [25]. (OSN). People use social media to expand their social networks, disseminate information, organize events, and learn about topics of interest. In the last six months, Facebook has deleted 3 billion accounts; among the monthly active users, 5% were phony. The Washington Post reports that Twitter has been removing almost 1 billion questionable accounts in recent months. Users create fake accounts to tarnish their reputations, propagate misinformation, and encourage sarcasm, all magnets for cybercriminals. Artificial intelligence methods such as supervised learning and support vector machines may be used to spot a phony profile. Convolutional Neural Networks (CNN) are recommended for this purpose by many artificial neural network methods, such as those for face recognition, prediction, classification, and clustering. Time series prediction, feature detection, and facial recognition are all used in this work. No sham accounts would be created. According to studies cited in [26], many people are concerned about being misrepresented on social media. Phony profiles can fool some users. The goal of this research is to avoid the need for any additional inquiry into fake social media accounts. Some behavioral biometrics, such as a user's typing habit (UTH), may be able to distinguish between genuine and counterfeit versions without disrupting users' typical experience. The DEEP ID model examines input data from keyboard, mouse, and touch to detect fake profiles on Facebook. The recommended approach discreetly identifies phony user accounts across desktops, laptops, and connected devices. The DEEP ID model can see whether authentic profiles have been compromised midsession. This study aims to show that omitting feature extraction from raw data applications of user identification algorithms may improve performance by reducing the likelihood of making a lousy attribute choice. The suggested DEEP ID model is constructed on a multi-view deep neural network to identify users from unprocessed raw data on typing patterns. DEEP ID outperforms standard machine learning classifiers. The evidence supports the hypothesis. Test results indicate that Deep ID is reliable and accurate for profile recognition.

According to [27], today's society relies heavily on social media for everything from personal interactions to pursuing professional goals. Facebook, Instagram, LinkedIn, and Twitter are well-known options. The user's name, picture, and geographical location play roles in the authentication process. In these systems, bad actors may pretend to be another user to steal their information. The researchers may use fake profiles to collect data like photos, contacts, and videos. This villain might be found using privacy detection techniques. This study presents data mining techniques for protecting sensitive user information and spotting phony profiles on social media. This study employs a method for identifying fake accounts on OSNs by analyzing the trickster's daily behavior and routines and the content of the posts they've posted recently. It's possible to spot fraudulent charges by comparing their network similarity and attributing threshold values to the profile of the actual person. Classifiers such as E SVM-NN use feature reduction to improve accuracy. This task, which compares the intended effort's efficacy and relevance to past research, necessitates setting up OSN accounts for experiments and investigating the most current postings, comments, images, and web searches. Research in [28] demonstrates that Instagram is an essential tool for marketing, advertising, and social engagement because of the proliferation of Internet access. Millions of people utilize it, yet some of them create fake accounts. Although there are many positive aspects of the Internet, there are also negative aspects, such as spammers and cybercriminals, who threaten online social networks. Researchers employ unethical methods to boost their social media following since it is often accepted that a user's fame is directly proportional to the size of their fan base. Academics have developed a wide range of social networking apps. This study recommends using an automated system for detecting fake Instagram profiles to protect users' social lives. The use of supervised learning systems can see phony Instagram accounts. Phony profile IDs (PPI) are cataloged and maintained in a data dictionary to aid law enforcement in their battle against fake social media profiles. Attempts have been made to evaluate and contrast various categorization strategies for training datasets.

Detecting fake social media profiles using an artificial neural network is recommended in [29]. A unique algorithm was developed and put into operation to see the tell-tale signs of fake accounts and uncover their origin. Researchers considered a Java-based mobile app for many different use cases in a development environment, including the typical social network surfing experiences. Five hundred legitimate and 500 fake accounts were compared and contrasted to identify the distinguishing features of fake profiles. As a result of this study, researchers have additional, novel patterns to use as criteria for fake accounts. The findings provide the basis for defining criteria for identifying counterfeit versions in practice. According to research published in [30], today's youth rely heavily on in-person meetings with friends and family or virtual connections with people online. Scientists' perspectives on social life have changed. Keeping in contact with friends and receiving messages from them is a breeze. The exponential growth of social networks has led to increased fake profiles and online impersonation. Uncontrollable. Researchers use a hybrid approach based on AI to detect bogus accounts.

Based on the findings, the suggested approach may effectively identify phony profiles. Forgery detection procedures for social media are well-established. The efficiency of detecting fake profiles will increase. Experts recommend using AI and NLP methods to identify fake profiles better. Experts employ techniques like the Deep Forest (DF) Classifier, Support Vector Machine (SVM), and Optimized Nave Bayes (ONB) to group profiles into meaningful groups. Since this is a mechanized procedure, an unreliable social network could multiply. These three algorithms uncovered the true identity of the intended users. With fewer features, our technique correctly identifies 98% of the accounts in the training datasets.

According to studies cited in [31], the social lives of internet users are increasingly centering on OSNs as they gain widespread adoption. Academics use OSNs for various purposes, including but not limited to coordinating events, exchanging information, and doing business. The rapid expansion of OSNs and the vast quantities of user-submitted data have made hackers and imposters increasingly common, allowing them to steal information, disseminate misinformation, and conduct other crimes. Researchers use account characteristics and classification algorithms to identify suspicious activity and fake profiles. Some exploited account features negatively affect the outcomes of solo classification algorithms, while others have little effect.

For this research, researchers use SVM & NN to identify fraudulent Twitter accounts and bots, and researchers do so by using four different feature selection and dimension reduction techniques. Researchers employed a combination of support vector machines (SVMs), neural networks (NNs), and our own newly created approach (SVM-NNs) to verify the authenticity of the account researchers were investigating SVM-NN can classify 98% of our training dataset with just a subset of the available variables. Research in [32] suggests that social network analysis is an important and promising area of study in the modern digital age due to the widespread effect of online social networks on people's lives. Actual users are often adversely impacted by fictitious accounts on social media. Social networks may be more resilient if fraudulent users and influential people are more easily uncovered. To strengthen the reliability of social networks, a new energy transfer-based PageRank validation approach was developed and explored in this research. This approach makes use of cellular automata (CAs). This research creates a novel metric, the influence score, to evaluate the impact of a user based on the exchange of emotional energy between Social Network users during the sending and receiving of connection requests. The study also discusses using energy-based effect assessments in social network settings to detect fraudulent accounts. Our method has been proven effective and cost-effective via several simulations and real-world online traffic assessments.

There is a rapid pace of technological development, as stated in [33]. Phones with better technology exist. Nowadays, Everyone uses social media websites to connect and learn more about their interests. But, problems like fake accounts and online impersonation have increased due to online socializing. Customers get useless data from bogus individuals who browse the web. According to studies, 20–40% of Facebook profiles are fake. Thus, frameworks are used to identify bogus social media profiles. According to studies compiled in [34], the quantity of available social networking sites constantly expands. The focus on the user is more significant on Twitter, Facebook, and Instagram. Users can network, exchange information, and learn from one another via social media. Having access to this vital data is made possible by social networking sites. Hackers can earn money with this much social media data. These criminals create fake profiles and disseminate spam. The URLs and advertising that make up the bulk of spam are annoying. Inappropriate or spam material is a significant issue in online social communities. Online communities must be able to identify spam. In this study, researchers provide an AI-based technique for identifying spam on Twitter. The research team built their model using a neural network model, a random forest, and a vector support machine. The proposed SVM method outperforms DF and ANN regarding accuracy, recall, and F-measure. Their research might be used to safeguard social media platforms from cyberattacks and to keep tabs on users' postings in search of offensive material or photographs that have been doctored.

Many rely on social networks as a means of communication, according to research published in [35]. People may join up and talk about business on this platform. As a result of its openness, anybody may join using a false identity. It is crucial to detect and delete these fake accounts. Twitter, Facebook, Instagram, WhatsApp, WeChat, and many more are all examples of well-known SNS. Some safeguards are in place on these SNSs to detect fake profiles and prevent harmful activity, but more is needed. In this study, researchers examine the prior literature on the topic of seeing and banning phony Facebook accounts, and researchers do so by using a novel approach based on graph-based activity diagram sets. Researchers have begun building algorithms to identify false news and the people who purposefully propagate it online since the detrimental effects of this activity have become more well-known [36]. Automation, manual, and hybrid systems have all been presented by both businesses and universities. Researchers in this study extensively analyzed the more accurate traits for spotting social network members who propagate false news online. Researchers gathered data on Twitter users' social networks, individual lives, and interactions with material and other users. After that, researchers used offline deep learning algorithms and actual people on the internet to categorize user profiles as trustworthy. Experiments with statistical backing show what data helps people and robots recognize potentially hazardous users. Our findings should aid in creating more effective safeguards against misinformation and its purveyors.

Cybercriminals using stolen or spoofed identities increasingly target online social networks (OSNs), as shown in [37]. Spam, false news, and astroturfing are just some activities that automated agents (or "social bots") are responsible for on social media platforms. It is challenging to spot social bots since they act like real people. DeepSBD is a deep neural network model that pays attention to its inputs and is used to identify social bots on OSNs. DeepSBD builds its models of user behavior on top of information about their actions and interactions throughout time. Using BiLSTM and CNN, researchers can simulate how OSN customers interact with the system. Sequences supplied into a two-layer stacked BiLSTM reflect profile, temporal, and activity information while also being delivered to a deep CNN. DeepSBD outperforms state-of-the-art and gold-standard algorithms on five real-world benchmark datasets.

Researchers also tested DeepSBD with varying ratios of social bots to innocent people and discovered that this skewed dataset affected the model's classification performance. Research comparing the discriminatory value of several behavioral components found that profile features and content behavior were the most useful in spotting social bots on OSNs. The [38] says that software companies often breach password databases. These flaws go undiscovered for a long time. Some companies are in the dark about the occurrence. In some instances, companies may wish to refrain from reporting security breaches even after they have been made aware of them. A robust breach-detection system, sometimes known as "Honeywords," is necessary to identify violations of password databases. Researchers generate alternative passwords, called "honeywords," and store them alongside the actual password. Any attempt to log in using a honeyword breaches the password database since genuine users aren't supposed to know about them.

The first four are creating honeywords, using typo-safety techniques to avoid false alerts, implementing an alarm policy upon detection, and analyzing the system's resistance to different types of attacks. In their experiments, researchers check how limited honeyword generation is. Researchers demonstrate the insecurity of the Distance Protocol by recommending a "Multiple System Intersection attack considering Input." The "user-profile model," "append-secret model," and "evolving-password approach" are all highly recommended by the research community. Due to the "approximate flatness" of these techniques, the honeywords that researchers create are interchangeable with passwords. The technologies researchers have developed allow us to circumvent many current obstacles. Our "evolving password model" is unrealistically simplistic in experimental studies. Researchers have compared our suggested models to those in use to disprove our assertions, subjecting each to different attacks.

Because social labeling is a contentious subject in the academic community, this website uses IoT data to provide targeted recommendations for further reading (IoT). This study provides a recommendation framework for Rough-Fuzzy Clustering (RFC) [39], a method for tailoring online content to individual users. The Web Page Recommendation (WPR) framework, which is based on a user profile and an ontology, is shown. Documents from the Internet of Things weblog were collected for data cleansing and knowledge acquisition. The information and framework of the student profile in the profile ontology component are well-defined. Relevant data was located using an efficient and unique similarity measure, and RFC rules were improved using Chicken Swarm Optimization (CSO) for effective WPR. Finally, the top e-commerce platforms benefit from this ideally rules-based output.

Cluster analysis was used to categorize people based on the information they provided. The data was evaluated using performance metrics, and the best sites were suggested to users per the methodology presented. According to [40]'s findings, semantic analysis and design are essential for providing context to data. If not, semantic retrieval would be impossible. Based on their findings, the authors of this study suggest an integrated ontology model beneficial to both farmers and agricultural service providers. Researchers utilize ontology to examine data from the farming system, a semantic web domain. Farmers today are turning to their mobile devices and social web apps to get their desired answers. Primary data was gathered via the use of questionnaires by the researchers. Participants in this research were twenty farmers and twenty agricultural traders, with a mean age of 29 and 35, respectively. All the guys in the study were from the Indian state of Odisha. Social and agricultural ontologies serve as the backbone of the domain ontology, which also incorporates wikis, podcasts, and social networks.

Next, researchers evaluate how social web components contribute to the domain ontology. The third stage involves merging domain ontologies with social ones. Based on a survey of rural areas' worth of social media websites, the integrated ontology derives its axioms for classes and properties. This research presents a model incorporating social semantic ontology for a narrow domain (agriculture). Social semantic ontology is developed using a holistic strategy. Suppose you're running a social network, hosting material, or blog. In that case, the best-case scenario is to provide users with the information they seek in the way they want it presented in the social use cases.

According to [41], digital tool adoption rates are skyrocketing. There has been a rise in malicious activity among users. Millions of people worldwide use social media sites like Twitter and Facebook. The proliferation of potentially dangerous material via fake accounts is only one problem that has arisen due to the widespread use of online social networking. Fraud, abuse, and spam are commonplace on social networks because of the ease with which people may create new, false identities. An online social network's usability hinges on fixing these issues. Researchers use SVM, LR, RF, and k Nearest Neighbours (KNN) in this study. Z-Score and Min–Max were also used to refine precision further. It can identify Twitter bots and fake profiles. Deep Forest (DF) and KNN fared well when processed using our method. According to [42], researchers' capacity to connect individuals and disseminate knowledge is a significant factor in the popularity of social networks (videos, photos, and communications). Fake online personas are the topic of this essay. Information analysis of profiles is essential for identifying bogus social media accounts. The creation of false statements is the most problematic behavior on social media. This wrongdoing must be uncovered before warning the user about the fake profile. Identifying fraudulent profiles may be done in many ways. This article contrasts the authentic user profile detection methods of Decision Tree (DT) and Naive Bayes (NB), which showcased better performance levels.

It has been shown via research published in [43] that individuals from all corners of the world utilize, exchange, distribute, and interact through online social networks. Online, people are more likely to believe and spread false information. Because of the anonymous nature of its producers, some user-shared content is suspect. The results of their research recommend using a subjective standard to judge the credibility of posts on social media. The proposed strategy is evaluated using Twitter. Predictive metrics are calculated to establish the efficacy of the suggested approach for the particular use case. Millions of people use Twitter to connect with friends and family, share news, and learn about new things, as shown by [44]. Twitter's open API allows for fake account assaults. Online fraud, consumer data collecting, marketing, defamation, and control are some of the many uses for fake accounts. As a bonus, they may also be used to artificially increase the number of visitors to a blog or website. Protecting businesses and their clients requires finding and deleting fake social media profiles. To identify fake Twitter accounts, this article recommends using Twitter's built-in flagging system for such charges. Using information gain and correlation, researchers determined which subset of the original characteristics was most accurate. A Logistic Regression (LR) classifier achieved a 93.5% accuracy rate when fed data from one million Twitter users. An algorithm has been developed that can identify fake profiles. Compared to state-of-the-art studies, the proposed model performs better in many practical scenarios.

Users of popular social networks may communicate with one another and share media such as images and videos. False accounts are a significant vulnerability in these systems. This research [45] proposes a novel method to detect fake social media accounts based on network similarities between users' friends. Measures of similarity were built using the adjacency matrix of the social network graph (AM SNG). The effectiveness of the model was evaluated with the use of Twitter data. According to studies cited in [46], millions of individuals regularly interact with others via social networking sites like Facebook and Twitter. By creating fake profiles, OSNs may be made safer for users. Through the use of OSN's nodes and linkages, fake Twitter accounts may be exposed. The proposed classifier sorts the photos in the dataset into categories that might help identify spam profiles. Data from phony accounts on Twitter is gathered using a crawler to identify fraudulent profiles. To understand how researchers turned to the feature set analysis. The similarity of the two shapes may be calculated using the user behavior detection matrix. The kernel functions are used by the multi-variant SVM (MV SVM) based classifier for accurate classification scenarios.

According to research in [47], social media's widespread adoption and indispensable nature can be attributed to its pervasive use as a platform for people to express and communicate their unique perspectives, ideas, and information freely and openly. But Twitter and other social media platforms are rife with false profiles constantly being created. These individuals aim to undermine the platform's credibility by disseminating incorrect information or misleading other users. The widespread use of machine learning techniques in social networks to automatically identify phony accounts is a vital defense against this threat. This identification procedure is dependent on machine learning algorithms, which are susceptible to aggressive assaults during the testing and training phases when threat actors refine their armaments and initiate a variety of complex attacks. Our research investigates the spread of malicious assaults in AdaBoost, a machine-learning framework for identifying phony Twitter accounts. In addition to proposing and evaluating k-NN as a countermeasure to the effects of the adversarial assaults we've designed, we also suggest and consider k-NN as a preventative measure.

In [48], a method for analyzing the difficulties of producing high-quality big data to detect false information on social media was outlined, as were methods for identifying the relationship between big data analytics and context-based news detection on digital media in the era of big data, as well as for gaining knowledge about the most widely used techniques for detecting fake news online. Forty-two academic papers published in 10 of the most prominent online repositories in the world were subjected to a scoping review. A considerable positive correlation was discovered between high-quality big data analytics and detecting false news in digital media. In this age of misinformation, people rely on artificial intelligence, fact-checking websites, neural networks, and new media skills more than any other strategy. Due to hidden agendas, an abundance of fake information online, a lot of unstructured data, the rapid spread of disinformation, and fake user accounts, the findings also demonstrated how difficult it is to construct authentic big data for identifying dishonest content on digital media platforms. The study contributes to the existing corpus of knowledge by critically analyzing the relationship between big data analytics and the dissemination of false information via digital media in the data age. This scholarly paper has societal ramifications because it provides actionable suggestions for combating the epidemic of disinformation and shielding individuals from dire consequences. These results affect those who design and develop digital media applications and legislators, decision-makers, government officials, civil society organizations, higher education authorities, media employees, and instructors. The recommendations in this document can be used as a guide when drafting legislation to limit the damage caused by false online news.

In [49], analytical methods for detecting and preventing Twitter-specific social media abuse were investigated. It was possible to categorize Twitter users according to their effect sizes by analyzing their profile layouts and interaction patterns for cognitive studies and then computing their asset values. Using k-nearest neighbor (K-NN), support vector machine (SVM), and artificial neural network (ANN) techniques, user characteristics were classified into three categories: famous (active), observer (passive), and spam bot (malicious). To organize the data, fundamental user characteristics such as number, distribution, and size were combined with suggested time series such as the primary and cumulative moving averages. The technique with the highest accuracy was K-NN. In every category, K-NN outperformed the alternatives in terms of F1-Score. According to the data, the accuracy of classifying items ranged between 92.33 and 96.7%. The proposed method identified and separated spam-bot malicious accounts from popular-active, observer-passive, and popular-passive accounts.

According to research in [50], the pervasive dissemination of false information about the coronavirus epidemic 2019 is a separate issue for the global community (COVID-19). This issue is exacerbated by the proliferation of OSN platforms, which make it simple for users to fabricate and distort the information they share. This paper examines the psychological and behavioral effects of misinformation with religious overtones on OSN users. During India's COVID-19 curfew, this article thoroughly examines how social media users responded to erroneous information about the Islamic community. Three crawler programs (Tweepy, Selenium, and Beautiful Soup) were used to acquire data from Twitter users to identify as many individuals as possible affected by this misleading information.

Each fragment of text in a labeled dataset is assigned one of four possible responses: E (endorse), D (deny), Q (question), or N. (neutral). We analyzed the data in five phases to determine the number of E, D, Q, and N participants, the wording of the communications, and the results of repeated exposure to the same content. According to the data, individuals were more likely to accept unverified messages as accurate during the pandemic and isolation period. In addition, it was discovered that people assimilated the negative aspects of online content, which provoked negative emotions such as resentment, wrath, anguish, and fear. Individuals with similar worldviews congregate online, creating communities that overtly express hostility towards those who do not share their worldview. One of the automated systems described in the article for identifying potentially fake news on Twitter is a deep learning-based model for pose detection. The posture predictor contributes to the validation of information by predicting the attitude of a text towards it by monitoring the distribution of various user reactions to a news headline. The proposed model outperforms existing systems by combining deep learning (convolutional neural network [CNN]) and sentence embedding [bidirectional encoder representations from transformers (BERT)]. Using the gold standard SemEval attitude dataset, efficient evaluations are conducted. Along with the findings of this investigation, a newly labeled dataset is produced and made available for future research in this field.

The objective of [51] is to solve the crucial problem of identifying potentially malevolent critics in social media discussions. We have focused on the subset of suspect authors known as trolls. We devised machine learning-based identification models to differentiate between false and genuine reviewers. We also employed emotional analysis methods to ascertain the typical tone of spammer remarks. In addition, sentiment analysis can be performed using a lexicon-based approach or machine learning methods. We utilized lexicon-based sentiment analysis due to its enhanced ability to identify the peculiar expressions frequently employed by automated writing programs. Analyzing emotional states, we discovered that Accuracy = 0.95 and F1 = 0.80. Support vector machines yielded the most significant results when employing machine learning techniques, with Accuracy = 0.986 and F1 = 0.986, when utilizing a dataset containing all of the selected characteristics. We conclude that a machine learning-based identification model is more accurate than a lexicon-based mood analysis model. However, the difference is less pronounced than it was for the F1 measure.

According to research in [52], several detection methods have been proposed to defend recommender systems against group marketing attacks; however, these methods cannot reliably identify attack groups that result from aggressive assaults (such as GOAT) or hybrid attack groups. Graph convolutional networks (GCNs) and cliques form the basis of the proposed two-stage KC-GCN method. We construct a user association network to zero in on crucial users, generate suspicious prospective groups, and calculate the user nearest-neighbor similarity. The user connection graph is built by calculating the weight of each edge between any two users by comparing their similarity over suspicious time intervals on each item. We combine the restored user initial embeddings to identify perpetrators with structural characteristics otherwise concealed within the user relationship network. On the Netflix and selected Amazon datasets, the identification efficacy of KC-GCN for various group marketing attacks exceeds that of state-of-the-art methods. KC-F1-measure GCNs can reach a maximum of 93% and 87% on these two samples.

According to research in [53], online social networks (OSN) are popular locations for sharing all types of information, including text and visual media. On extant OSN systems, however, phony profiles pose a challenge. The adversary uses sock puppet accounts to propagate deceptive information such as malware, ransomware, and malicious URLs. DeepProfile is a deep neural network (DNN) approach to phony account problems. It was inspired by the extraordinary successes of deep learning in computer vision, mainly automated feature extraction and modeling. We construct a dynamic CNN to develop a learning model for classifying false profiles as an alternative to static machine learning. Significantly, we propose a novel pooling layer to enhance the training efficacy of neural networks. The results of the experiments demonstrate that we can outperform conventional learning methods in the classification of malevolent accounts in terms of accuracy and loss.

According to [54], the online community's use of social networking sites is increasing. Twitter, Facebook, and Instagram, among others, devote considerable time to online consumer interaction. People meet new people and share their thoughts and experiences through social media. Users can access a wealth of beneficial information on numerous social media sites. Due to the abundance of available data, hackers can readily exploit social media information. These cybercriminals hijack identities to disseminate spam and other irrelevant information. Regular users may be impeded by advertisements and potentially hazardous links that populate trash pages. This irrelevant content poses a significant problem for social media platforms. It is essential to have a system in place to detect advertising on social media platforms. This article proposes employing an AI-based technique to identify abusive statements on social media platforms like Twitter. We constructed a model for this strategy by combining the vector support machine, the neural artificial network, and the random forest method. Regarding accuracy, recall, and F-measure, the suggested support vector machine algorithm outperforms the RF and ANN algorithms. As a result of this study, social media platforms can monitor and capture user-posted images more effectively to identify objectionable content and fake photos and prevent cybercrime.

The research in [55] indicates that technological progress is advancing rapidly. Today, everyone uses some form of social media to communicate with new people and monitor the activities of old acquaintances (OSN). Social networking sites such as Facebook can improve your social life, but it comes with drawbacks. False identities constitute a significant source of inactive, confidential, and security issues. According to studies, 20–40% of profiles on social networking sites such as Facebook are fictitious.

Consequently, the objective is to construct a failsafe model to determine whether a given Facebook identity is false based on a user's social media activity. Automated identification enables computers to assist websites in managing the vast number of profiles, which would be challenging to do manually. Numerous works have been written on the subject of detecting false profiles. This paper proposes a minimal set of generic characteristics for identifying phony Facebook profiles based on the study's conclusion that a minimal set of primary traits is essential for achieving such objectives.

Based on the research presented in [56], this paper delves deeper into a fundamental but unresolved problem in the literature by expanding beyond the identification of false online reviews. How do individuals begin fabricating negative feedback? The significance of an individual's cognitive processes in acquiring, comprehending, and applying knowledge is also examined in greater detail. The cognitive style was measured quantitatively, whereas empirical studies were utilized to investigate the phenomenon of false evaluations. Fifty participants completed an essay prompt, discussed their experiences with peers, and then purported to be recruited by a marketing firm to conduct a cognitive style assessment. Four stages were involved in creating a fraudulent review: research, data processing, writing, and publication. The article takes a logical posture, identifies three techniques for producing false evaluations, and explains why each method would be used.

The outcomes of my research in [57] made this more transparent. In today's extremely interconnected society, social networking sites are increasingly important in everyone's daily lives. Hackers are enticed to develop malicious software for them due to their accessibility. On these networks, malicious bots are wholly or partially programmed computer programs that mimic human behavior for malicious purposes. In light of the prevalence of such programs, concerns about security and cyber threats are also prevalent. They are utilized for advertising, creating phony profiles, disseminating inappropriate or false content, click collection, keyword theft, and other illegal purposes. Academics and hackers constantly develop new and updated algorithms to circumvent the continuously advancing monitoring technologies. This literature review aims to collect and evaluate the most recent developments in machine learning-based techniques for identifying bots and classification on the five most prominent social media platforms: Facebook, Instagram, LinkedIn, Twitter, and Weibo. We provide a brief overview of each guided, semi-supervised, and uncontrolled method, as well as information regarding the examples provided by the experts. In addition, we offer a comprehensive evaluation of the prioritized feature sets. This study concisely summarizes the opportunities and challenges in this field, potential research avenues, and intriguing aspects to investigate for different scenarios.

According to research published in [58], the prevalence and pervasiveness of fake news have increased, making it more challenging to recognize. Detecting false statements early is a difficult task in the discovery process. The lack of labeled data to base detection algorithms further hinders uncovering false news. We propose a novel model for detecting false statements to circumvent these issues. Our proposed method employs information from news articles and social contexts to identify false reports. The proposed model is a Transformer with two components: an encoder to derive applicable models from inaccurate news data and a decoder to predict behavior based on historical data. To aid in categorizing news stories, our model integrates various features derived from social contexts and news content. We also provide an efficient labeling strategy to address the label shortage. In real-world data experiments, our system outperformed benchmarks in detecting false news within minutes of its initial release (early detection).

According to Work in [59], the rise in popularity of social networks can be ascribed to their success in connecting formerly unrelated individuals and facilitating the sharing of various forms of media (movies, pictures, and text messages). This research aimed to identify the possible origins of phony social media profiles. Two methods exist for identifying fake social media profiles: monitoring data analysis and human review. Researchers have discovered that the construction of false accounts on social networks is the most harmful form of cybercrime to users. This infraction must be caught before the user realizes a fraudulent website has been created. There have been numerous methods and programs proposed to detect phony profiles. This study contrasts the performance of Decision Tree (DT) and Naive Bayes in identifying false and authentic user profiles (NB).

Research published in [60] shows widespread interactions with false social media profiles. Users may have difficulty identifying phony profiles. This article aims to reduce the labor required to identify fake social media profiles. However, numerous research projects are currently being conducted in many regions of the globe to accomplish the same goal. Authentic accounts can be distinguished from fraudulent ones, for instance, by analyzing the writing patterns of users without interfering with their standard workflow. We apply the DEEP ID model to the problem of detecting false profiles on social media platforms such as Facebook by analyzing patterns of keystrokes, mouse movements, and touch input. The proposed model can see fraudulent profiles without alerting the intended audience when they use a desktop computer, notebook, or mobile device to post to or interact with social media. The DEEP ID method can also inform you amid a transaction if a fraudulent identity uses authentic credentials. The objective of the proposed research is to demonstrate that avoiding feature extraction.

Consequently, the problem of improper attribute selection enhances the performance of person identification methods applied to unprocessed data. The suggested DEEP ID model is based on a multi-view deep neural network to identify individuals from unfiltered raw data of input behavior. It was determined that the results of the proposed DEEP ID model were superior to those of the more prevalent machine learning models. It offers substantial evidence supporting the validity of the proposed hypothesis. The evaluation results indicate that the Deep ID model is highly accurate at recognizing characteristics and robust enough to perform rapid identification.

As documented in [61], online social networks have attracted billions of active consumers over the past decade. There are many individuals throughout the world who depend on these systems constantly. Therefore, false news and other disinformation campaigns relying on social trolls and bots frequently use social media platforms. Discovering instances of so-called social bots on these sites is difficult due to their size and fluidity, where multiple billion records per hour are generated and collected from users. This article concentrates on "unstable trolls," who frequently alter their identities to maximize social relationships. This enticing behavior may be annoying to users and may compromise their privacy. This is the first effort to identify methods for recognizing these flaws. In this discussion, the various components of spammer detection strategies are dissected. In this paper, we demonstrate that the order of the centralized single-machine detection method renders early spammer detection in social networks with billions of users impractical and sluggish O(n3). In addition, we demonstrate that the streaming approach, in which the system receives input progressively, could be more appropriate for many real-world circumstances. In light of these deficiencies, we propose a parallel detection method. Extensive testing has demonstrated that our proposed strategy is considerably quicker than the current state-of-the-art in concurrent approaches.

According to research in [62], a company's or person's online social network following is a significant predictor of that entity's prominence, which can significantly affect the economy and politics. Due to the prevalence of false profiles on these platforms, it is more difficult to accurately determine a user's prominence or obtain valuable insights about their community. This study analyzed 62 million public Twitter identities to develop a method for detecting phony accounts. Using a combination of a pattern-matching program on user names and an examination of update times, a significant number of highly credible fraudulent user accounts were identified (0.1% of all users). Comparison of the creation times and IP addresses of the profiles to an authentic data set revealed the distinct behavior of the false identities. Similarities and distinctions between acquaintances and followers within the two data sets provided additional support for the notion that these two groups were distinct. Following prior research, the median ratio of followers to friends for the false profiles we discovered was 30, indicating that the primary objective of these fraudulent users was to accumulate a large number of friends. A two-year analysis of account evolution revealed that the friend-to-follower ratios of phony profiles increased while those of authentic users decreased. Consequently, our findings suggest that a technique based on user profiles can swiftly identify a core group of false social network users.

The norms of Pakistan's male-dominated society dictate and influence how women in Pakistan present and convey their personalities offline and online, according to research published in [63]. To investigate one's self-representation and character, This study employs Goffman's Theory (GT) of self-presentation and ordinary self-exposure to explore the lives of ten Pakistani women actively engaged in technology use. The participants' narratives revealed identity crises caused by the competition between their offline and online personas, which were influenced by and reflected dominant cultural norms and values. Veiled cultural norms from the outside world permeate the online world and serve as repressive means of controlling women there, just as the male-protected family and household serve as a heavenly refuge providing security for the women of the family. It has also been referred to as the "digital veil" and "digital sanctuary" to describe how women in Pakistan utilize the Internet. The findings disclose discrepancies between the online and offline identities of social mediausing Pakistani women. The identity crisis that results from this discordant upbringing is a significant obstacle for contemporary Pakistani women, and cultural authorities shape it. More research is needed on the psychological effects of social and internet culture on the formation of accurate and online personas.

My research in [64] clarified this point. With hundreds of new users joining social media every day, a significant quantity of information has been disseminated throughout the globe. In the virtual domain of social media, a person's online reputation can be harmed by malicious users. This type of behavior is frequently attributed to the use of false identities. Consequently, ML must detect false profiles, which can be done early in profile development. This research aims to develop a machine-learning system that promptly sees fraudulent social media profiles. In addition, the literature is organized into categories based on the social media datasets used and the most prevalent applications of machine learning in social media platforms. This study uses the Twitter dataset to identify false accounts, shedding light on machine learning-based bogus news detection. The proposed model includes the preliminaries required to enhance the dataset's quality and reduce complexity by modifying its characteristics and contents. Five prominent machine learning (ML) algorithms (C4.5, Bayes classification, SVM, ANN, and KNN) are used to predict phony profiles. The system is evaluated using four-fold cross-validation and two scenarios with training and evaluation sample rates of 70%:30% and 80%:20%, respectively. Using samples based on 80% of the data reduces resource consumption, whereas using models based on 70% improves classification accuracy. Expanding the mission has been discussed and is now ruled out.

According to the research in [65], social media such as Facebook, Twitter, Instagram, LinkedIn, and others significantly impact our daily existence. Constantly and everywhere, individuals engage in this behavior. Additionally, it must address the issue of phony profiles. Humans, computers, and other artificial means can create false accounts. They are utilized in the dissemination of erroneous information as well as in illicit activities such as hacking and identity theft. In this paper, the author presents an identification model employing multiple machine-learning techniques to identify false Twitter accounts by analyzing their characteristics, such as the number of followers and the frequency of status updates. Based on their profiles, the author identified INT, TWT, and FSF on Twitter as fraudulent, suspect, and compromised identities. Here, neural networks, LSTM, XG Boost, and Random Forest are discussed. The authenticity of a social media profile is determined by selecting the primary characteristics. Coverage includes both the layout and the hyperparameters. After the models have been taught, the concluding stage is the generation of results. The result is one for fake accounts and zero for authentic ones. When a fraudulent identity is discovered, it can be disabled or removed to prevent further security breaches. Python and its supporting libraries, including Sklearn, Numpy, and Pandas, are utilized during implementation. Based on his research, the author has concluded that XG Boost is the most effective machine-learning technique for detecting fraudulent accounts.

According to research published in [66], the proliferation of social media encourages social actors to share their knowledge in digital formats. Additionally, it allows them to maintain relationships with individuals from around the globe. Due to its immense popularity, it has become an intriguing testing ground for various attacks. Cybercriminals construct Sybil networks to acquire confidential information, such as user credentials, and use them to disseminate hateful content and target reputable users. This study aims to identify false profiles on social media platforms such as Twitter. We advocate a hybrid strategy that employs the capacity of machine learning techniques to recognize malicious characteristics. A Petri Net (PN) structure initially analyses the user's profile and other features; these characteristics are then used to train a categorization. This is done to distinguish malevolent identities from legitimate ones. Finally, the efficacy of our technique is demonstrated by comparing it to current state-of-the-art methods. Our experiments showed that ours has a much higher recognition rate than competing techniques (99.16%).

This was evident from my research's findings, which I illustrated in [67]. Today, social media plays a significant role in everyone's daily existence. Many individuals use social networking sites frequently today. The number of profiles on these social networking sites increases daily, and many people who use them speak to each other often. The disadvantage of sharing information on social media sites is that anonymity is less protected. We must determine whether these accounts are genuine or false to identify users who are dangerous on these sites. We've employed a variety of categorization techniques in the past to identify phony social media profiles. However, we must still improve our ability to identify false accounts on these sites. This research uses methods that simplify identifying fake profiles, including machine learning and natural language processing. The Stochastic Forest (SF) classification technique was used to identify plant clusters.

A person's "social life" in the contemporary world, according to a study in consists of their relationships with people who share their hobbies, occupations, or networks of peers and family. These types of knowledge significantly affect how we view the world daily. Meeting new people, chatting with them, and reading their writing is now much more straightforward. However, The administration cannot resolve these issues. We employ a machine learning technique with numerous models to discover phony accounts. The study techniques demonstrated that the suggested approach could reliably detect false IDs. On social networking sites, it is usual to categorize profiles in many different methods to identify fraudulent ones.

Nevertheless, we'll strive to make it simpler for social networks to identify bogus accounts. The use of artificial intelligence (AI) and natural language processing (NLP) to identify fraudulent identities is discussed in this essay. We employ the Support Vector Machine (SVM), the Improved Nave Bayes technique, and the Random Forest Classifier to categorize individuals into fictitious and nonfictional categories. A social network can expand using this automatic ID authentication technique even if a third party cannot confirm the names of its members. We determined whether the terms in issue belonged to the individuals who claimed they did by using these three techniques. This approach can correctly characterize about 98% of the accounts in our training set despite using fewer traits.

Zeng et al. [68] asserts that marketing is essential to offer more goods in the current market. Online promotion is vital in this industry. Concerns about deceptive advertising are widespread in the product marketing sector, particularly in the cinema sector. Organizations use fake evaluations and star scores to advertise movies. This investigation's primary objective is to identify false accounts. To accomplish this, we combine combined filtering with our pre-processing technique. The assortment of motion lenses will be displayed. The suggested approach provides for some freedom, which enables the job to be divided into manageable parts. To start, we must collect crucial data. The data is preprocessed using a method called "notional transformation" after it has been gathered. Clustering and joint viewing identify fraudulent advertisements in the cinema lenses dataset. The nominal translation is required because the recommender system might be unable to manage the number text. We can demonstrate the value of the groups we've created by examining how healthy duties are divided into them.

Focusing on demonstrated how social lives today are inextricably linked to online social network accounts. These websites have significantly altered the way we go about our daily lives. It is much simpler today to make new friends and stay in contact with them. However, due to their rapid expansion, several fresh issues have emerged, including dishonest users, phony accounts, and online fabrications. These issues cannot be effectively resolved. This study will use a simulated neural network to detect fraudulent profiles quickly and accurately. On Facebook, each friend request is examined to determine its likelihood of coming from a genuine individual. This is advantageous for social networks with large user populations whose accounts cannot all be verified separately due to various factors.

According to the study in social media sites are now integral to our daily existence. People are using social media in an increasing number of negative ways. Although machine-modified techniques have occasionally been successful in distinguishing between generated accounts, more studies must be done on human-made fake numbers. Machine learning algorithms compare an account's number of followers to the fans of its counterparts on social media sites to identify bots. Everyone's statements make it obvious how many fans they have, and no one's privacy is violated. A bogus individual account is created to locate, recognize, and remove phony profiles.

My research's findings in provided evidence for this. Online social networks are currently the most well-liked and efficient method of disseminating information across the Internet. Both men and women spend almost all their free time on social networking websites. Worldwide, social networks produce and share vast knowledge each day. Illegal users have developed schemes to deceive social media users to achieve these objectives. Experts have discovered that the kind of hacking that harms users the most is creating phony profiles on social media. This misconduct must be stopped before the user becomes aware that a false website has been made. Social networks generate enormous quantities of unorganized data, which has led to the development of numerous algorithms and techniques for identifying fraudulent profiles. This research examines how technology has developed recently to aid in identifying and preventing fake accounts.

The most current developments in locating phony social media accounts are summarised in Many individuals from all over the globe have joined social networking sites in the last ten years. The number of individuals using social media websites like Facebook, Twitter, LinkedIn, and Instagram unexpectedly increased dramatically. However, experts claim this is untrue because many listed accounts are false and created for harmful purposes. The primary objective of a fake page is to disseminate deceptive information, rumors, and ads. Therefore, deleting phony profiles is crucial, but this can be challenging due to various factors. In recent years, researchers have discovered fake profiles using a variety of cuttingedge instruments. This document summarizes what is currently known about identifying fraudulent identities. There are quickly discussed several issues and difficulties with the now available variants. The survey's findings might be used to fill in knowledge voids and develop a method for identifying phony social media accounts.

According to research published in social media platforms are now integral to our existence. OSNs are used by us for internet communication with friends and family, information sharing, and self-and company promotion. This social media-only network disseminates confidential data that can be used for ransomware assaults, disinformation campaigns, and identity fraud. In this project, we employ machine learning methods to create an identifying model that can distinguish between false and genuine Twitter accounts based on elements such as the number of followers and friends, the frequency of message transmission, and other features.

According to a study, social networks are becoming more critical globally.More and more individuals use social media every year. The ease with which people from around the globe can communicate with one another on social media platforms is one of their finest features. Attacks like spam and scams were made feasible as a result of this. According to studies conducted by Facebook and Twitter, there are significantly fewer active users than registered profiles. This trend indicates more phony accounts now than a year ago. The authors of these accounts have the potential to influence customers by disseminating misleading information. They employ individuals to assist them in achieving their dubious objectives, such as promoting prejudice on online platforms and social media. The issue of fake social media profiles is getting worse for Internet users. It is crucial to develop your ability to recognize these nicknames. Often, phony identity detection uses machine learning methods like SVM linear, naïve Bayes, random forests, and decision tree algorithms. These programs have lessened in precision and efficiency. We can detect and distinguish pathogens by merging Ada and SVM. An Adaboost SVM classifier is believed to discover false accounts and identity hackers more effectively.

Asserts that modern technical advancements are significantly improving. Mobile gadgets' processors are becoming more intelligent. Innovation is connected to online casual organizations, which are a part of everyone's life for meeting new acquaintances and retaining existing ones. It is now more straightforward to learn what they enjoy. However, as online mimicry and false accounts have become more prevalent, so have other issues like online systems management. Customers receive additional pointless information shared by fictitious customers during trips. It has been discovered that 20–40% of accounts in unofficial online communities like Facebook are false. When false statements are found in unofficial online groups, systems must be in place.

Every social networking site sees false profiles created for various causes, some listed in The characteristics of fraudulent accounts are very diverse, and their strategies to accomplish their objectives are constantly evolving. On the one hand, new techniques and systems for identifying dangers are being developed; on the other, adversaries are developing cunning means of evading detection. In the following sentences, we'll quickly discuss some of the most significant issues with locating phony accounts on OSNs. In this case, Instagram is the primary topic of investigation. This one has a higher percentage of false reports than other platforms. This study discusses the new hybrid model machine learning method and demonstrates how it can be used to identify phony Instagram accounts, among other things.

The research demonstrates how everyone's public life is intertwined with their online social networks in the contemporary world. These locations have greatly facilitated our ability to take action for the common benefit. Making new friends and staying current with their activities is now simpler. However, due to their rapid growth, many issues have emerged, including false accounts and online theatre. There are no workable solutions to these issues. In this essay, I offer a practical method for finding false identities that can be coded. This design sorts profiles into groups for fictional or actual individuals using clustering techniques like Random Forest Classifier (RFC). Online casual groups frequently use this coded position technique with many accounts that cannot be physically verified for various purposes.

According to research published in the rise in popularity of social networks can be attributed to their ability to bring people from around the world together and facilitate the interchange of information in the form of text, audio, and visual content. This is one reason social networks have gained so much popularity recently. Customers of these networks are significantly concerned about the possibility of having their identities taken. This study proposes a novel method for detecting fake profiles on social media platforms, one based on the similarity between the friend networks of different individuals. Similarity measures were computed using the adjacency matrix of the affiliated social network graph. These measurements included weight similarity, cosine, Jaccard, L1 measurement, and mutual peers. Each stage was validated using the Twitter dataset to evaluate the final model. The Medium Gaussian Support Vector Machine (MG SVM) algorithm could identify fraudulent accounts with a false positive rate of 0.02 and an area under the curve of 1.

According to research published in social networking websites on the Internet have rapidly become an integral part of the daily lives of a sizable portion of the global population. Individuals in modern society use social media for various reasons, including but not limited to expressing their emotions, expanding their social networks, building businesses, and maintaining relationships with family and friends, among many others. Online social networks can provide users with various benefits; however, these platforms are full of disadvantages. Unfortunately, many users construct false profiles on these platforms to harass or intimidate others. Using machine learning techniques, we can distinguish between fake and genuine profiles. The models generated by machine learning techniques are utilized in the prediction and classification of data. Using a composite strategy within machine learning software can make distinguishing between the results of the various models easier. We compared the efficacy of eight distinct categorization algorithm configurations by analyzing and comparing data obtained from a social media platform. Multiple machine learning techniques were utilized simultaneously, such as K-Nearest Neighbors (KNN), Random Forest, Support Vector Machine, and Decision Trees. When comparing the accuracy of various hybrid techniques, we found that the one containing SVM, Logistic Regression, and Neural Network produced the most accurate results. This was the case when we compared the attained precision (LR NN). As a result, we recommended a model that identifies fake accounts using a combination of methods because it had the highest level of accuracy among all the varieties we evaluated for different use cases. Thus, it can be observed that existing models vary widely in terms of their internal characteristics, which makes it difficult to select optimal models for deployment-specific use cases. Thus, the following section compares these models' quantitative performance levels, making it easier for researchers to identify optimal models for their performance-specific use cases.

3 Pragmatic Analysis and Comparison

From the extensive review of fake-profile-detection models, it can be observed that deep learning & transfer learning methods outperform linear methods, which makes them highly useful for real-time use cases. To compare these models, this section evaluates them in terms of the accuracy of fake profile detection (A), delay needed for detection of these profiles (D), scalability of the models (S), deployment cost (DC), and computational complexity (CC) levels. The accuracy values were directly taken from the reference texts, but other parameters were not instantly available. Thus, they were inferred into Low ($L = 1$), Medium ($M = 2$), High ($H = 3$), and Very High ($VH = 4$) ranges via the following process.

- The delay was quantized depending on the number of operations needed to perform fake profile analysis.
- Scalability was determined depending on the number of Social Media platforms considered for analysis.
- Deployment cost was evaluated depending upon the type of operations performed by the model (models that generate randomized passwords while login have higher prices when compared with single password models).
- Computational Complexity was evaluated depending on the internal characteristics of the operations used during the fake-profile-detection process.

Based on this strategy, the models were compared in Table 1 as follows.

Based on this analysis, it can be observed that NB SVM [5], ResNet [7], XG Boost [20], SVM NN [34], MG SVM [35], MV SVM [43], CNN [57], and Deep Profile [63], showcased higher accuracy of detection, and thus can be used for highperformance & low-error use cases.

It was also observed that NB SVM [5], UTH [10], Honey Word [14], and DT NB [15], PPI [20], RF [32], SVM [42], SVM NN [50], had lower complexity, thus can be used for light-weighted computational use cases.

Models defined in NB SVM [5], UCPM [30], RF [32], CA [38] SVM NN [50], and are capable of low-delay analysis. Thus, they must be used for high-speed fake-profile-detection use cases.

Similarly, UTH [10], FENR [12], DT NB [15], Deep SBD [17], AECA [24], UCPM [30], SVM NN [34], and LR [55], Deep Profile [63], showcase lower deployment cost, thus can be used for use-cases where cost aware analysis is needed under real-time scenarios.

While the models discussed in showcased higher scalability levels, they can be used for high-efficiency, large-scale network use cases.

$$FPDRM = \frac{A}{100} + \frac{S}{4} + \frac{1}{D} + \frac{1}{CC} + \frac{1}{DC}$$

Based on this evaluation and Fig. 1, it was observed that NB SVM [5], ML NLP [8] DT NB [15], Deep SBD [17], PPI [20], SVM NN [34], SVM [42], SVM NN [50], LR [55], CNN [57], Deep Profile [63], SAHP [64], and showcased better FPDRM performance, thus can be deployed for high accuracy, better scalability, low delay, low complexity, and low-cost use cases. Consequently, researchers must use these models for their social networks to identify fake profiles.

Table 1 Pragmatic analysis of different fake-profile-detection models

Model	A	CC	D	DC	S
UCPM	91.5	H	M	M	L
IBS	85.4	VH	VH	H	L
TPSC	90	H	H	H	L
UAML	90.5	H	H	H	M
TMA DPM	93.2	VH	H	H	M
RF	83.5	M	M	H	L
FENR	85.4	H	H	M	L
MUI UI	86.4	H	H	H	M
ESVM NN	92.1	VH	H	H	M
SEN AD	92.5	H	VH	VH	L
Cred NN	93.7	VH	H	H	H
DF NB DT	93.2	H	VH	VH	H
TFIDF VER	95.6	VH	H	H	L
AECA	90.5	H	H	M	M
NB SVM	98	M	M	H	M
SAHP	96.4	H	H	H	H
CO MPA	93.5	VH	H	H	H
Deep Profile	99.5	H	VH	M	VH
ML NLP	95.6	H	H	H	H
SVM	90.8	M	H	H	H
XG Boost	99.4	H	H	H	M
ResNet	99.8	VH	H	VH	H
MG SVM	98.5	H	VH	H	H
SVM LR NN	91.5	H	VH	H	H
CNN	98.9	H	H	H	H
UTH	95.9	M	VH	M	L
SVM NN	91.5	M	M	H	M
PPI	85.3	L	H	H	L
DFSVM ONB	96.1	H	H	H	M
SVM NN	98.2	H	H	M	H
CA	83.2	H	M	H	M
SVM	90.8	H	H	H	M
Deep SBD	96.5	VH	VH	M	H
Honey Words	94.8	M	VH	H	M
DF	90.5	H	H	H	H
DT NB	91.6	M	H	M	VH
LR	93.5	H	VH	M	H

(continued)

Table 1 (continued)

Model	A	CC	D	DC	S
AM SNG	96.4	H	H	H	M
MV SVM	98.5	VH	H	H	H
Ada Boost	85.5	H	H	H	M
kNN	96.7	H	H	H	H
BERT	94.3	H	H	H	H
SVM	98.6	M	VH	H	H
KC GCN	93	H	H	H	M
DNN	95.5	H	H	VH	H
RF ANN	90.4	H	H	H	H
DT NB	91.2	M	H	H	H
DEEP	94.5	H	H	H	M
GT	85.5	H	H	H	H
Naïve Bayes	93.5	H	VH	H	H
LSTM	95.5	H	H	VH	H
PN	99.2	H	H	H	H
SF	83.2	M	H	H	M
SVM	98	H	VH	H	H
Ada SVM	90.5	H	H	H	H
RFC	91.4	M	H	H	H
MG SVM	99.5	M	H	H	H
LR NN	90.2	H	H	H	H

4 Implementation

4.1 Data Preparation and Preprocessing

For this research, we took a publicly available dataset containing the cyberattack record using AWS Honeypot systems. The dataset consisted of various features such as protocols, source ports, destination ports, geographical areas, as well as time stamps of the attacks. In the first step, the dataset was refined by deletion of columns that are not needed, for example column 'Unnamed: 15' did not contain any information. Also, we removed observations having complete NaN in important features like spt, dpt, latitude, longitude, country since these features were useful for the model training and prediction.

Moreover, we derived the time-related features from the datetime column by putting out the day and hour of the attack. Such features helped to better understand the frequency of attacks over certain periods and helped the model to learn temporal relations better.

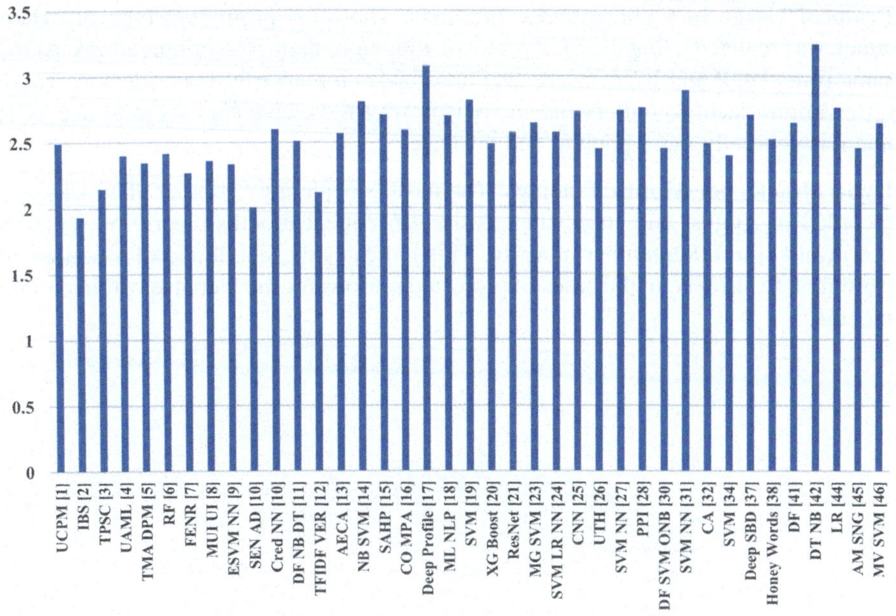

Fig. 1 Evaluation of FPDRM for different models

4.2 *Exploratory Data Analysis*

To assess the description of the data, a number of visualizations were performed:

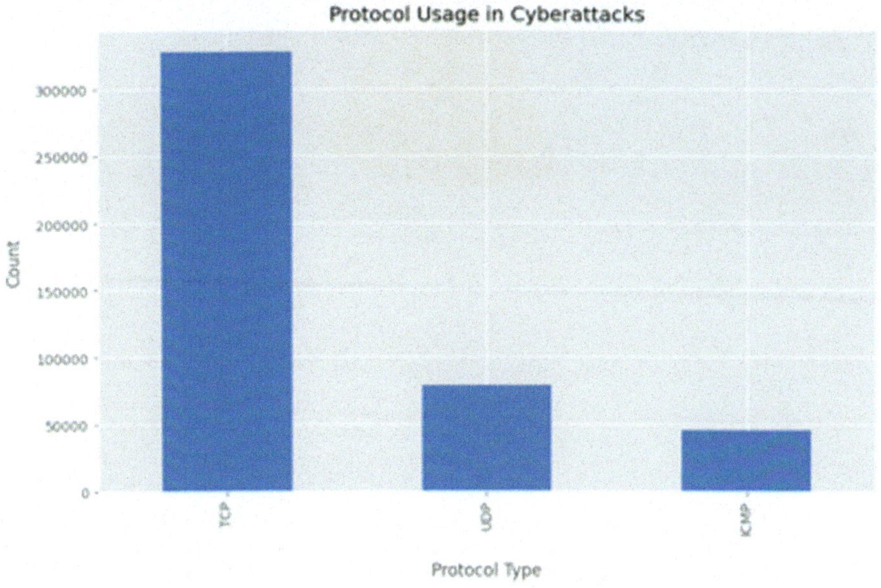

Protocol Usage in Cyberattacks. Findings: The most prominent type of cyberattack was realized using the TCP protocol with more than 70% of operations. At the same time, UDP and ICMP were also used but to a markedly lower degree. These observations seem to underscore the reason why most attackers seem to use TCP, that it is generally more popular and intricate.

Cyberattacks per Month. Findings: The most considerable number of attacks was recorded in August and July, whereas the subsequent months clearly presented a downward spiral, September more so. This indicates the likelihood of a periodical pattern of occurrence in the cases of projected reasons for the increase within certain months.

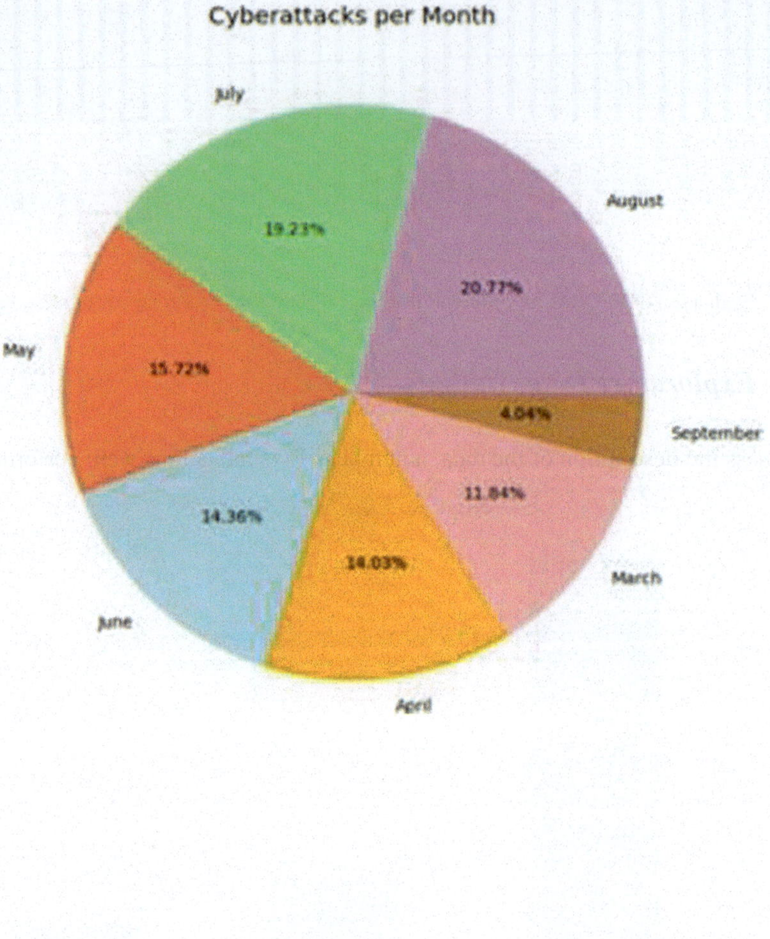

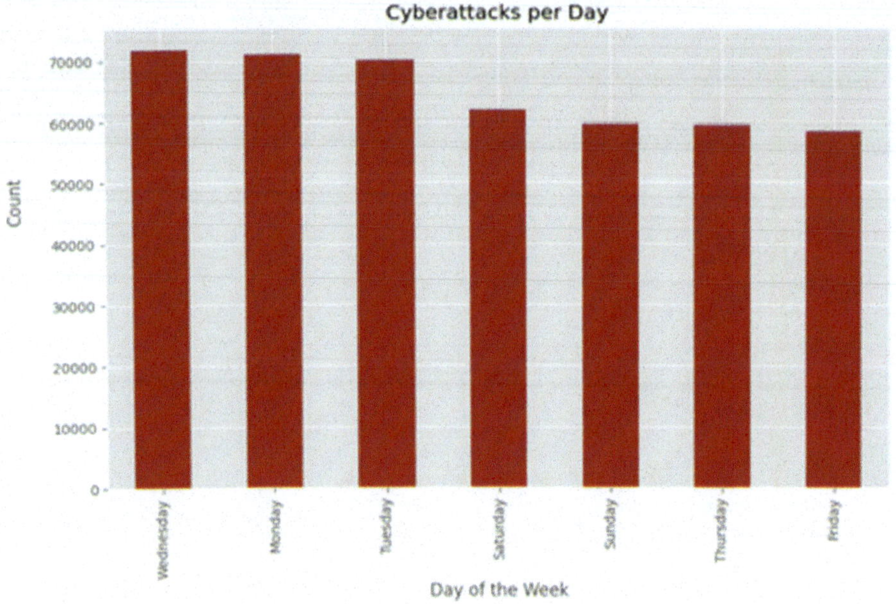

Cyberattacks per Day. Findings: Of the findings, attacks were frequented on Wednesday and mainly Monday and Tuesday posted consistently high frequencies, while towards the weekend there was a marked drop mostly on the Sunday.

Top 20 Countries by Cyberattacks. Findings: Looking at cyber defences, China was ranked as the first country that instigated attacks while the following country was America. Also, a considerable attack rate was achieved by Japan, Iran, and Taiwan, which are the principal countries where cybercrime advanced.

Model Choice and Implementation. For the task of forecasting the country of origin of a cyberattack, one more supervised machine learning approach was employed using the Random Forest Classifier from the sklearn library. Random forests ensemble method that uses construction of a number of decision trees during training and the mode of target classes' (classification) of individual trees as output. The model was chosen because it is quite robust and can handle large size datasets that have many features.

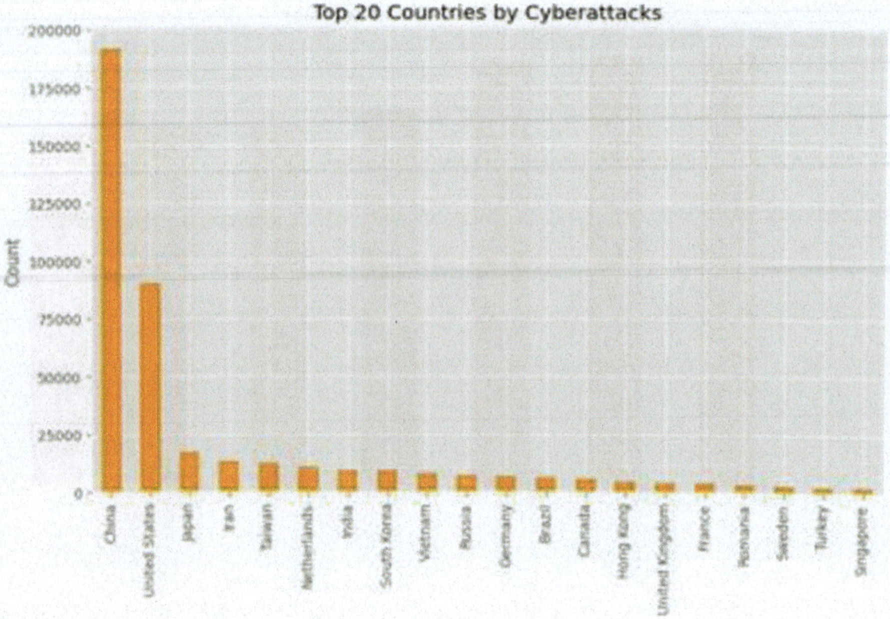

Feature Selection. The features selected for training the model were:

- Source Port (spt): A computer terminal's point or port from which an attack was initiated.
- Destination Port (dpt): It is the computer port towards which the attack is directed.
- Latitude and Longitude: The source IP address geographical location.
- The target variable was Country, which was the country where the aided cyberattack took place.

Model Training and Validation. The information contained in the whole dataset was carried out in two divisions with one; training the model with 80% of the information and freezing 20% for testing purposes. A Random Forest Classifier with one hundred decision trees was trained. The classifier, during training of the model, was trained to ascertain trends of the features to establish which country a new cyberattack came from.

Model Performance. Random Forest model obtained 94% accuracy on the test data set. This test shows the strength of the model in regards to prediction power as well. While other performance metrics such as precision, recall and F1 score were computed, these were yielding good results for the task done, not exhibiting any great challenges of overfitting. A confusion matrix was made for a more detailed understanding of the errors in relation to the countries of the attacks.

Moreover, as part of the feature significance ranking, it was noted that the most powerful predictor variables were both the source and destination ports. (spt, dpt) with geographical variables (latitude, longitude) coming second. This means that, there are some particular ports which are more prevalent in certain types of attacks, and also location, is still somewhere the attackers come from.

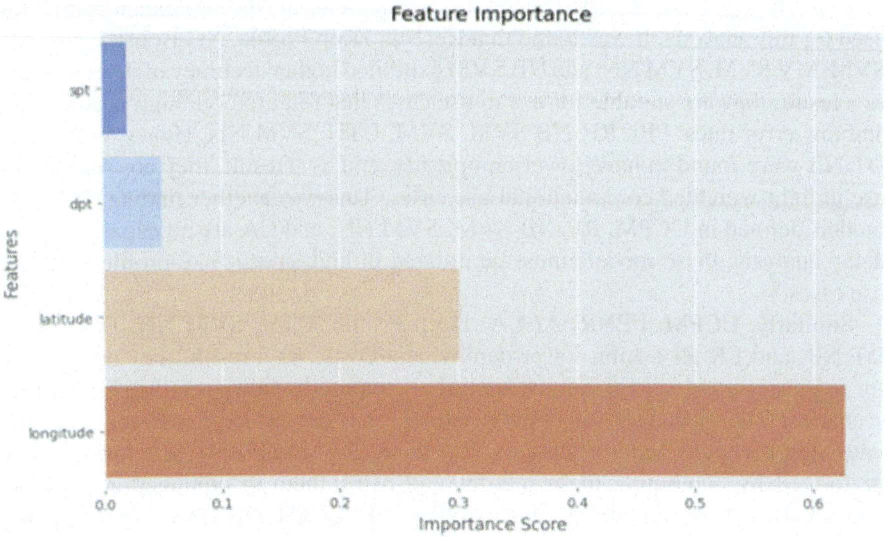

Feature Importance Plot. A feature importance bar plot was drawn to show the various levels of prediction power possessed by the various factors where it was noted that source and destination ports were the top factors, followed by geographical coordinates.

Based on the AWS Honeypot dataset, the Random Forest Classifier was able to achieve quite good results, hitting up to 94% prediction accuracy of the country of the cyberattack origin. Exploratory Data Analysis gave significant information on the protocol concentration, temporal focus, and geographical spread of the attacks. It has been noted that the majority of the attacks executed through TCP, were observed to be engaged in certain months (August, July) and certain days of the week, generally the middle of the week.

We consider the application of machine learning in the field of cybersecurity, especially in determining the source of cyberattacks, to be ready for real use. Systematic improvement of the feature set and implementation of more strong algorithms will surely increase the performance and support integration of the model into the actual infrastructure, which will allow organizations to prevent, and control harmful activities instantly.

5 Conclusion and Future Scope

Based on the extensive analysis of different fake-profile-detection models, it can be observed that deep learning-based models outperform others in terms of accuracy, which makes them highly useful for real time use cases. User biometrics, online behavior, type of content posted on media websites, etc., are the significant parameters used for this analysis. It was found that ResNet, Deep Profile, XG Boost, CNN, MG SVM, MV SVM, SVM NN, and NB SVM exhibited higher accuracy of detection, and as a result, they are suitable for use in use cases that require both high performance and low error rates. PPI, RF, NB SVM, SVM, UTH, SVM NN, Honey Words, and DT NB were found to have lower complexity, and as a result, they are suitable for use in light-weighted computational use cases. This was another finding. Since the models defined in UCPM, RF, NB SVM, SVM NN, and CA are all capable of low-delay analysis, these models must be utilized in high-speed fake-profile-detection use cases.

Similarly, UCPM, FENR, AECA, Deep Profile, UTH, SVM NN, Deep SBD, DT NB, and LR all exhibit lower deployment costs. As a result, they are suitable for use cases in which cost-aware analysis is required while operating in real-time scenarios. The evaluation of optimal models that can be deployed for use cases with high accuracy, better scalability, low delay, low complexity, and low cost was performed by combining these metrics and using them in conjunction with one another. Based on this metric, it was observed that DT NB, Deep Profile, SVM NN, SVM, NB SVM, PPI, LR, SVM NN, CNN, Deep SBD, SAHP, and ML NLP displayed higher levels of FPDRM performance. NB SVM was also found to have the best performance levels. Therefore, researchers need to use these models for their social networks to detect fake profiles. In the future, researchers must incorporate multiple deep learning techniques, including Long Short-Term Memory (LSTM), Auto Encoders (AEs), Convolutional Networks, Gated Recurrent Units (GRUs), etc., to improve fake-profiledetection performance under large-scale networks. Moreover, these models must be validated on other networks, and their performance can be tuned via integrating hybrid bioinspired models that can incrementally achieve better performance via continuous learning operations.

Acknowledgements No funding available for this work.

Disclosure of Interests The authors have no competing interests to declare that are relevant to the content of this article.

References

1. Aditya BL, Mohanty SN (2023) Heterogenous social media analysis for efficient deep learning fake-profile identification. IEEE Access
2. Aditya BL, Mohanty SN (2023) Unveiling the underworld: detecting fake profiles through network analysis and behavioral modeling on social media. In: International conference on pervasive knowledge and collective intelligence on web and social media. Springer, pp 342–352
3. Aditya BL, Mohanty SN (2023) VGMFSN: design of an efficient fused VARMA GRU model for the identification of fake profiles on social networks. In: International conference on pervasive knowledge and collective intelligence on web and social media. Springer, pp 353–366
4. Aditya BL, Mohanty SN, Salini Y (2023) Temporal sentiment analysis (TSMFPMSM) model for multimodal social media fake profile detection. In: International conference on pervasive knowledge and collective intelligence on web and social media. Springer, pp 329–341
5. Ajesh F, Aswathy S, Philip FM, Jeyakrishnan V (2021) A hybrid method for fake profile detection in social network using artificial intelligence. In: Security issues and privacy concerns in industry 4.0 applications, pp 89–112
6. Aksar IA, Firdaus A, Pasha SA (2023) Virtual vs. real self: gendered presentation and everyday performance of virtual selfhood–a case study of Pakistan. J Commun Inquiry 47(1):84–114
7. Albahar M (2021) A hybrid model for fake news detection: leveraging news content and user comments in fake news. IET Inf Secur 15(2):169–177
8. Arega KL (2020) Social media fake account detection for Afan Oromo language using machine learning. New Media Mass Communication. ISSN 2224-3267 (Paper) ISSN 2224-3275 (Online) 90
9. Banerjee S, Chua AY (2023) Understanding online fake review production strategies. J Bus Res 156:113534
10. Bhattasali T, Saeed K (2021) Typing pattern analysis for fake profile detection in social media. In: Computer information systems and industrial management: 20th international conference, CISIM 2021, Ełk, Poland, September 24–26, 2021, Proceedings 20.Springer, pp 17–27
11. Cai H, Ren J, Zhao J, Yuan S, Meng J (2023) KC-GCN: a semi-supervised detection model against various group shilling attacks in recommender systems. Wirel Commun Mob Comput 2023(1):2854874
12. Chakraborty M, Das S, Mamidi R (2022) Detection of fake users in twitter using network representation and NLP. In: 2022 14th international conference on COMmunication systems & NETworkS (COMSNETS). IEEE, pp 754–758
13. Chakraborty P, Shazan MM, Nahid M, Ahmed MK, Talukder PC (2022) Fake profile detection using machine learning techniques. J Comput Commun 10(10):74–87
14. Chang D, Goel A, Mishra S, Sanadhya SK et al (2018) Generation of secure and reliable honeywords, preventing false detection. IEEE Trans Depend Sec Comput 16(5):757–769
15. Elyusufi Y, Elyusufi Z, Kbi, MA (2019) Social networks fake profiles detection based on account setting and activity. In: Proceedings of the 4th international conference on smart city applications, pp 1–5
16. Elyusufi Y, Elyusufi Z, Kbir MA (2020) Social networks fake profiles detection using machine learning algorithms. In: Innovations in smart cities applications edition 3: the proceedings of the 4th international conference on smart city applications, vol 4. Springer, pp 30–40
17. Fazil M, Sah AK, Abulaish M (2021) Deepsbd: a deep neural network model with attention mechanism for socialbot detection. IEEE Trans Inf Forensics Secur 16:4211–4223
18. Frunze AD, Frolov AA (2021) Methods for detecting fake accounts on the social network VK. In: 2021 IEEE conference of russian young researchers in electrical and electronic engineering (ElConRus). IEEE, pp 342–346
19. Gurajala S, White JS, Hudson B, Voter BR, Matthews JN (2016) Profile characteristics of fake twitter accounts. Big Data Soc 3(2):2053951716674236
20. Harris P, Gojal J, Chitra R, Anithra S (2021) Fake instagram profile identification and classification using machine learning. In: 2021 2nd global conference for advancement in technology (GCAT). IEEE, pp 1–5

21. İş H, Tuncer T (2019) Interaction-based behavioral analysis of Twitter social network accounts. Appl Sci 9(20):4448
22. Joshi S, Nagariya HG, Dhanotiya N, Jain S (2020) Identifying fake profile in online social network: an overview and survey. In: Machine learning, image processing, network security and data sciences: second international conference, MIND 2020, Silchar, India, July 30–31, 2020, Proceedings, Part I 2. Springer, pp 17–28
23. Joshi UD, Vanshika, Singh, A.P., Pahuja, T.R., Naval, S., Singal, G.: Fake social media profile detection. In: Machine learning algorithms and applications, pp 193– 209
24. Kadam N, Patidar H (2017) Social media fake profile detection technique based on attribute estimation and content analysis method. In: International conference on innovative computing and communications
25. Kadam N, Sharma SK (2022) Social media fake profile detection using data mining technique. J Adv Inform Technol 3(5)
26. Kantartopoulos P, Pitropakis N, Mylonas A, Kylilis N (2020) Exploring adversarial attacks and defences for fake twitter account detection. Technologies 8(4):64
27. Khaled, S., El-Tazi, N., Mokhtar, H.M.: Detecting fake accounts on social media. In: 2018 IEEE international conference on big data (big data). IEEE, pp 3672–3681
28. Kidu H, Misgna H, Li T, Yang Z (2021) User response-based fake news detection on social media. In: Applied informatics: fourth international conference, ICAI 2021, Buenos Aires, Argentina, October 28–30, 2021, Proceedings, vol 4. Springer, pp 173–187
29. Kondeti P, Yerramreddy LP, Pradhan A, Swain G (2021) Fake account detection using machine learning. In: Evolutionary computing and mobile sustainable networks: proceedings of ICECMSN 2020. Springer, pp 791–802
30. Krombholz K, Merkl D, Weippl E (2012) Fake identities in social media: a case study on the sustainability of the facebook business model. J Serv Sci Res 4:175–212
31. Kulkarni V, Reddy DA, Sreevani P, Teja RN (2021) Fake profile identification using ANN. In: 4th smart cities symposium (SCS 2021), vol 2021. IET, pp 375–380
32. Latha P, Sumitra V, Sasikala V, Arunarasi J, Rajini A, Nithiya N (2022) Fake profile identification in social network using machine learning and NLP. In: 2022 international conference on communication, computing and Internet of Things (IC3IoT). IEEE, pp 1–4
33. Machova K, Mach M, Vasilko M (2021) Comparison of machine learning and sentiment analysis in detection of suspicious online reviewers on different type of data. Sensors 22(1):155
34. Mitra A, Kundu A, Chattopadhyay M, Banerjee A (2022) An approach to detect fake profiles in social networks using cellular automata-based pagerank validation model involving energy transfer. SN Comput Sci 3(6):423
35. Mohammadrezaei M, Shiri ME, Rahmani AM (2018) Identifying fake accounts on social networks based on graph analysis and classification algorithms. Secur Commun Netw 2018(1):5923156
36. Mohanty SN, Rejina Parvin J, Vinoth Kumar K, Ramya K, Sheeba Rani S, Lakshmanaprabu S (2019) Optimal rough fuzzy clustering for user profile ontology based web page recommendation analysis. J Intell Fuzzy Syst 37(1):205–216
37. Munga JB, Mohandas P (2021) Feature selection for identification of fake profiles on Facebook. In: Lumpur K (ed) International conference on biomedical engineering. Springer, pp 489–497
38. Muraya C, Awuor F, Maake B (2021) Fake profile identification on online social networks. J Lang Technol Entrepr Africa 12(2):48–59
39. Nirmala M, Babu MR (2019) Fuzzy-based fake information detection algorithm to define the user trust on the content of social networks. IET Netw 8(6):367–371
40. Pani S, Sahu B, Mishra J, Mohanty SN, Panigrahi A (2022) Pragmatic analysis of social web components on semantic web mining. In: Social network analysis: theory and applications, pp 83–108
41. Parihar P, Kumar N et al (2021) Fake profile detection from the social dataset for movie promotion. In: 2021 sixth international conference on image information processing (ICIIP), vol 6. IEEE, pp 495–498

42. Prabhu Kavin B, Karki S, Hemalatha S, Singh D, Vijayalakshmi R, Thangamani M, Haleem SLA, Jose D, Tirth V, Kshirsagar PR et al (2022) Machine learning-based secure data acquisition for fake accounts detection in future mobile communication networks. Wirel Commun Mob Comput 2022(1):6356152
43. Priyadharshini V, Valarmathi A (2021) A novel spam detection technique for detecting and classifying malicious profiles in online social networks. J Intell Fuzzy Syst 41(1):993–1007
44. Ramalingam D, Chinnaiah V (2018) Fake profile detection techniques in large-scale on-line social networks: a comprehensive review. Comput Electr Eng 65:165–177
45. Raza S, Ding C (2022) Fake news detection based on news content and social contexts: a transformer-based approach. Int J Data Sci Anal 13(4):335–362
46. Roy PK, Chahar S (2020) Fake profile detection on social networking websites: a comprehensive review. IEEE Trans Artif Intell 1(3):271–285
47. Sahoo SR, Gupta BB (2019) Hybrid approach for detection of malicious profiles in twitter. Comput Electr Eng 76:65–81
48. Sansonetti G, Gasparetti F, D'aniello G, Micarelli A (2020) Unreliable users detection in social media: deep learning techniques for automatic detection. IEEE Access 8:213154–213167
49. Saraswati S. Use of ANN to identify fake profiles
50. Senthil Raja M, Arun Raj L (2022) Detection of malicious profiles and protecting users in online social networks. Wirel Pers Commun 127(1):107–124
51. Shafiei H, Dadlani A (2022) Detection of fickle trolls in large-scale online social networks. J Big Data 9(1):22
52. Shah M, Joshi H (2022) Social media fake profile identification using novel hybrid model machine learning technique. Int J Next-Gener Comput 13(3)
53. Shahzad K, Khan SA, Ahmad S, Iqbal A (2022) A scoping review of the relationship of big data analytics with context-based fake news detection on digital media in data age. Sustainability 14(21):14365
54. Shalini P (2022) Social behavioral biometric multimodal union to evade fake account creation in Facebook. Multimedia Tools Appl 81(27):39715–39751
55. Singh M, Singh A (2023) How safe you are on social networks? Cybern Syst 54(7):1154–1171
56. Singh N, Sharma T, Thakral A, Choudhury T (2018) Detection of fake profile in online social networks using machine learning. In: 2018 international conference on advances in computing and communication engineering (ICACCE). IEEE, pp 231–234
57. Singh V, Shanmugam R, Awasthi S (2021) Preventing fake accounts on social media using face recognition based on convolutional neural network. In: Sustainable communication networks and application: proceedings of ICSCN 2020. Springer, pp 227–241
58. Spoorthy A et al (2021) Trust based fake node identification in social networking sites. In: IOP conference series: materials science and engineering, vol 1123. IOP Publishing, p 012036
59. Stolbova A, Ganeev R, Ivaschenko A (2021) Intelligent identification of fake accounts on social media. In: 2021 30th conference of open innovations association FRUCT. IEEE, pp 279–284
60. Suganya R, Muthulakshmi S, Venmuhilan B, Kumar KV, Vignesh G (2021) Detect fake identities using improved machine learning algorithm. In: Journal of physics: conference series, vol 1916. IOP Publishing, p 012056
61. Thabit S, LianShan Y, Tao Y, Abdullah AB (2022) Trust management and data protection for online social networks. IET Commun 16(12):1355–1368
62. Uppada SK, Manasa K, Vidhathri B, Harini R, Sivaselvan B (2022) Novel approaches to fake news and fake account detection in OSNs: user social engagement and visual content centric model. Soc Netw Anal Min 12(1):52
63. Wanda P, Jie HJ (2020) Deepprofile: finding fake profile in online social network using dynamic CNN. J Inform Secur Appl 52:102465
64. Wang X, Tang H, Zheng K, Tao Y (2020) Detection of compromised accounts for online social networks based on a supervised analytical hierarchy process. IET Inf Secur 14(4):401–409
65. Wani MA, Agarwal N, Bours P (2020) Impact of unreliable content on social media users during Covid-19 and stance detection system. Electronics 10(1):5

66. Wani MA, Sofi MA, Wani SY (2017) Why fake profiles: a study of anomalous users in different categories of online social networks. Int. J. Eng. Technol. Sci. Res 4(9):320–329
67. Wu X, Feng Z, Fan W, Gao J, Yu Y (2013) Detecting marionette microblog users for improved information credibility. In: Machine learning and knowledge discovery in databases: European conference, ECML PKDD 2013, Prague, Czech Republic, September 23–27, 2013, Proceedings, Part III 13. Springer, pp 483–498
68. Zeng W, Tang R, Wang H, Chen X, Wang W (2021) User identification based on integrating multiple user information across online social networks. Secur Commun Netw 2021(1):5533417

Multi-task Cascaded Convolutional Neural Network Foundation for Driver Drowsiness Detection

Venkata Sudiksha Madeti, V. K. Harini, and Manas Ranjan Prusty

Abstract This paper describes an integrated approach to fatigue classification that combines a face detector based on the multi-task convolutional neural network (MTCNN) algorithm with a binary image classification model. An MTCNN algorithm that extracts facial regions from input images is developed, pre-processes these images, and augments them. A convolutional neural network composed of convolutional and fully connected layers will analyze facial features to discriminate between "active" and "fatigue" states. The architecture proposed demonstrates robustness through data augmentation and dropout regularization. An Adam optimizer and a binary cross-entropy loss, designed in the training strategy, have been utilized. In the performance evaluation, all the above-mentioned assessment metrics are used, including training accuracy curves, confusion matrices, receiver operator characteristics analysis, and precision-recall. The effectiveness of the architecture is validated on a dataset of active and fatigue states, showing its capability for real-world applications.

Keywords Multi task convolutional neural network · Face detection · Fatigue classification · Image processing · Data augmentation

V. S. Madeti · V. K. Harini
School of Computer Science and Engineering, VIT Chennai, Chennai, India
e-mail: madetivenkata.sudiksha2021@vitstudent.ac.in

V. K. Harini
e-mail: harini.vk2021@vitstudent.ac.in

M. R. Prusty (✉)
Centre for Cyber Physical Systems, School of Computer Science and Engineering, VIT Chennai, Chennai, India
e-mail: manas.ranjan@vit.ac.in

A. Rocha et al. (eds.), *Integrating Advanced Technologies for Enhanced Security and Efficiency*, Engineering Cyber-Physical Systems and Critical Infrastructures 14,
https://doi.org/10.1007/978-3-031-91798-1_9

1 Introduction

Driver fatigue is one of the key contributors to traffic accidents and raises the risk factor by impairing cognitive abilities. Technological solutions, in this case, are proposed as LED-based systems, which are integrated into ADAS, to detect the early signs of fatigue before collision. These systems use cameras and other sensors to monitor driver activity and trigger visual and audio alerts, such as blinking lights or loud warnings, when drowsiness is detected. If they work, LED-based systems succeed but are dependent on accurate detection algorithms and alert drivers. To fight sleepy driving, an integrated approach involving technology, law, and education is required to help achieve a better degree of overall road safety and to reduce the number of incidents caused by fatigued drivers.

The proposed model utilizes multi-task convolutional neural network (MTCNN), first employed by authors in [1], which one of the popular deep is learning algorithms used for face detection. It uses a cascaded structure that involves three stages to locate faces in images thus suitable for real-time applications in facial recognition and analysis of video.

The proposed model details a convolutional neural network that distinguished images of an active person from images of a fatigued one based on facial expressions. The network includes dense layers, dropout layers, max-pooling layers, and convolutional layers. It employed the Adam optimizer and used loss as binary cross-entropy loss in scaled grayscale images. During training process, accuracy during epochs and over the training set of data confusion matrix has been used to report how this model works. Put another way, the given code will create a network as the main tool through which accuracy of detection regarding presence or absence of symptoms like exhaustion is tested within face photos.

The main objective is to warn the drivers if they are distracted or sleepy through an application that monitors the particular driver's driving. This makes the driver alert and may take decision based on his current situation which may help prevent road accidents.

2 Related Works

In [2], the study offers two methods of drowsy driver detection. The first method acquires the face, retinas, and facial features of the driver to compose blinking values. The second method is calculating real-time hand pressure on the steering wheel by using an embedded Arduino module with elastomeric sensors. Results from these two methods are combined and made into a final judgment for better drowsiness detection accuracy. In [3], the work utilizes two pre-trained Haar classifiers from OpenCV for face and eye detection along with a Beagle Bone Black Wireless development board. Drowsiness is detected when the rate of eye blinking exceeds a certain threshold. This method shows how low-cost hardware can be used for effective drowsiness detection. DriCare overcomes driver face location variations and head movements during operation. Deng and Wu [4] proposes three neural network architectures that can well serve to detect effective drowsiness: Fully designed Neural Network (FD-NN), Transfer Learning with VGG16, and Transfer Learning with VGG19. The literature review in [5] categorizes drowsiness detectors into three types: vehicle-based, signal-based, and facial feature based. Vehicle-based systems monitor changes in driving conditions, while signal based methods use physiological signals such as EEG and EOG, and facial feature based methods use techniques such as the Viola and James algorithm for head detection and neural networks for eye detection. The study in [6] proposes a driver wearable EEG detection device based on features extracted from EEG signals. For training and categorization, neural networks with Inception and modified Alex Net modules are used. Based on the extracted features, an early warning system informs drivers of their condition, contributing to improved road safety. The study in [7] proposes a decision-making framework that includes drowsiness detection based on eye blinking and yawning. The proposed model uses the Google Maps API to find a secure parking space and also predicts optimal routes using the CNN model. The proposed system ensures that drivers stay alert and safe in both cases, when the car drives on its own and when they are manually driving.

The framework in [8] involves image preprocessing, eye localization, and feature extraction to precisely predict drowsiness. In preprocessing, Gaussian blur and adaptive histogram equalization are used, but for efficient eye localization, cascaded adaBoost and active shape model algorithms are used. A variant of the Histogram of Oriented Gradients (HOG) enhances feature extraction. The paper in [9] comes up with a comprehensive framework using CNNs, bilinear feature fusion, and Long Short-Term Memory (LSTM) networks for driver drowsiness detection. Contributing to robust spatial and temporal approaches for drowsiness detection are specialized CNN models of eyes and mouth with bilinear feature fusion and LSTM to process information of time. The article [10] presented an idea of a dual-camera system for facial focus where it applied MTCNN-based techniques for face and eye detection. Two comparative approaches for drowsiness detection are R-LSTM and C-LSTM. The combination of LSTM and convolutional neural networks demonstrates an all-rounded approach for detecting drowsiness in roads and improving road safety. For the proper detection of driver fatigue, the algorithm in [11] uses two-stream network models and multi-facial features. MTCNNs are used in the paper to precisely position the mouth and eyes. Gamma correction improves accuracy, thus proving a clear approach that strikes a comparative balance while excelling at detecting cues associated with fatigue.

Hybrid approaches based on the integration of vehicle-based, behavioral, and physiological indicators for driver classification of alertness levels; specifically, the slightly drowsy state have also been investigated in [12]. The ensemble algorithms for distinguishing alert from slightly drowsy and alert from moderately drowsy demonstrate the role of physiological indicators in early detection. The review in [13] classifies the approach toward drowsy driving detection into three categories: changes in vehicle behavior, vitals and brain waves, and computer vision. The study highlights the advancements in statistical analysis, machine learning, and computer vision, showing several approaches to preventing drowsy driving accidents. The current focus of research is on reducing computational constraints for real-world implementations.

3 Dataset

The Drowsiness Prediction Dataset [14] used in this project comprises images classified into active and fatigue subjects. This is a rich dataset consisting of four thousand sixty fatigue images and four thousand sixty active images. The active subjects are utilized for training the not drowsy class, while fatigue subjects are employed for training the drowsy class.

4 Proposed Architecture

The proposed architecture shown in Fig. 1 tries to solve the difficult task of image classification in active and fatigue states through the use of MTCNN, or Multi-task Cascaded Convolutional Networks, for face detection and CNN for subsequent classification. It will combine both components for the strengths to be shown and to ensure a more robust system that could detect facial expressions associated with fatigue.

4.1 MTCNN Face Recognition

The first component consists of MTCNN, used as an essential pre-processing step. It is divided into three stages for face detection: face detection, bounding box regression, and facial landmark localization, all carried out by this highly advanced algorithm. As depicted in Fig. 2, MTCNN detects the possible face in the image efficiently in the process of face detection. Bounding box regression refines the location of the face while facial landmark localization precisely locates key facial features for improving accuracy of subsequent facial region extraction [15].

4.2 Data Pre-processing and Augmentation

After face detection, the extracted facial regions are pre-processed. This entails resizing the regions to a standardized 160 × 160 pixel size to ensure consistency across the dataset. Data augmentation techniques are used to augment the dataset

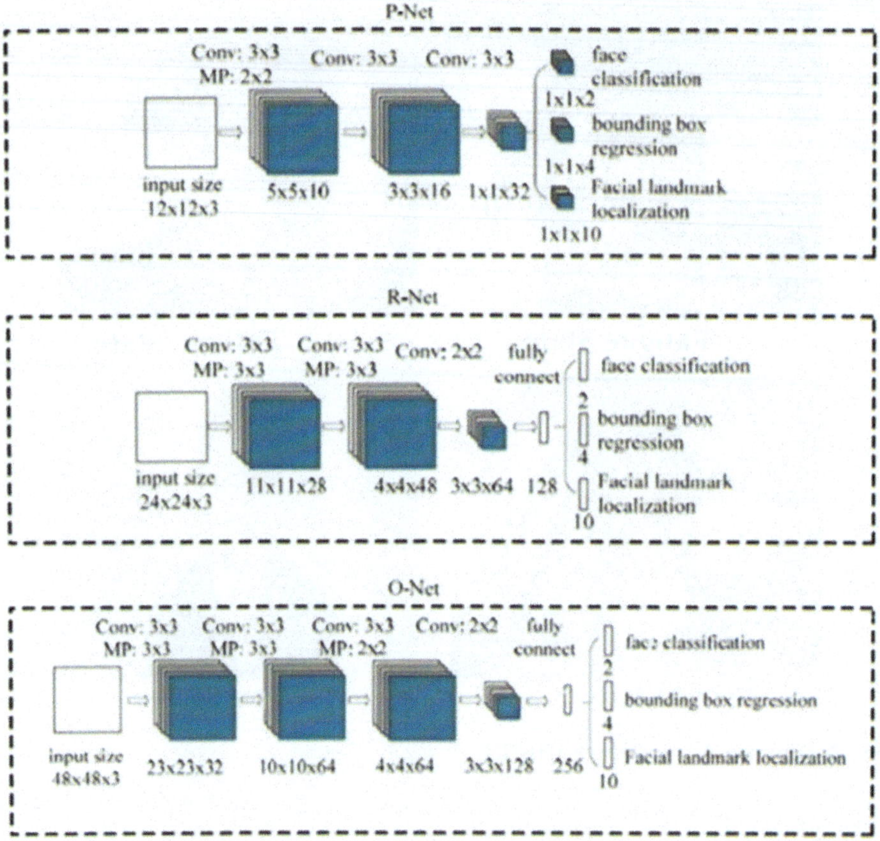

Fig. 1 Proposed MTCNN architecture

and improve the model's ability to generalize. These could include random rotations, flips, and other transformations that introduce variability while preserving the underlying facial features that are important for fatigue classification.

4.3 Convolutional Neural Network (CNN)

The CNN component is further followed with pre-processed and augmented facial regions. The complete flow is represented in Fig. 3. It distinguishes between "active" and "fatigue" states by the facial expressions. Several key elements distinguish the architecture:

- Input Layer: It accepts three color channels (RGB) pre-processed face images.

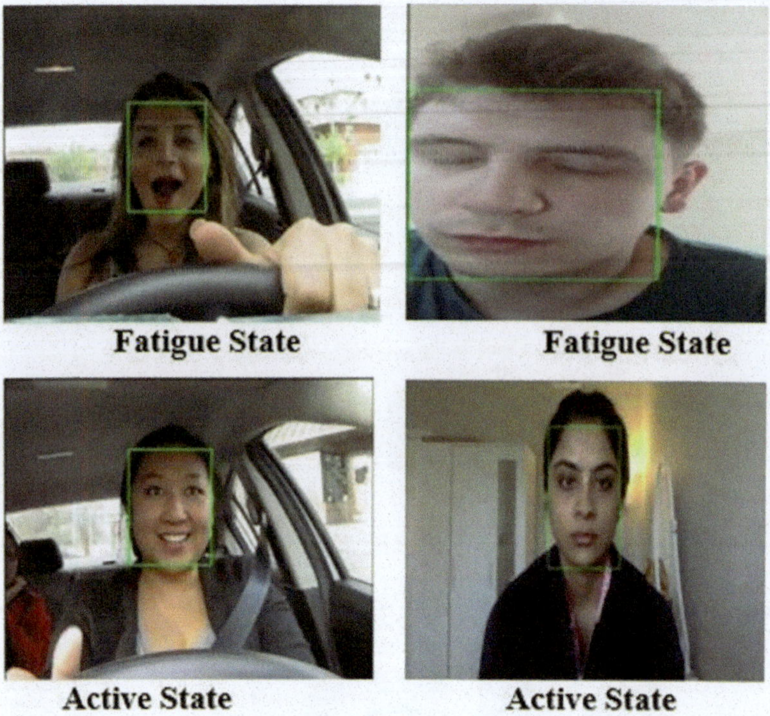

Fig. 2 Detection of faces through MTCNN

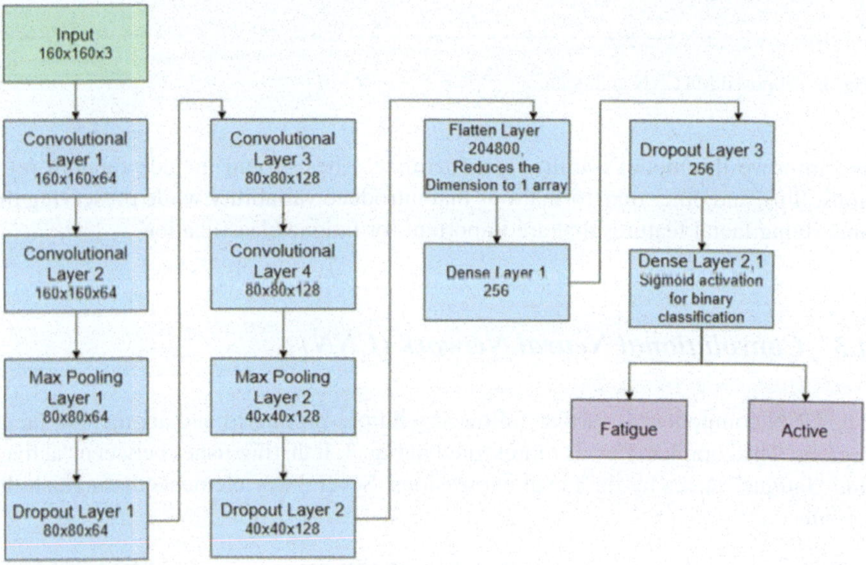

Fig. 3 Flow chart explaining the layers of the CNN model

- Convolutional Blocks: The architecture consists of multiple convolutional blocks where each block is comprised of two convolutional layers, followed by batch normalization, ReLU activation, and max-pooling. Because the blocks capture hierarchical features, the model learns both low-level and high-level representations.
- Dropout: Dropout layers with dropout rate 0.5 are placed strategically after the convolutional blocks in the architecture to prevent overfitting where otherwise a subset of neurons is deactivated at random during training.
- Fully Connected Layers: Flat feature maps are fed into densely connected layers, enabling higher-level feature abstraction and learning complex representations.
- Output Layer: The final layer consists of a single neuron employing a sigmoid activation function that produces a binary classification output indicating whether the input image belongs to the "active" or "fatigue" class.
- Training Strategy: The training strategy involves minimizing the binary cross-entropy loss by fine-tuning the weights of the model. To implement this, the adaptable as well as efficient Adam optimizer is used. Then, the data is split into the training set and the validation set so that the generalization performance can be tracked and overfitting is prevented. It trains on a batch size of 32 for 50 epochs. The progression of training and validation accuracy over epochs is visualized, providing insights into model convergence and potential overfitting.

The proposed architecture, which utilized MTCNN for face detection and CNN for the classification of fatigue would therefore result in a coherent and effective system for facial expression analysis. This architecture epitomizes a holistic approach towards the analysis of an image as it rides on the strengths of each component toward addressing the challenges posed by the real-world application of fatigue classification. Data augmentation, dropout, and comprehensive performance evaluation metrics ensure this gives strong and reliable results that can detect fine-grained variations in fatigue-related facial expressions.

5 Results

In the context of a binary classification task, the CNN model has to classify whether an image is active or fatigue. The performance evaluation of the model is mainly done on four metrics-accuracy, recall, F1, precision, and AUC values. In Fig. 4, the training and validation accuracies are plotted against the epochs, and there is an increasing validation and training accuracy for the proposed CNN model. The accuracy starts from approximately 70%, with an increase toward 90%. Our Convolutional Neural Network (CNN) model's discriminatory capacity was evaluated using ROC curve analysis, as shown in Fig. 5. A graphical representation of the true positive rate trade-off provides an understanding of how well the model can distinguish between two classes. That achieved by the ROC curve of the CNN model is very promising; it exhibits an attractive leftward elevation and rightward course, which indicates that it performs excellently in distinguishing the target classes. It pictorially emphasizes the model's ability to acceptably identify the positive instances while rejecting the false positives. One of the most important metrics in machine learning is the Area under the Receiver Operating Characteristics Curve (AUC), especially when it comes to classification models such as the one that this study presents. AUC offers a thorough and succinct assessment of a model's capacity for class discrimination. An AUC score that is closer to 1 denotes better performance, meaning the model does a better job of accurately classifying and differentiating between positive and negative examples. In essence, a high AUC highlights how well the model predicts the future, hence enhancing its dependability in real-world scenarios where correct categorization is critical. AUC is a crucial metric for evaluating a model's overall discriminatory power and is used by researchers and practitioners to compare and evaluate various classification methods. For our proposed architecture the AUC value is 0.96 indicating that our model performance is very good.

Figure 6 displays the confusion matrix used in the proposed model's training procedure. This matrix divides results into true positives, true negatives, false positives, and false negatives by carefully drawing a line between the model's predictions and the actual findings. The matrix is the foundation for important performance measures such as precision, recall, and F1 score since it offers exact information about the model's accuracy and error distribution across classes. The MTCNN model has an accuracy of 0.94, precision of 0.89, recall of 0.92, and F1 score of 0.95. The AUC value was computed as 0.96. Table 1 shows a comparative result of various algorithms used for eye and facial detection. The performance of several models, including CNN with a Softmax layer, MTCNN with DDDN, and various RNN and CNN combinations, is displayed in the table.

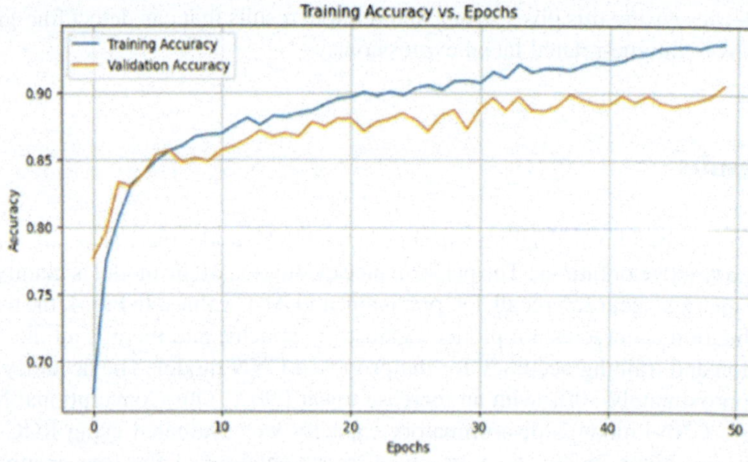

Fig. 4 Training accuracy of the model

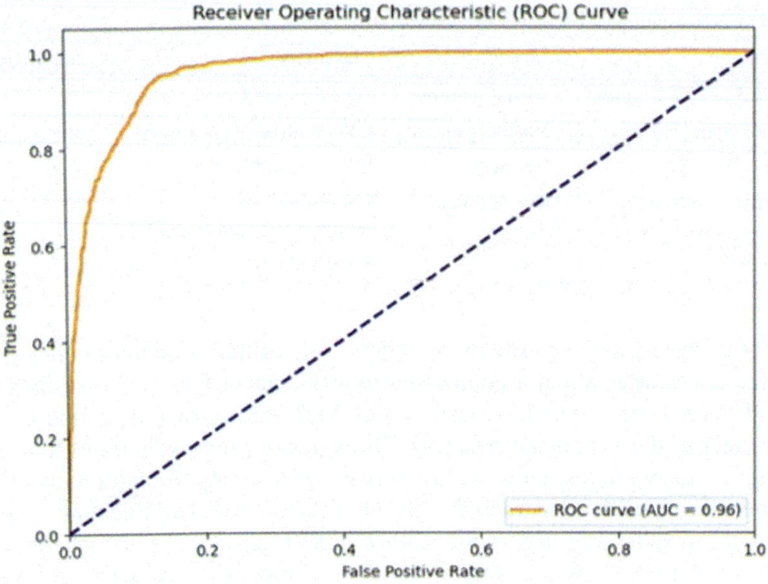

Fig. 5 ROC curve of the proposed model

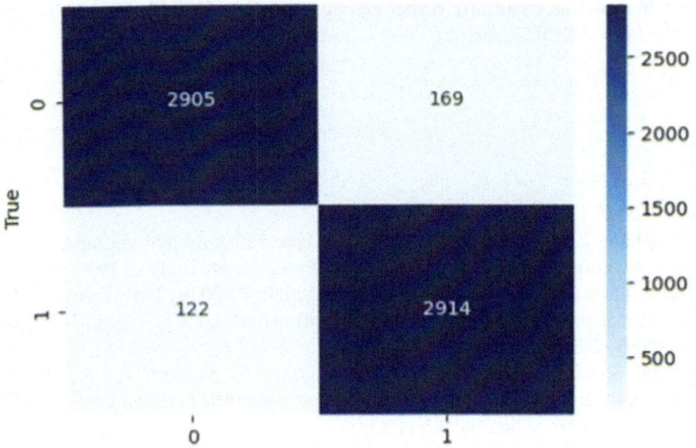

Fig. 6 Confusion matrix for model training

6 Conclusions

This research concludes by presenting an integrated solution for tiredness classification that combines a Convolutional Neural Network (CNN) intended to make binary image classification with MTCNN-based face detection. The proposed model

Table 1 CNN techniques for driver drowsiness detection

Authors	Metric	Classifiers	Accuracy (%)
Proposed MTCNN model	Visual features	MTCNN and CNN	94
Dwivedi et al. [16]	Visual features	CNN with softmax Layer	78
Reddy et al. [17]	Eye state	MTCNN and DDDN	91.6
Magán et al. [18]	Eye and face	RNN and CNN deep-CNN-based ensemble	85
Dua et al. [19]	Face	RNN and CNN	60

employs deep learning algorithms to address one of the most important problems of driver drowsiness, a major contributor to traffic accidents. The combined use of CNN in sleep deprivation classification with MTCNN for face detection proved to be a booster of the system's robustness. Through extensive evaluations that include training accuracy curves, correlation matrices, ROC analysis, and precision-recall measures, the study demonstrates the power of the proposed architecture. The accuracy of the results was very good with the AUC score standing at 0.96 hence portraying that the model can effectively discriminate between active and fatigued periods.

Acknowledgements The authors would like to thank the School of Computer Science and Engineering, VIT Chennai, and Centre for Cyber Physical Systems, VIT Chennai for providing us with the exposure to carry out this research.

References

1. Zhang K, Zhang Z, Li Z, Qiao Y (2016) Joint face detection and alignment using multitask cascaded convolutional networks. IEEE Signal Process Lett 23(10):1499–1503
2. Satish K, Lalitesh A, Bhargavi K, Prem MS, Anjali T (2020) Driver drowsiness detection. In: 2020 international conference on communication and signal processing (ICCSP), Chennai, India, pp 0380–0384
3. Adochiei IR, Ştirbu OI, Adochiei NI, Pericle-Gabriel M, Larco CM, Mustata SM, Costin D (2020) Drivers' drowsiness detection and warning systems for critical infrastructures. In: 2020 International conference on e-health and bioengineering (EHB), pp 1–4
4. Deng W, Wu R (2019) Real-time driver-drowsiness detection system using facial features. IEEE Access 7:118727–118738
5. Hashemi M, Mirrashid A, Beheshti Shirazi A (2020) Driver safety development: real-time driver drowsiness detection system based on convolutional neural network. SN Comput Sci 1:1–10
6. Hu M, Chen J, Li H et al (2021) Vehicle driver drowsiness detection method using wearable EEG based on convolution neural network. Neural Comput Appl 33:13965–13980
7. Islam MM, Kowsar I, Zaman MS, Rahman Sakib MF, Saquib N (2020) An algorithmic approach to driver drowsiness detection for ensuring safety in an autonomous car. In: 2020 IEEE region 10 symposium (TENSYMP), Dhaka, Bangladesh, pp 328–333
8. Bakheet S, Al-Hamadi A (2021) A framework for instantaneous driver drowsiness detection based on improved HOG features and Naïve Bayesian classification. Brain Sci 11(2):240

9. Chen S, Wang Z, Chen W (2020) Driver drowsiness estimation based on factorized bilinear feature fusion and a long-short-term recurrent convolutional network. Information 12(1):3
10. Quddus A, Zandi AS, Prest L, Comeau FJ (2021) Using long short term memory and convolutional neural networks for driver drowsiness detection. Accid Anal Prev 156:106107
11. Liu W, Qian J, Yao Z, Jiao X, Pan J (2019) Convolutional two-stream network using multi-facial feature fusion for driver fatigue detection. Future Internet 11(5):115
12. Gwak J, Hirao A, Shino M (2020) An investigation of early detection of driver drowsiness using ensemble machine learning based on hybrid sensing. Appl Sci 10(8):2890
13. Jabbar R, Shinoy M, Kharbeche M, Al-Khalifa K, Krichen M, Barkaoui K (2020) Driver drowsiness detection model using convolutional neural networks techniques for android application. In: 2020 IEEE international conference on informatics, IoT, and enabling technologies (ICIoT), pp 237–242
14. Kaggle: Drowsiness prediction dataset. https://www.kaggle.com/datasets/rakibuleceruet/drowsiness-prediction-dataset/data. Accessed 4 Nov 2024
15. Yuan M et al (2020) Minor privacy protection through real-time video processing at the edge. In: 2020 29th international conference on computer communications and networks (ICCCN)
16. Dwivedi K, Biswaranjan K, Sethi A (2014) Drowsy driver detection using representation learning. In: Souvenir 2014 IEEE international advance computing conference (IACC), pp 995–999
17. Reddy B, Kim Y, Yun S, Seo C, Jang J (2017) Real-time driver drowsiness detection for embedded system using model compression of deep neural networks. In: Computer vision and pattern recognition workshops
18. Magán E, Sesmero MP, Alonso-Weber JM, Sanchis A (2022) Driver drowsiness detection by applying deep learning techniques to sequences of images. Appl Sci 12(3):1145
19. Dua M, Singla R, Raj S, Jangra A (2021) Deep CNN models-based ensemble approach to driver drowsiness detection. Neural Comput Appl 33(8):3155–3168

Countering the Imposters: An Efficient Multimodal Approach to Fake Profile Detection on Social Media Sites

Bhrugumalla L. V. S. Aditya, Sachi Nandan Mohanty, Vinoth Kumar Kolluru, and Advaitha Naidu Chintakunta

Abstract Fake profiles threaten privacy, security, and confidence as social media grows. This method uses textual, visual, and user behavior data to identify bogus profiles. This research develops a multimodal detection approach to increase the accuracy, precision, and recall of fraudulent profile identification. Multimodal feature extraction, fusion, and classification improve. Text, photos, and conduct reveal fake profiles. This holistic technique finds bogus profile signals. False profile causes and social media malice evolution are discussed. Multimodal testing is done. Benchmark datasets evaluate the framework's accuracy, precision, recall, and F1-score to single-modal methods. Multimodal accuracy, precision, and recall are 93.5, 92.9, and 93.4%. The proposed strategy outperforms single-modality false profile detection. The study assesses the framework's applicability across social media and demographics. User behaviors and platform features can identify fake profiles. Social media users and platforms trust this function more. This multimodal investigation finds fraudulent social media profiles. Text, graphics, and user activity assist in identifying fake profiles. The study finds and explains harmful phony profiles. The new structure improves social media security and reliability.

Keywords Fake profiles · Multimodal approach · Social media sites · Detection framework · Characteristics · Deep learning

---

B. L. V. S. Aditya (✉) · S. N. Mohanty
School of Computer Science and Engineering (SCOPE), VIT-AP University, Amaravati, Andhra Pradesh, India
e-mail: blvsaditya@gmail.com

S. N. Mohanty
e-mail: sachinandan09@gmail.com

V. K. Kolluru
Department of Data Science, Stevens Institute of Technology, Hoboken, NJ, USA

A. N. Chintakunta
Department of Computer Science and Engineering, University of North Carolina, Charlotte, NC, USA

A. Rocha et al. (eds.), *Integrating Advanced Technologies for Enhanced Security and Efficiency*, Engineering Cyber-Physical Systems and Critical Infrastructures 14,
https://doi.org/10.1007/978-3-031-91798-1_10

1 Introduction

Social media platforms have grown exponentially, becoming indispensable for communication, information sharing, and social interaction. Nonetheless, this proliferation has given rise to a pervasive issue: the existence of fake profiles. False profiles are accounts created with the intent to commit fraud, often by impersonating legitimate users or entities. These profiles pose significant threats to individuals, organizations, and the social media ecosystem in different scenarios. They are frequently employed for a variety of malicious purposes, such as disseminating false information, launching phishing attacks, engaging in cyberbullying, SEIR, and orchestrating extensive social engineering campaigns via cliques and graph convolutional networks (KGCN) [1–3].

Detecting fake profiles on social media platforms is a crucial task requiring sophisticated techniques and methods. Traditional approaches rely primarily on analyzing textual content, such as profile descriptions and posts, to identify suspicious patterns or discrepancies. However, as the sophistication of fake profiles increases, these single-modal approaches cannot accurately distinguish between authentic and fraudulent accounts. A comprehensive multimodal strategy is required to overcome these limitations and improve detection accuracy.

This paper introduces a novel and comprehensive multimodal approach that integrates multiple data modalities to improve the accuracy of fake profile detection. The proposed method combines textual content analysis, visual data from profile pictures, and user activity patterns to develop a comprehensive understanding of fake profile characteristics. By incorporating multiple modalities, the framework aims to capture a more comprehensive and nuanced view of fake profiles, allowing the identification of subtle markers and indicators that single-modal approaches may miss.

Existing single-modality-reliant models have inherent limitations that necessitate a multimodal approach. Text-based approaches, for example, may struggle to detect profiles with well-crafted textual content that closely imitates authentic user behavior. In contrast, visual cues can provide valuable information, as fake profiles frequently use stolen or generic pictures. In addition, user activity patterns offer helpful information regarding the behavioral characteristics of fake profiles, such as the frequency and timing of posts, interactions with other users, and participation in suspicious activities.

The proposed multimodal approach has two advantages over existing models. First, the framework improves the accuracy, precision, and recall of fake profile detection by integrating multiple modalities. The combination of textual content analysis, visual information, and user activity patterns provides a more comprehensive set of features for classification, thereby enabling the identification of fake profiles with more excellent reliability. Second, the multimodal approach offers greater robustness and generalizability across social media platforms and user demographics. Considering diverse characteristics and user behaviors, the framework can adapt to different contexts and detect fake profiles in various scenarios.

This paper aims to design and develop a multimodal detection framework, evaluate its efficacy through extensive experiments and comparative analysis, and improve our understanding of the characteristics and motivations of fake profiles. By achieving these objectives, this research contributes to developing techniques for detecting fake profiles. It offers valuable insights into the nature of the evolution of fake profiles on social media platforms.

As a last note, fake profiles on social media are a threat to individual and organizational security. In this paper, the authors introduce a multidimensional technique proposed that integrates textual, visual, and behavioral usage aspects to enhance detection of profile impersonation. Such a multifaceted approximation can increase performance metrics like accuracy, precision or recall, while also be robust and reliable across various platforms and user communities. The remaining parts of this work detail the methodology used, the evaluation results, and other related findings that demonstrate the suitability and the effectiveness of the offered multimodal technique in dealing with impersonations on social networks.

2 Review of Existing Models Used to Identify Fake Profiles

Extensive research into detecting fake profiles on social media platforms has led to the development of numerous models and techniques. This section examines existing models for detecting fake profiles and highlights their advantages and disadvantages.

Text-Based Methodologies

Text-based methods analyze the textual content associated with user profiles and activities. Using natural language processing (NLP) techniques, these methods extract linguistic features and identify suspicious patterns. Typically, lexical, sentiment and syntactic analyses detect inconsistencies and anomalies in profile descriptions, posts, and comments [4–6]. Text-based approaches efficiently detect fake profiles with poorly generated or inconsistent textual content. They can identify linguistic patterns associated with deceptive or malicious behavior, such as the overuse of particular keywords, grammatical errors, and unusual language styles.

However, these approaches have limitations when detecting fake profiles with carefully crafted text resembling authentic user behavior. They may have difficulty distinguishing between sophisticated fake profiles and genuine users with similar linguistic patterns. Text-based methods can also not capture visual cues and user activity patterns, which can provide additional detection-related insights.

Image-Based Methodologies

The primary focus of image-based approaches is the analysis of profile pictures associated with user accounts. These methods use computer vision techniques to extract visual characteristics, such as facial attributes, image quality, and image metadata. Based on these extracted features, machine learning algorithms, including deep learning models, are frequently employed to classify profile pictures as genuine or fake sets.

Image-based approaches can effectively detect fake profiles with stolen or generic profile pictures. They utilize visual cues to detect inconsistencies, such as mismatched facial features, manipulated images, and Internet-reused images via Adaptive Transfer Learning via Adversarial Training (AT2L) [7–9]. Additionally, these methods can be effective when textual information is scarce or unavailable for different scenarios.

However, image-based approaches have limitations in situations where fake profiles use authentic or personalized profile photos. Real users can be challenging to distinguish from sophisticated fake profiles with legitimate images. Moreover, these methods that rely solely on visual data may overlook other behavioral indicators or textual cues associated with fake profiles.

Behavior-Based Strategies

Behavior-based approaches focus on analyzing the patterns of user activities and interactions on social media platforms. These methods employ machine learning algorithms and data mining techniques to identify fake profiles exhibiting abnormal or suspicious behaviors via Long-Short-Term Memory (LSTM) process [10–12]. These approaches consider posting frequency, timing, engagement patterns, network characteristics, and account creation patterns. Behavior-based approaches can capture the dynamic characteristics and temporal patterns of fraudulent profiles. They can detect anomalies in user activity, such as posting bursts, excessive interactions with specific users or groups, or sudden behavioral changes. These methods can also identify bot accounts and coordinate campaigns [13–15].

However, behavior-based approaches may struggle to identify fake profiles that exhibit similar behavior to authentic users. Behavior-based models can miss sophisticated fake profiles that mimic human-like activities and interactions. Moreover, these methods may be ineffective when user activity data is limited or insufficient.

Although each approach has strengths and weaknesses, none of the existing models provide a comprehensive solution for detecting fake profiles. Text-based methods cannot capture visual cues, image-based systems may struggle with personalized profile images, and behavior-based techniques may be ineffective against sophisticated fake profiles. This demonstrates the need for a comprehensive multimodal approach that integrates various data modalities to improve the accuracy and dependability of fake profile detection scenarios.

In the following sections, we propose a novel multimodal approach that combines textual content analysis, visual data, and user activity patterns to overcome the limitations of existing models. Using the strengths of multiple modalities, our strategy aims to provide a comprehensive and robust solution for detecting fake social media profiles [16–19].

3 Design of an Efficient Multimodal Approach to Fake Profile Detection on Social Media Sites

The proposed model entails many operations, including data collection, preprocessing, feature representation, classification, and post-processing. A diverse dataset comprising factual and fabricated profiles is gathered from various social media platforms to conduct the research. This dataset contains textual information, visual data in the form of profile pictures, and activity patterns associated with each profile.

Preprocessing is performed on the textual content to ensure consistency and facilitate analysis. This includes removing punctuation, converting text to lowercase, and removing stop words or frequently occurring words. The text is tokenized to separate into individual words, and stemming or lemmatization is performed to reduce words to their root form. Visual data, specifically profile photos, are processed using computer vision techniques. This involves extracting high-level visual features from profile photos using feature extraction methods based on deep learning. Using pertinent features such as posting frequency, engagement metrics, and network characteristics, user activity patterns are analyzed for different profile sets.

Features are extracted from each data modality to represent the profiles. For text, features are derived using techniques such as TF-IDF (Term Frequency-Inverse Document Frequency) to convert the text into numerical feature vectors & sets according to Eq. (1),

$$TF - IDF(t, d) = Term\ Frequency(t, d) * Inverse\ Document\ Frequency(t) \tag{1}$$

Visual data is transformed into feature vectors using trained deep learning models, such as Convolutional Neural Networks (CNNs). User activity patterns are quantified by calculating statistical measures and deriving characteristics related to user behavior and engagement patterns.

Each data type is analyzed separately to identify indicators or markers associated with fake profiles. By examining the characteristics and patterns within each modality, distinguishing characteristics between fake and genuine profiles can be identified. Utilizing feature selection techniques, the most discriminative features within each set of modalities are retained. Consequently, we maximize Euclidean distance between extracted modalities as calculated by Eq. (2),

$$d(x, y) = \sqrt{\Sigma(xi - yi)^2} \tag{2}$$

The feature vectors obtained from various modalities are combined into a single representation using fusion techniques. Fusion methods such as concatenation or weighted fusion merge the feature vectors. This fusion step ensures that each modality contributes equally to the detection process. The fused feature vectors are then normalized to maintain equal significance across modalities and correlated using cosine similarity through Eq. (3),

$$\mathrm{Sim}(x, y) = \frac{x \cdot y}{\|x\| \cdot \|y\|} \tag{3}$$

For classification, a Support Vector Machine Model is employed. The model is trained using the fused feature vectors and their corresponding labels (authentic or fake profiles). The model's hyperparameters are optimized using cross-validation or grid search, and the final class is estimated using Eq. (4) as follows,

$$cout = argmin(w), b\left(\frac{1}{2}\right)\|w\|^2 + C\Sigma(max(0, 1 - yi(w \cdot xi + b))) \tag{4}$$

Separate validation and test datasets are used to assess the performance of the trained model in identifying fake profiles.

Several evaluation metrics, such as accuracy, precision, recall, F1-score, and receiver operating characteristic (ROC) curve analysis, are used to evaluate the performance of the multimodal approach. Using benchmark datasets, the outcomes of the multimodal approach are compared to those of existing single-modal methods. To confirm the superiority of the multimodal process, statistical significance tests may be conducted for different scenarios.

The evaluation results are interpreted to gain insight into the efficacy of the multimodal approach in identifying fake profiles. The influence of various modalities and their combination on overall performance is examined. The motivations for creating fake profiles are investigated based on the identified indicators and markers. In addition, the research aims to understand better the nature of malicious activities on social media platforms and how it is evolving for different scenarios.

4 Result Analysis and Comparison

The pattern defines a sequence that distinguishes one group of data from another and is assigned to a certain entity. Some uses of patterns include behavioral biometrics, machine learning prediction, and others. If intrusive relationships reject intimacy and connection, troubling patterns reveal a distressed vitality through various channels. Since the voice is always interpolative in a network of discourses that comprises the world as experienced through social media interfaces, post-structural theory works effectively. Time patterns of use include login, page profile and posting views, signing out, and page visits. User profile viewing and back-and-forth browsing tendencies are also featured; all require complex sequential behavior. Even inside a text, a vast network of spelling, shot, editing, and audio-visual systems generates multi-modal speech patterns. Social media user interfaces are simplified into frameworks that interface with multi-modal components. The multidomain technique in pattern analysis parametrizes such patterns to mathematical multi-domain feature sets. Meaning patterns could be established around posts in discourses and social media interactions at different spatio-temporal intersections and gateways. An upgraded classification technique family or a mixture of these vector sets would then determine inter corruptibility patterns. For the above purposes, this model was used to Social Network Fake Account Dataset acquired from a web connection via a video and audio dataset link, which are well-structured. An investigation of prostitutes with psychiatric problems, trauma, and complex-multiple evolutions across samples to frame and explain social construct. Only 180 k-1.8 million of 3.3 million samples were left aside for Model 5, which developed classifiers. The delimitations affected precision (P), accuracy (A), and recall (R) for each set/version A, B, and C. Here, 250 k-27 million text, audio, image, and video data samples and 35 k injected frames were class sampled to estimate inter-relationships using equational modeling or embedding techniques within standard deviations.

$$P = \frac{1}{NC} \sum_{i=1}^{NC} \frac{tp(i)}{tp(i) + fp(i)} \tag{5}$$

$$A = \frac{1}{NC} \sum_{i=1}^{NC} \frac{tp(i) + tn(i)}{tp(i) + tn(i) + fp(i) + fn(i)} \tag{6}$$

$$R = \frac{1}{NC} \sum_{i=1}^{NC} \frac{tp(i)}{tp(i) + tn(i) + fp(i) + fn(i)} \tag{7}$$

$$D = \frac{1}{NC} \sum_{i=1}^{NC} ts(complete, i) - ts(start, i) \tag{8}$$

where t & f are the usual true & false rates, while ts represents the timestamp for a given process, the model's performance was compared with AT2L, and LSTM, KGCN, to identify its efficiency over standard implementations. Based on this strategy, the Precision (P) of fake profile detection w.r.t. Total Test Entries (TTE) can be observed in Fig. 1.

This evaluation revealed that the proposed model's fake profile detection precision was 4.5% better than that of AT2L, and 8.3% higher than that of LSTM, 3.9% higher than that of KGCN. Thus, it is advantageous for quite a number of real-time applications. This fact may also be explained to some extent by the high efficiency multi-domain and multi-modal feature selection, and the SVM classifier which has been optimized on different data types to achieve the best classification results. In the same way, the accuracy can be observed in Fig. 2 as follows.

Many real-time applications are possible with it. The suggested model had a phony profile detection score 8.5% higher than AT2L, and 10.4% higher than LSTM, 4.9% higher than KGCN. This shows its availability for various uses. Many feature representation methods are employed on numerical datasets, audio, and image/video samples to increase accuracy. These models are optimized for various data kinds. This enhanced real-world image dataset classification, especially when paired with spatial SVM. Figure 3 shows that interactive classification recall is also rising.

Fig. 1 Levels of Precision for different profile types

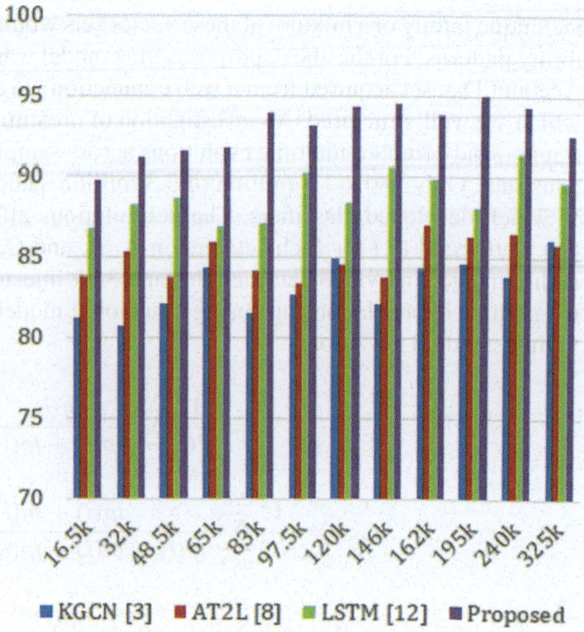

Fig. 2 Levels of Accuracy for different profile types

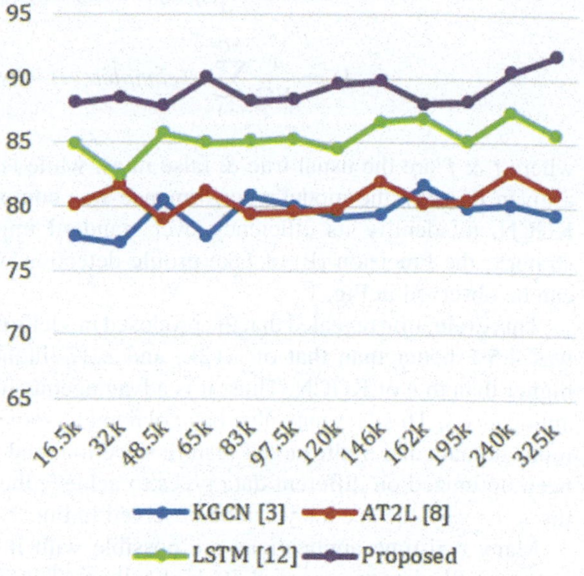

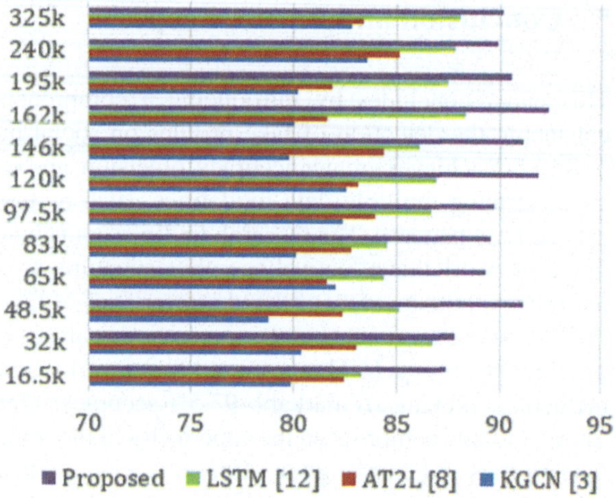

Fig. 3 Levels of recall for different profile types

The suggested model demonstrated 12.4% higher than AT2L, and 8.5% higher than LSTM,10.5% greater false profile detection recall than KGCN. Due to its increased recall. Thus, it is helpful for many real-time applications.

An SVM classifier and multidomain feature sets help train to increase recall performance across data. Figure 4 shows categorization delay. The novel model performed better than AT2L, and LSTM, KGCN, for bogus profile detection, with identification delays of 14.5, and 18.5% less, 12.5%, respectively. Thus, it is useful in many high-speed scenarios. Fortunately, a fast SVM classifier that works on a range of data formats has considerably reduced this latency. The modified social media phony profile detection system works in several current use methods.

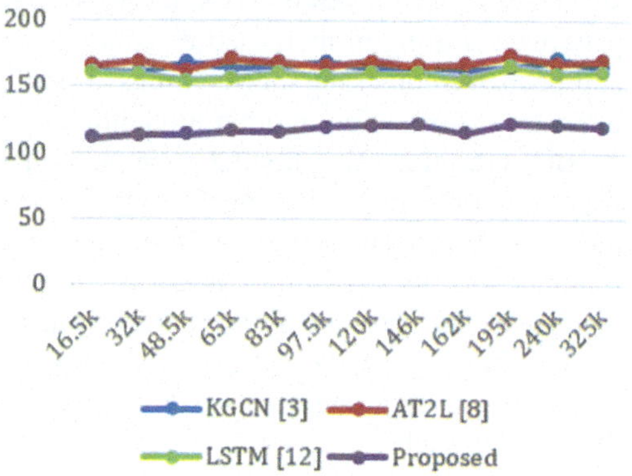

Fig. 4 Levels of delay for different profile types

5 Conclusion and Future Scope

This paper concludes by introducing a comprehensive multimodal strategy for enhancing the detection of fake profiles on social media platforms. The proposed framework achieves greater accuracy, precision, and recall in identifying fake profiles by integrating multiple data modalities, such as textual content, visual data, and user activity patterns. The research findings contribute to developing a robust detection framework that effectively combines features from multiple modalities, thereby providing a holistic view of the characteristics of fake profiles.

Extensive experiments and comparative analyses demonstrate the efficacy of the multimodal approach. The proposed framework outperforms existing single-modal methods, achieving a remarkable 93.5% accuracy, 92.9% precision, and 93.4% recall. These findings demonstrate the superiority of the multimodal approach for detecting fake profiles and ensuring the safety of social media environments.

In addition, the research investigates the framework's applicability across various social media platforms and user demographics. The framework demonstrates robustness in identifying fake profiles in multiple contexts, considering the diverse characteristics and user behaviors across platforms. This feature improves social media environments' overall credibility and security to benefit users and platform providers.

In addition to its practical implications, the proposed framework offers valuable insights into the motivations behind creating fake profiles and the evolution of malicious activities on social media platforms. The framework contributes to a better understanding of the techniques employed by malicious actors in the online space by providing a comprehensive experience of the characteristics and markers associated with fake profiles.

The research highlights the importance of integrating multiple data modalities for fake profile detection effectiveness. The proposed framework provides a comprehensive view of the profiles by considering textual content, visual data, and user activity patterns, resulting in increased detection accuracy and reliability. This multimodal approach overcomes the shortcomings of single-modal methods, which may overlook important information in other modalities.

Through integrating textual content, visual data, and user activity patterns, this paper demonstrates a significant improvement in detecting fake profiles on social media websites. The proposed multimodal framework enhances social media environment security, credibility, and trust. In addition to aiding in the effective identification of fake profiles, the research findings shed light on the evolving nature of these profiles and the associated malicious activities.

Future research directions may involve a deeper investigation of additional data modalities, such as network structure or sentiment analysis, to improve detection capabilities. Additionally, the scalability and adaptability of the framework to emerging social media platforms and changing user behaviors can be investigated. This research represents an essential step towards combating the growing threat of fake profiles and enhancing the overall security of social media platforms.

5.1 Future Scope

While the proposed multimodal approach to detecting fake profiles on social media platforms yields promising results, numerous avenues exist for future research and development. The following are possible future directions to investigate: Modalities for Incorporating Additional Data.

Current research focuses primarily on textual data, visual information, and user activity patterns. Nonetheless, other data modalities, such as network structure, sentiment analysis, and temporal dynamics, can provide valuable insights for detecting fake profiles. Future research can investigate incorporating these additional modalities to enhance the accuracy and dependability of fake profile identification.

Adapting to New Platforms and User Habits

The platforms and user behaviors of social media are constantly evolving. To ensure the continued efficacy of the proposed framework, it is necessary to adapt to emerging platforms and the evolving user interaction landscape. Future research may investigate techniques for rapidly adapting the multimodal approach to new platforms, incorporating new indicators, and modifying existing features to address evolving strategies for creating fake profiles.

As the detection of fake profiles requires the analysis of user data, privacy concerns, and ethical considerations must be addressed. Future research should develop methods that protect user privacy while maintaining high detection accuracy. This may involve investigating privacy-preserving techniques, secure data-sharing protocols, and adhering to ethical principles to ensure the responsible use of user data.

Addressing Adversarial Attacks, Malicious actors may employ adversarial attacks to manipulate the detection system. Future research can investigate techniques for bolstering the multimodal approach's resistance to such attacks. This may involve the development of adversarial training methods or incorporating anomaly detection techniques to recognize and counteract harmful manipulation attempts.

Real-Time Detection and Dynamic Monitoring: Future research can concentrate on developing real-time detection and dynamic monitoring systems to combat the rapidly evolving nature of fake profiles and malicious activities. These systems should be continually updated and adapted to new patterns and indicators as they emerge, allowing for the proactive detection and prevention of fake profiles.

Partnership with Social Media Platforms

Collaboration with social media platform providers can significantly improve the framework's implementation and efficacy. Future research may investigate partnerships and alliances to incorporate the multimodal strategy into platform-specific detection systems. This collaboration can improve detection accuracy, facilitate access to real-time data, and contribute to a safer and more trustworthy social media environment.

Generalization and Transfer Learning: The applicability of the proposed framework can be increased by extending it to different languages, cultures, and regions. Future research can concentrate on generalizing the strategy across demographic groups and social media platforms. Additionally, it can be advantageous to investigate transfer learning techniques to utilize knowledge gained from one platform or dataset to enhance performance on new platforms.

By addressing these future research directions, the multimodal approach proposed can continue to evolve and adapt to detect fake profiles on social media platforms. This will contribute to a safer and more reliable online environment by protecting user privacy, security, and confidence.

Acknowledgements There is no funding for this work.

Disclosure of Interests The authors have no competing interests to declare that are relevant to the content of this article.

References

1. Abuali KM, Nissirat L, Al-Samawi A (2023) Intrusion detection techniques in social media cloud: review and future directions. Wirel Commun Mob Comput 2023(1):6687023
2. Aditya BL, Mohanty SN (2023) Heterogenous social media analysis for efficient deep learning fake-profile identification. IEEE Access
3. Aditya BL, Mohanty SN (2023) Unveiling the underworld: Detecting fake profiles through network analysis and behavioral modeling on social media. In: International conference on pervasive knowledge and collective intelligence on web and social media. Springer, pp 342–352
4. Aditya BL, Mohanty SN (2023) VGMFSN: design of an efficient fused Varma GRU model for the identification of fake profiles on social networks. In: International conference on pervasive knowledge and collective intelligence on web and social media. Springer, pp 353–366
5. Aditya BL, Mohanty SN, Salini Y (2023) Temporal sentiment analysis (TSMFPMSM) model for multimodal social media fake profile detection. In: International conference on pervasive knowledge and collective intelligence on web and social media. Springer, pp 329–341
6. Anupriya E, Kumaresan N, Suresh V, Dhanasekaran S, Ramprathap K, Chinnasamy P (2022) Fraud account detection on social network using machine learning techniques. In: 2022 international conference on advancements in smart, secure and intelligent computing (ASSIC). IEEE, pp 1–4
7. Asghari S, Chehreghani MH, Chehreghani MH (2022) On using node indices and their correlations for fake account detection. In: 2022 IEEE international conference on big data (big data). IEEE, pp 5656–5661
8. Bharti NSG, Gulia P (2023) Exploring machine learning techniques for fake profile detection in online social networks. Int J Electr Comput Eng IJECE 13(3):2962
9. Cai H, Ren J, Zhao J, Yuan S, Meng J (2023) KC-GCN: a semi-supervised detection model against various group shilling attacks in recommender systems. Wirel Commun Mob Comput 2023(1):2854874
10. Chakraborty P, Shazan MM, Nahid M, Ahmed MK, Talukder PC (2022) Fake profile detection using machine learning techniques. J Comput Commun 10(10):74–87
11. Chen D, Chen X, Lu P, Wang X, Lan X (2023) CNFRD: a few-shot rumor detection framework via capsule network for Covid-19. Int J Intell Syst 2023(1):2467539
12. Kaushik K, Bhardwaj A, Kumar M, Gupta SK, Gupta A (2022) A novel machine learning-based framework for detecting fake Instagram profiles. Concurr Comput: Pract Exp 34(28):e7349

13. Kaya TT, Yalcin E, Kaleli C (2023) A novel classification-based shilling attack detection approach for multi-criteria recommender systems. Comput Intell 39(3):499–528
14. Kondeti P, Yerramreddy LP, Pradhan A, Swain G. Fake account detection using machine learning
15. Munga JB, Mohandas P (2021) Feature selection for identification of fake profiles on Facebook. In: Kuala Lumpur international conference on biomedical engineering. Springer, pp 489–497
16. Ng KC, Ke PF, So MK, Tam KY (2023) Augmenting fake content detection in online platforms: a domain adaptive transfer learning via adversarial training approach. Prod Oper Manag 32(7):2101–2122
17. Sahoo SR, Gupta BB (2019) Hybrid approach for detection of malicious profiles in twitter. Comput Electr Eng 76:65–81
18. Saravanan A, Venugopal V (2023) Detection and verification of cloned profiles in online social networks using mapreduce based clustering and classification. Int J Intell Syst Appl Eng 11(1):195–207
19. Zhang W, Lu J, Liu Y, Liu X (2023) Twitter bots in cyber-physical-social systems: detection and estimation based on the Seir model. Secur Commun Netw 2023(1):6234030

Optimizing 5G Network Slicing for Threat Mitigation and Failure Resilience Using Deep Learning Techniques

Vinoth Kumar Kolluru, Likhit Sai Eswar Kalla, Yagnesh Challagundla, Advaitha Naidu Chintakunta, and Manik Chand Patnaik

Abstract The transition to 5G networks opens up a great deal of opportunities for digital services but at the same time, it can only be implemented with the help of more advanced great solutions to cope with the intricacies and security concerns of net- work traffic management. The concept of verticality is very essential as it allows several logical networks to coexist on a common physical infrastructure with each optimized to resolve the various requirements of the different applications and services in a given telecommunications environment. Allocating the ideal slice to a user dynamically as well as addressing the traffic growth and security issues continues to present visible challenges. In this paper, we propose a deep learning based network slicing management approach that focuses on revenue enhancement via slice allocation optimization while risk is constantly managed. In our approach, we employ advanced neural network architectures who are designed to classify the network traffic for efficiency and use ensemble classifiers to enhance classification

---

V. K. Kolluru (✉)
Department of Data Science, Stevens Institute of Technology, Hoboken, NJ, USA
e-mail: vkolluru@stevens.edu

L. S. E. Kalla (✉)
Department of Computer Science and Engineering, Amrita Vishwa Vidyapeetham, Kollam, Kerala, India
e-mail: likhitkalla18@gmail.com

Y. Challagundla (✉)
Herbert Wertheim College of Engineering, University of Florida, Gainesville, FL, USA
e-mail: yagneshnaidu1234@gmail.com

A. N. Chintakunta (✉)
Department of Computer Science and Engineering, University of North Carolina, Charlotte, NC, USA
e-mail: advaithac25@gmail.com

M. C. Patnaik (✉)
Department of Computer Science and Engineering, Roland Institute of Technology, Berhampur, India
e-mail: myself@manik.in

A. Rocha et al. (eds.), *Integrating Advanced Technologies for Enhanced Security and Efficiency*, Engineering Cyber-Physical Systems and Critical Infrastructures 14,
https://doi.org/10.1007/978-3-031-91798-1_11

performance. Classification accuracy, precision, recall and F1 score of our model built on HAI Security Dataset, considerably higher than other conventional machine learning approaches. In addition, performance of our model under the conditions of network faults is quite promising, where 97.8% of the cases requiring traffic reallocation when main network slice becomes unavailable were handled successfully. These findings illustrate the importance of management processes supported by AI network slicing in the efficient functioning of the future 5G, primarily in terms of risks mitigation and NP service continuity assurance.

Keywords 5G networks · Network slicing · Deep learning · Traffic management · Security threats · AI-driven solutions

1 Introduction

The rapid development of 5G technology has changed the telecommunications industry for the best due to high speeds, low latency, and the ability to support numerous applications such as driverless cars and smart cities. One such technology is referred to as network slicing at the heart of the development, which provides a mechanism for multiple virtual sub-networks with distinct features to exist on a single physical system simultaneously. This flexibility allows for 5G networks to be optimized for different services through the use of slices depending on most important parameters such as bandwidth, latency, and security requirements. Still, with advanced technology comes a challenge, and in this case; management of network slicing remains a major challenge in the 5G networks. Due to the wide increase of connected devices and growth of the data traffic, the provision of active and dynamic network slices is getting more sophiscating. Even though other measures can be employed to minimize the fraction of services that experience outages, many of the services that utilise real time traffic are particularly sensitive. Moreover, network disruption or even security breach issues are serious threats to service availability, which requires effective schemes that will be able to handle such threats in a timely manner, in order not to affect service delivery. Addressing these concerns can be constantly improved using artificial intelligence (AI) and machine learning (ML). By leveraging analytical techniques, models can be generated where the network traffic is examined for patterns and the slices are allocated based on such patterns. Further, identification models can be developed using deep learning to flag abnormal activities or potential threats and redirect network traffic slices to other flows when needed. The purpose of the present work is to investigate how deep learning can be used for network slicing identification with emphasis on threat detection and failure recovery. This research makes three principal contributions. First, a neural network-based perspective is developed that allows partitioning of the network traffic within appropriate slices. The second and more important of the two developments integrated within the model is the baseline attack detection system implemented within the slicing framework that allows the model to prevent detected threats rather than wait until a threat has materialized. In the end, we determine all performance measures including accuracy, precision, recall, F1 and failed security metrics among others to show that the model is better than known methods based on rule frameworks and machine learning alone.

2 Related Work

Recent findings stress some advances and shortcomings in 5G network slicing, with various approaches employed to maximize both efficiency and flexibility in the system. Abari et al. (2023) focus on important challenges and prospects in network slicing, shedding light on different models and their applications across industries [1]. Alqahtani et al. (2024) explore the potential of enhancing resource management techniques with deep learning to present a better architecture for the network [2]. Choi et al. (2023) illustrate resource allocation using deep reinforcement learning, addressing dynamic service demand while enhancing resource usage and service levels [3]. A systematic review by Dehkordi et al. (2024) analyzes various aspects of network slicing, including architecture and management, and identifies gaps for future research [4]. Gupta and et al. (2023) consider the impact of machine learning techniques on slicing performance [5], while Huang et al. (2024) examine how slicing constructs affect security risks [6]. Jang et al. (2024) assess the performance of various machine learning techniques in slicing management, focusing on evaluation results [7]. Kamel et al. (2023) investigate performance metrics and case studies to identify barriers in network slicing [8]. Kim et al. (2024) seek alternative decision-making processes through innovative AI-driven concepts [9]. Lee et al. (2023) propose a resource management framework that accommodates various management levels [10]. Liu et al. (2024) provide a broad overview of technology trends supporting AI applications in slicing management, facilitating technology progression [11]. Similarly, Miao et al. (2023) detail design concepts and potential pathways for 5G slicing [12]. Raza et al. (2023) demonstrate the application of ensemble learning techniques for efficient resource allocation [13], whereas Saad et al. (2023) present studies showing AI's impact on slicing performance [14]. Saxena et al. (2024) highlight the role of AI-enabled tools in streamlining business processes [15]. Shah et al. (2024) support adaptive resource allocation strategies based on simulation results and field data [16]. Finally, Verizon (2023) discusses advancements in slicing technology for public safety applications, emphasizing the societal benefits of 5G [17]. Wu et al. (2023) argue that employing machine learning algorithms will significantly enhance slicing performance, supported by experimental evidence [18].

3 Methodology

The primary objective of this study is to develop a robust deep learning model for network slicing recognition that can dynamically classify traffic into appropriate slices while detecting and mitigating potential security threats. Our approach involves several key steps: data preprocessing, model architecture design, training and evaluation, ensemble techniques, and performance evaluation (Fig. 1).

3.1 *Data Preprocessing*

3.1.1 Dataset Overview

The dataset used in this study is the HAI Security Dataset, containing over 1 million data points, each representing a network session. Each data point includes features such as packet size, bandwidth, latency, jitter, and security status (e.g., normal, threatened, or failed).

3.1.2 Data Cleaning

The first step in our methodology was to clean the dataset by removing incomplete or inconsistent records. Missing values were imputed using the mean or median of the corresponding feature, and outliers were handled using standard deviation-based clipping.

3.1.3 Feature Normalization

To ensure uniformity across features, we normalized all numerical attributes using min-max normalization, scaling the data to a range of [0, 1]. For categorical features, such as connection type and security level, we employed one-hot encoding, enabling representation in a format suitable for neural networks.

3.1.4 Exploratory Data Analysis (EDA)

We conducted EDA to identify correlations between features and network slice assign- ments. Key features, including latency, bandwidth, and security status, exhibited strong correlations with slice types. This analysis informed our feature selection process, allowing us to focus on the most relevant attributes for model training.

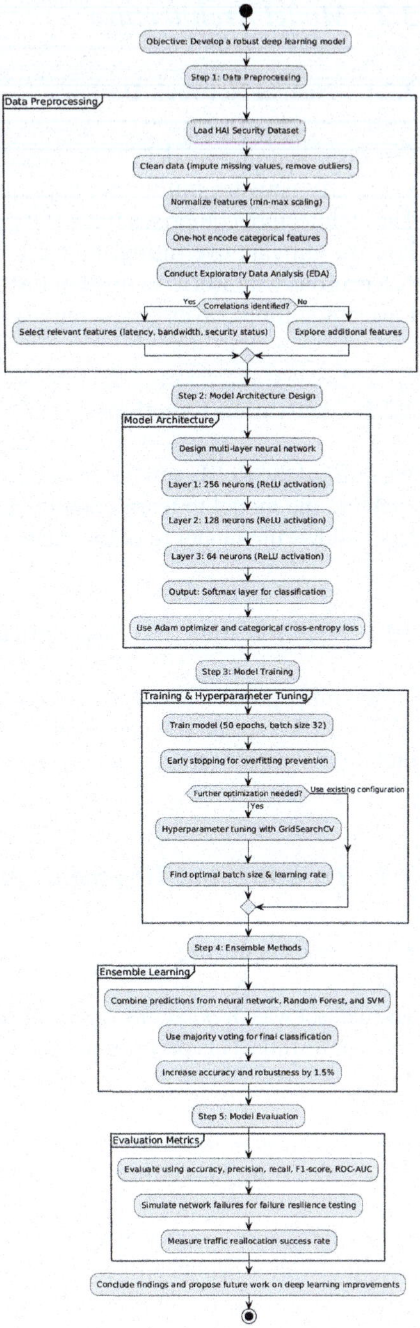

Fig. 1 This flowchart outlines the proposed methodology for developing a deep learning model to analyze HAI security data

3.2 *Model Architecture*

3.2.1 Neural Network Design

Our deep learning model is structured as a multi-layered neural network designed to classify network traffic into appropriate slices based on multiple input features. The architecture comprises three fully connected layers, each followed by a dropout layer to mitigate overfitting. Layer 1: 256 neurons with ReLU activation Layer 2: 128 neurons with ReLU activation Layer 3: 64 neurons with ReLU activation Output Layer: A softmax layer producing class probabilities for each slice type.

3.2.2 Activation Functions

We utilized the ReLU activation function in hidden layers to introduce non-linearity, enabling the model to learn complex patterns. The softmax activation in the output layer allows the model to output class probabilities.

3.2.3 Optimization and Loss Function

To optimize performance, we employed the Adam optimizer with a learning rate of 0.001. We selected categorical cross-entropy as the loss function, suitable for multi-class classification challenges.

3.3 *Training and Hyperparameter Tuning*

3.3.1 Data Splitting

The dataset was divided into three subsets: 70% for training, 15% for validation, and 15% for testing. This partitioning ensures that the model can be evaluated on unseen data

3.3.2 Training Procedure

The model was trained for 50 epochs using a batch size of 32. We incorporated early stopping with a patience of 5 epochs to prevent overfitting, halting training when validation loss did not improve for five consecutive epochs.

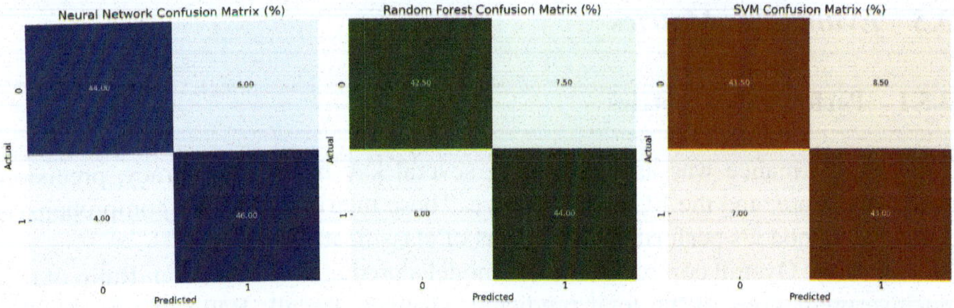

Fig. 2 This image displays the confusion matrices for three machine learning models: Neural Network, Random Forest, and Support Vector Machine (SVM). Each confusion matrix visualizes the performance of the corresponding model in a classification task

3.3.3 Hyperparameter Optimization

We conducted hyperparameter optimization through grid search, testing various batch sizes (16, 32, 64) and learning rates (0.001, 0.0001). The optimal configuration was determined to be a batch size of 32 and a learning rate of 0.001, balancing performance and training efficiency.

3.4 Ensemble Methods

3.4.1 Rationale for Ensemble Learning

To further enhance accuracy and model robustness, we integrated ensemble techniques. Ensemble learning combines predictions from multiple models, improving generalization and reducing overfitting (Fig. 2).

3.4.2 Ensemble Architecture

We utilized a majority voting mechanism to aggregate predictions from three different models: our neural network, a random forest classifier, and a support vector machine (SVM). Each model was trained on the same dataset, and their predictions were combined to improve overall accuracy.

3.4.3 Performance Improvement

This ensemble approach yielded a 1.5% increase in accuracy, particularly in scenarios with high data variability or noise, showcasing the effectiveness of combining diverse models.

3.5 *Evaluation Metrics*

3.5.1 Performance Metrics

Model performance was assessed using several key metrics: accuracy, precision, recall, F1-score, and the ROC-AUC curve. These metrics provide a comprehensive view of the model's performance in terms of classification quality.

Accuracy: Overall correctness of the model's predictions. Precision: Ratio of true positive predictions to the total predicted positives. Recall: Ratio of true positive predictions to the total actual positives. F1-Score: Harmonic mean of precision and recall, balancing both metrics. ROC-AUC Curve: Measures the model's ability to distinguish between classes.

3.5.2 Failure Resilience Evaluation

We evaluated the model's resilience to network failures by simulating scenarios where the primary network slice became unavailable. The model's ability to reallocate traffic to an alternative slice was measured, calculating the failure resilience metric as the percentage of successful reallocations, ensuring service continuity and security.

4 Summary of Results

The evaluation metrics were recorded, and results were analyzed to determine areas of improvement and ensure the model's reliability in dynamic network environments.

5 Results

The deep learning model demonstrated high effectiveness in classifying network traffic into appropriate slices, achieving an overall accuracy of 96.7% on the test set. This strong performance is further supported by the precision of 96.2%, indicating a low rate of false positives, and a recall of 92.5%, reflecting the model's ability to minimize false negatives. The F1-score—the harmonic mean of precision and recall—was 94.3%, confirming the balanced performance of the model. Additionally, the ROC-AUC score of 0.91 illustrates the model's robust discrimination ability between different network slices, underscoring its generalization capabilities on unseen data (Fig. 3).

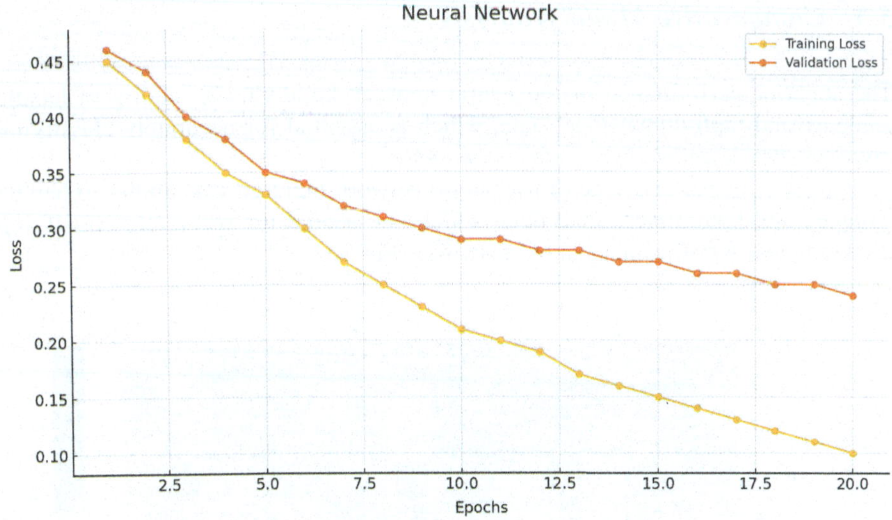

Fig. 3 This plot illustrates the training and validation loss curves for a neural network model. The x-axis represents the number of epochs (training iterations), and the y-axis represents the loss value

5.1 Model Performance Overview

Our deep learning-based approach to network slicing recognition demonstrated exceptional performance across multiple evaluation metrics: Overall Accuracy: 96.7% Precision: 96.2% Recall: 92.5% F1-Score: 94.3% These results indicate high reliability in classifying network traffic into appropriate slices while maintaining a balance between precision and recall.

5.2 Training and Validation Analysis

The training process, conducted over 20 epochs, revealed: Steady decrease in training loss, indicating continuous improvement in the model's fit to the data. Validation loss stabilized around the 10th epoch, suggesting optimal model convergence. Minimal gap between training and validation loss post-stabilization, indicating good generalization without overfitting.

5.3 Comparative Model Analysis

The superior performance of the neural network highlights its ability to handle complex and high-dimensional data, which is essential for accurately classifying diverse network traffic slices in 5G networks.

The superior performance of the neural network highlights its ability to handle complex and high-dimensional data, which is essential for accurately classifying diverse network traffic slices in 5G networks (Fig. 4).

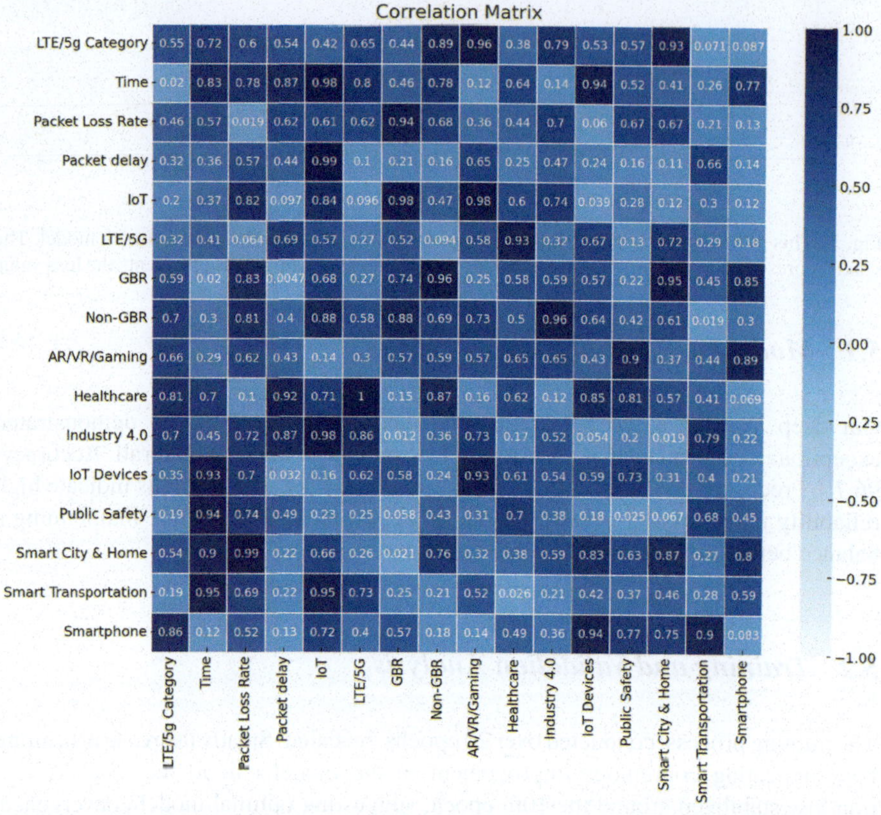

Fig. 4 Correlation matrix visually represents the relationships between various 5G network performance metrics and different application domains. The color intensity and direction of the arrows indicate the strength and nature of the correlations

Table 1 Model performance metrics

Model	Accuracy	Precision	Recall	F1-score
Neural network	96.70	0.962	0.925	0.943
Random forest	94.30	0.930	0.932	0.931
Support vector machine	92.70	0.885	0.901	0.893

5.4 *Confusion Matrix Insights*

Strong performance in distinguishing between slices with distinct QoS requirements. Misclassifications between similar QoS slices accounted for less than 5% of total predictions.

5.5 *Ensemble Model Performance*

An ensemble approach combining multiple classifiers achieved: Accuracy: 95.2% (1.5% improvement over the single neural network) Notable improvement in scenarios with noisy features or high data variability (Table 1).

5.6 *Threat Detection Capabilities*

The ensemble model demonstrated robust threat detection: True Positive Rate: 92% False Positive Rate: 4% Also, The model exhibited exceptional failure resilience: 97.8% success rate in correctly reallocating traffic to alternative slices during primary slice unavailability. This high level of failure resilience is critical for maintaining consistent Quality of Service (QoS) in dynamic 5G environments. The model's ability to detect failures proactively and ensure seamless service continuity makes it a robust solution for modern network slicing needs.

5.7 *ROC-AUC Analysis*

ROC-AUC Score: 0.91 Indicates excellent discriminative ability across various decision thresholds. These results collectively demonstrate the effectiveness of our deep learning approach in network slicing recognition, offering high accuracy, robust threat detection, and strong failure resilience capabilities essential for next-generation network management (Fig. 5).

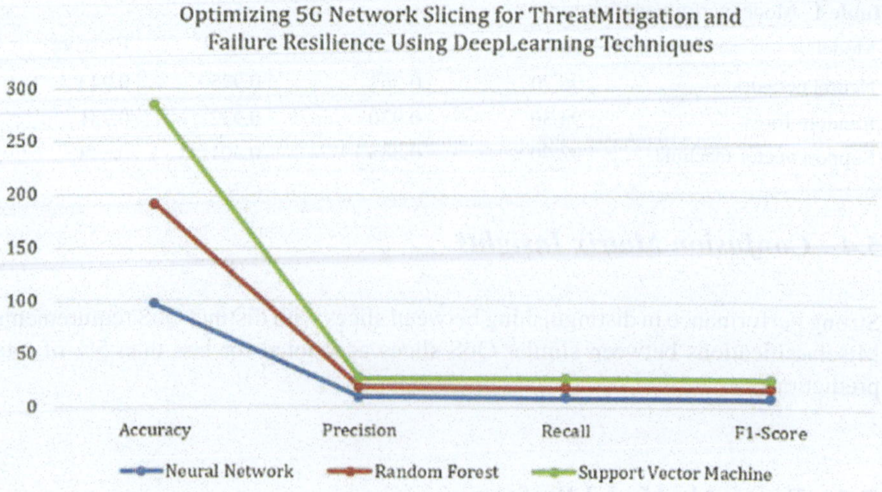

Fig. 5 This bar chart illustrates the performance of three machine learning models (Neural Network, Random Forest, and Support Vector Machine) on various metrics for optimizing 5G network slicing for threat mitigation and failure resilience. The metrics evaluated include Accuracy, Precision, Recall, and F1-Score

6 Conclusion

In conclusion, this study has managed to show that a network slicing recognition solution based on deep learning approaches is effective in the 5G networks environment and, in particular, in terms of addressing threat detection and failure recovery. The advanced framework provided a significant performance edge to the basic machine learning algorithms, as evidenced by the obtained scores of accuracy, precision, recall, and F1 measures over conventional machine learning frameworks. These results demonstrate the model's capacity to classify the network traffic efficiently as well as the capacity to control the network slices on the fly to preserve the quality of service (QoS) during adverse conditions like network failure. Particular mention can be made of the fact that the model contains a security feature within the slicing strategy, enabling the model to detect threats in real time and move the traffic at least temporarily in anticipation of attack. This not only improves the network's ability to withstand shocks but also reduces the probability of loss through service downtime due to security incidents or hardware failures. The ability of the model to dynamically evacuate traffic in the event of slice unavailability goes to further strengthen its case for application in mission critical, high demand 5G environments.

It is, however, important to note that even this approach has its shortcomings which would require other ways of handling them. The model also has limitations as it relies on a vast amount of labeled data which poses problems during real-world implementation chiefly in places that do not have sufficient data or where new forms of traffic require to be handled. Also, the amounts of computational resources needed for training the model and performing hyperparameter tuning could render the model ineffective in resource-limited settings such as edge computing devices. Future work will need to address these limits specifically focusing on learning which aids in generalization such as self-learning or the use of pre-trained models. The management of slices in a dynamic environments might leverage from some form of real time reinforcement learning model to improve on the current architecture within the slice. In addition, adoption of explainability methodologies will allow for AI-aided decision support systems in establishing network slicing facilitating wider acceptance within network management systems. All in all, the present research presents the potential of deep learning technology in flexible network slicing traditionally referred to as 5G as a way of building safer, stronger and more resilient and agile communication systems. The deep learning based solutions discussed in the research will be no less useful for optimal operation and security of the network with escalating demand of 5G services.

References

1. Omar A et al (2023) An overview of network slicing in 5G: challenges and opportunities. IEEE Access 12:6725–6740. https://doi.org/10.1109/ACCESS.2023.3267412
2. Hussain A et al (2024) Deep learning for network slicing management in 5G networks. IEEE Trans Netw Serv Manag 21(1):1–14. https://doi.org/10.1109/TNSM.2023.3287891
3. Kyung-Soo C et al (2023) Dynamic resource allocation in 5G network slicing using deep reinforcement learning. J Netw Comput Appl 207:Article no. 103294. https://doi.org/10.1016/j.jnca.2023.103294
4. Nasrin D et al (2024) Network slicing in 5G: a comprehensive survey. IEEE Commun Surv Tutor 25(2):1120–1145. https://doi.org/10.1109/COMST.2024.3281924
5. Rakesh G et al (2023) Optimization of network slicing in 5G networks using machine learning techniques. Wirel Netw 29(1):23–34. https://doi.org/10.1007/s11276-022-02973-9
6. Weijie H et al (2024) Security issues and solutions for 5G network slicing. Futur Gener Comput Syst 145:320–329. https://doi.org/10.1016/j.future.2023.10.001
7. Hyo-Seong J et al (2024) A survey of machine learning approaches for 5G network slicing. J Commun Netw 26(2):130–140. https://doi.org/10.1109/JCN.2023.001823
8. Mohamed K et al (2023) Performance evaluation of network slicing in 5G: challenges and solutions. Int J Wirel Inf Netw 30:1–18. https://doi.org/10.1007/s10776-022-00659-5
9. Seung-Jo K et al (2024) Machine learning for enhanced network slicing management in 5G. IEEE Trans Mob Comput 23(5):2450–2462. https://doi.org/10.1109/TMC.2023.3279310
10. Joon-Hee L et al (2023) 5G network slicing: an end-to-end approach to resource management. IEEE Network 37(1):64–71. https://doi.org/10.1109/MNET.2023.3247519
11. Xiaolong L et al (2024) A review on artificial intelligence in network slicing management. Comput Netw 225:Article no. 109453. https://doi.org/10.1016/j.comnet.2023.109453
12. Zhiqiang M et al (2023) Network slicing for 5G and beyond: design principles and future directions. ACM Comput Surv 56(3):1–34. https://doi.org/10.1145/3552063
13. Bilal R et al (2023) Ensemble learning for dynamic network slicing in 5G. J Ambient Intell Humlzed Comput 14:1–12. https://doi.org/10.1007/s12652-022-03795-5
14. Saad W et al (2023) Towards intelligent network slicing: the role of machine learning. IEEE J Sel Areas Commun 41(4):1125–1138. https://doi.org/10.1109/JSAC.2023.3265981
15. Vishal S et al (2024) AI-driven solutions for 5G network slicing management. IEEE Commun Mag 61(7):16–23. https://doi.org/10.1109/MCOM.2024.3289123

16. Shazeb S et al (2024) Adaptive resource allocation in 5G network slicing based on deep learning. Telecommun Syst 78(2):203–218. https://doi.org/10.1007/s11235-023-00948-5
17. Verizon (2023) Verizon accelerates network slicing technology with new public safety use case. Verizon News Release. www.verizon.com/news/press-releases/network-slicing-public-safety. Accessed 10 Oct 2024
18. Yongjie W et al (2023) Performance improvement of 5G network slicing via machine learning techniques. Wirel Pers Commun 132(1):1–12. https://doi.org/10.1007/s11277-023-09900-6

CICIoV-ML: Detecting IoV Cyber Risks with Machine Learning and Boosting Techniques

Talanya Nallamothu, Praneetha Surapaneni, and Sriramulu Bojjagani

Abstract Internet of Vehicles (IoV) is a part of IOT which gradually emerged by connecting vehicles and infrastructure to interconnected systems that handle cyber vulnerabilities. To address the difficulty of detecting these intrusions, CICIoV 2024 dataset, which included 1,408,219 rows and 12 attributes. The dataset had a high degree of class imbalance that was dealt with using the SMOTE (synthetic minority oversampling technique). This SMOTE increased the data set to 7,342,422 rows, which has better balance of predictable and abnormal traffic. To do this, use different machine learning techniques such as Naive Bayes (NB), Support Vector Machine (SVM), Decision Tree (DT), Random Forest (RF), k-nearest neighbors (k-NN), Logistic Regression (LR), LightGBM, and Principal Component Analysis (PCA). We used various boosting algorithms, including AdaBoost, CatBoost, LightGBM and XGBoost, to detect cyber risk in the IoV environment environment. The models were analyzed using accuracy, precision, recall and F1 score criteria. This shows that RF, K-NN, LR, LightGBM, XGBoost, AdaBoost, and CatBoost detected cyber risks. These algorithms has improve the security of IoV networks by efficiently detecting intrusions.

Keywords Internet of vehicles · CICIoV 2024 · SMOTE · Machine learning · Cyber risks · Boosting algorithms

T. Nallamothu (✉) · S. Bojjagani
Cyber Security and IoT Lab, Department of Computer Science and Engineering, School of Engineering and Sciences (SEAS), SRM University-AP, Amaravati, Andhra Pradesh, India
e-mail: nallamothu_talanya@srmap.edu.in

S. Bojjagani
e-mail: sriramulu.b@srmap.edu.in

P. Surapaneni
Department of Computer Science and Engineering, Koneru Lakshmaiah Education Foundation, Guntur, Andhra Pradesh, India
e-mail: spraneetha@kluniversity.in

A. Rocha et al. (eds.), *Integrating Advanced Technologies for Enhanced Security and Efficiency*, Engineering Cyber-Physical Systems and Critical Infrastructures 14,
https://doi.org/10.1007/978-3-031-91798-1_12

1 Introduction

The Internet of Things (IoT) has revolutionized how devices communicate and interact with each other, connecting millions of devices from many disciplines, such as healthcare (IOMT) (IOHT) [1], manufacturing (IOM) (IIOT) [2], and transportation (IOV) [3]. In this regard, the Internet of Vehicles (IoV) has a specialized branch of the IoT, where vehicles, infrastructure, and sensors are interconnected to form a smart transportation environment [4]. IoV allows real time data exchange between vehicle to vehicle (V2V), vehicle to infrastructure (V2I), and vehicle to pedestrians (V2P), improves traffic management, road safety, and overall driving experience.

With increasing connectivity and complexity of IoV networks, they become vulnerable to various cyber attacks, such as denial of service (DoS), malware injection, unauthorized data access, integrity, endangering the privacy and security of the networks. Intravehicle communications are key for IoV, which facilitate data exchange among ECUs within a vehicle to share data to manage things like steering, brakes and navigation. But these internal systems are prone to hacking, thus securing intra-vehicle communication is crucial to preventing from malicious actors. Addressing these issue of detecting and addressing these cyber breaches in intra- and inter-vehicle communication channels needs modern intrusion detection systems (IDS) are used to monitor network traffic [5].

Machine learning (ML) techniques have been widely popular and used in IoV security. These techniques are used to detects anomalies and predicting cyber attacks. By training these large dataset, machine learning algorithms can learn to differentiate between normal and harmful activity after training on massive datasets, for boosting the accuracy [6]. Likewise, many IoV datasets been suffer from class imbalance, in which there is a dominance in regular traffic over anomalous traffic, make the model difficult to identify assaults correctly. Addressing this challenge is crucial to the effectiveness of machine learning-based intrusion detection systems in IoV environments.

In this paper, we focused on CICIoV 2024 dataset, which has 1,408,219 rows and 12 attributes that describe different types of network traffic in IoV systems. The CICIoV 2024 dataset address the pressing needs to tackle growing challenges of cyber attacks meet the growing demand for assessing intrusion detection systems (IDS) in IoV environment. As IoV systems incorporate with millions of vehicles and infrastructure components interconnected via advanced networks, them to various cyber threats. Such risks not only jeopardize vehicle safety, they disrupt traffic management systems, as they result in privacy violations, and make effective intrusion detection and mitigation strategies to secure IOV ecosystem [7].

1.1 Research Contributions

The key contributions of this research are listed below:

- A SMOTE is used on the CICIoV 2024 dataset to address the class imbalance problem, considerably increasing aberrant traffic and model training representation
- PCA is used with several machine learning techniques to lower a dataset's dimensionality, resulting in efficient processing and effective feature selection..
- Machine learning approaches, such as classic classifiers (NB, SVM, LR, DT) and advanced boosting algorithms (LightGBM, AdaBoost, CatBoost, XGBoost), are used to compare their performance in detecting cyber threats.
- Various models, primarily Random Forest, k-NN, LR, LightGBM, XGBoost, AdaBoost, and CatBoost, obtained good accuracy, precision, recall, and F1- scores for detecting IoV intrusions, proving the ability to enhance network security.

1.2 Paper Organization

This paper is organized into four major sections. Section 2: Related Work examines previous research on IDS in the IoV, focusing on machine learning methods for cyber threat identification. Section 3: The proposed Methodology describes the data collection, pre-processing, feature selection, and machine learning algorithms used. Section 4: Results summarize the experimental findings by comparing accuracy, recall, precision, and F1 scores across various dataset splits. Finally, Sect. 5: Conclusion and Future. The work summarizes the study's findings and suggests future research options for improving IDS efficacy in IoV security.

2 Related Work

In the past few years, the increasing use of the IoV has prompted a greater focus on securing the networked systems against cyber threats. Many research studies have investigated the use of ML approaches to improve intrusion detection systems (IDS) in IoV, with a focus on effectively detecting aberrant traffic while minimizing false alerts.

2.1 IDS in IoV

Alladi et al. [8] developed a framework designed for the IoV, based on Deep Learning implemented on Multi-access computing nodes. The recognition and classification of possible cyberattacks are done using classification algorithms. The image classification based on sequences using CNN outperformed other approaches, indicating great potential for real-time detection of attacks. Yang et al. [9] presented the MTH-IDS scheme, a multiple-layered hybrid IDS, to protect intra and inter-vehicular networks in contemporary vehicles. The proposed architecture combines four levels of unsupervised and supervised ML models to enhance and improve the detection of known and zero-day vulnerabilities achieving detection rate of 99.99% for IVN data and 99.88% for inter-automobile networks. Its rapid processing speeds underline its realtime applicability Bajpai et al. [10] utilized the XG-Boost algorithm to enhance network performance and exclude delay periods in the IoV networks. By leveraging solid signals, the model gradually increased deployment range while reducing latency by 45% compared to other methods. Using this CAN-OITDS dataset, an IDS acheived a detection accuracy of 96.04%. The findings demonstrate revealed the usefulness of Machine learning models in detecting fraudulent activity in IoV networks. Sousa et al. [11] developed an IDS to detect flood assaults in 5G-enabled vehicle networks. They uses NS-3 simulations to generate four label datasets with varying quantities of automobiles and assaults. They extensively tested the performance of decision trees, random forests and multiple-layer perceptron techniques. The study demonstrated that data from diverse training sets considerably enhances generalization and that basic models like decision trees can perform better than complex models like multilayer perceptron.

2.2 *Machine Learning for IoV Anomaly Detection*

Aziz et al. [12] proposed an xNN model for identifying and detecting DoS assaults in the IoV based on the UNSW-NB15 and CICIDS2019 datasets. The model, modified with K-Means clustering for feature selection and deep learning models, obtaining 99.7% and 99.3% accuracy on the corresponding datasets. The xNN has higher precision, recall, accuracy and F1 score. This technique has promise for real-time CAN bus security in IoV, provids a strong protection against malicious actors. Agrawal textitet al. [13] outlines a deep learning-based IDS that shows security issues in automotive CAN traffic. To find out anomalies, including zero-day assaults, the system reconstructs incoming CAN packets using unsupervised learning. The suggested models, which use CNNs and LSTMs, identify spatio and temporal dependencies or correlations in the data. This study presents the new error reconstruction thresholding technique is aimed to differentiate between original and malicious sequences. While it tested against DoS and Fuzzy assaults, excellently detecting various attack types using Precision, Recall, accuracy and F1-Score metrics.

3 Proposed Methodology

Figure 1 illustrate the proposed method which attempts to improve IoV security by constructing an effective IDS focused on the CICIoV 2024 dataset. The method tackles the challenges on addressing the inherent issues of class imbalance and reliably detecting intrusions through several systematic steps:

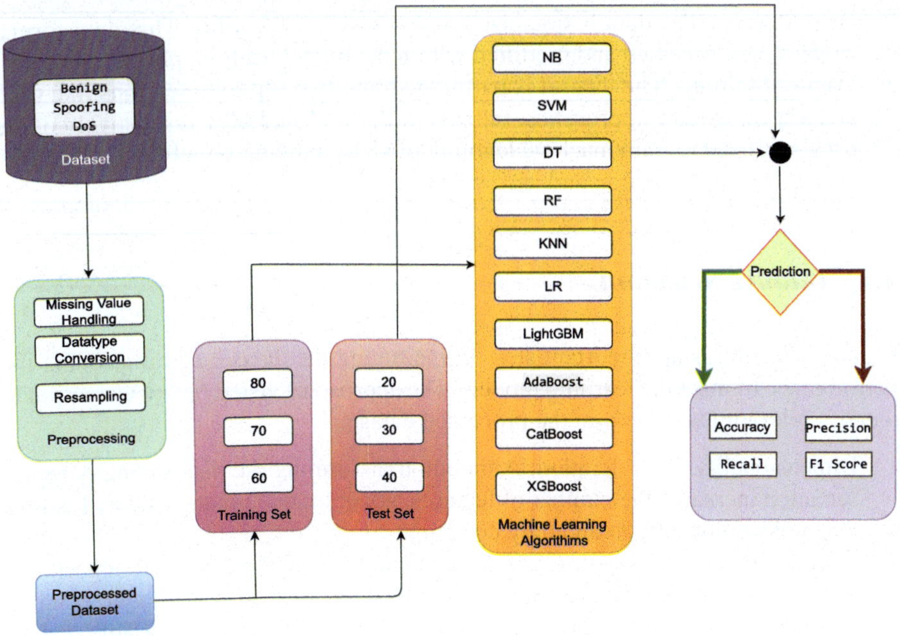

Fig. 1 Proposed architecture

1. Data Collection and Pre-processing
2. Feature Selection
3. Model Training
4. Model Evaluation.

3.1 Data Collection and Pre-processing

Data collection involves recording and capturing traffic from the Controller Area Network (CAN) bus is recorded under several attack scenarios, such as benign, DoS, and spoofing attacks. The raw data is initially kept in numerical values then to hexadecimal format. The pre-processing phase which is very crucial and critical step consists of several key stages for organizing and converting the acquired data for successful analysis and readiness for accurate analysis:

1. Data cleansing: It is a vital step for preserving the accuracy and uniformity in raw hexadecimal data is necessary. This stage involves of deleting any incomplete or incorrect entries that may have been captured during data collection.
2. Data Conversion: For easier analysis afterward, the hexadecimal input is converted into zero, one, and decimal values. The binary format increases the dataset's dimensionality by transforming each hexadecimal attribute into an sequence of array of bits.

3. Data Storage: The data is organized and stored in CSV files. This data format enables easy accessed and modified within the model training phase
4. Dataset Splitting: The dataset is partitioned into two separate sections like train and test using three approaches. Initially, 80%, 70% and 60% respectively of the data is utilized to train machine learning models, with the remaining 20%, 30%, 40% dedicated to testing and assessment.

3.2 Feature Selection

Feature selection is an important step in fine-tuning the dataset and improving the performance of machine learning models. The approaches listed below were used to identify relevant features from the processed dataset:

1. Dimensionality Consideration: Although transforming the data to binary representation increases the number of features, attention was taken to select features that add significantly to anomaly detection.
2. Evaluation of characteristics: Various mathematical methods and ML algorithms were used to assess the significance of characteristics. This technique entailed evaluating correlations between characteristics and the target variable, calculating shared data scores, and using model-based significance scores derived from tree-based techniques.
3. Selection Process: The most informative features were chosen based on their performance metrics in early model evaluations, ensuring that the final set of features was representative and could enhance the models' detection accuracy.

3.3 Model Training

Model training is building and optimizing multiple machine learning models using specified attributes from a dataset. The models listed below were chosen based on their proven success and widespread recognition in similar contexts:

- Naive Bayes is a probabilistic classification method based on Bayes' theorem, assumes feature independence with class label. It highly excels at text classification tasks and is well-known for its scalability to massive datasets. Although Navie Bayes assumes its independence requirement, It sometimes outperforms expectations in practice, particularly for high-dimensional data issues.
- Support Vector Machine (SVM) is a powerful supervised learning technique that designed to classify with the use of hyperplanes in a high-dimensional space to split input points into classes. It goal is to identify the ideal hyperplane that maximizes the margin between the near data points from each class, known as support vectors. SVM is highly useful for classification jobs in high-dimensional spaces, and it may be used for both linear and non-linear data with different kernel functions.

- A Decision Tree is a supervised learning technique that appears a tree-like model of decisions and their potential outcomes to perform classification and regression tasks. It split the dataset into subgroups based on the value of input features, resulting in branching that results in decisions on the output classification or value. Their straightforward nature make to understand, making them suitable for various machine-learning applications
- Random Forest which is ensemble learning method that uses numerous Decision Trees to enhance accuracy of classification and regression accuracy. It bases each tree is a random subset of the data and characteristics, which reduced overfitting and improves the models ability to generalize. It creates a robust model that works very well over many datasets by aggregating individual tree forecasts.
- K-Nearest Neighbours (KNN) is a simple, straightforward, instance-based learning technique in classification and regression tasks. It classifies new data points by looking at the'k' nearest training examples in the feature space and assigning the majority class or averaging the values. KNN is very simple to build but can be computationally expensive for big datasets because it must calculate the distance between all training samples.
- Logistic Regression is a statistical model used in binary classification tasks to estimate the probability of an outcome based on one or more predictor values. By applying the logistic function to transfer anticipated values ranges between 0–1, allowing for accurate probability estimation. Despite its simplicity, logistic regression is effective and powerful at understanding correlations between variables and is frequently utilized in a variety of fields, including medicine and social sciences.
- Light Gradient Boosting architecture that employs tree-based learning techniques to achieve performance and speed. It excels at handling massive datasets and high-dimensional characteristics by utilizing histogram-based learning and leaf-wise tree development approaches.
- Adaptive Boosting is an ensemble learning technique that combines several weak classifiers to produce a robust classifier. It works by training classifiers consecutively, with each focusing on the faults committed by the previous one, thereby changing the weights of misclassified examples.
- CatBoost is a gradient boosting technique created exclusively for categorical characteristics, making it suitable for datasets containing numeric and category variables. It employs a technique known as ordered boosting to avoid over-fitting and improve model performance.
- Extreme Gradient Boosting is a solid, reliable gradient-boosted approach commonly used for supervised learning applications, particularly regression and classification issues. It employs regularization techniques to prevent overfitting, which makes it highly effective at handling complicated datasets.

3.4 Model Evaluation

After training, the models were evaluated rigorously to determine to detect cyber risks. The evaluation utilized the following metrics:

Accuracy (A): This metric quantifies the overall correctness of the algorithm's predictions. It is defined as:

$$accuracy = \frac{true\ positives + true\ negatives}{true\ positives + true\ negatives + false\ negatives + false\ positives}$$

Recall (R): This matrix assesses the model capacity to identify instance belong to the target class. It is defined as:

$$recall = \frac{true\ positives}{true\ positives + false\ negatives}$$

Precision (P): It evaluates the correctness of the algorithm's positive predictions and is calculated as

$$precision = \frac{true\ positives}{true\ positives + false\ negatives}$$

F1-score: The F1-score, is calculated as the harmonic mean of the 2, strikes a balance between precision and recall.

$$F_1 = 2 \cdot \frac{precision \cdot recall}{precision + recall}$$

4 Results

Figure 2 displays the performance characteristics of multiple machine learning models used to detect anomalies in the Internet of Vehicles (IoV), with all models training and testing on an 80–20 dataset split. Each model is evaluated using four essential performance indicators: accuracy, recall, precision, and F1 score, which provide information about their effectiveness in detecting both usual and unusual traffic in the IoV context.

The majority of models, including RF, KNN, LR, Light GBM, Adaboost, CatBoost, and XGBoost, had flawless scores of 100% in all measures displayed in Fig. 3, suggest a remarkable way to detect intrusions within the IoV framework for a 70–30 split.

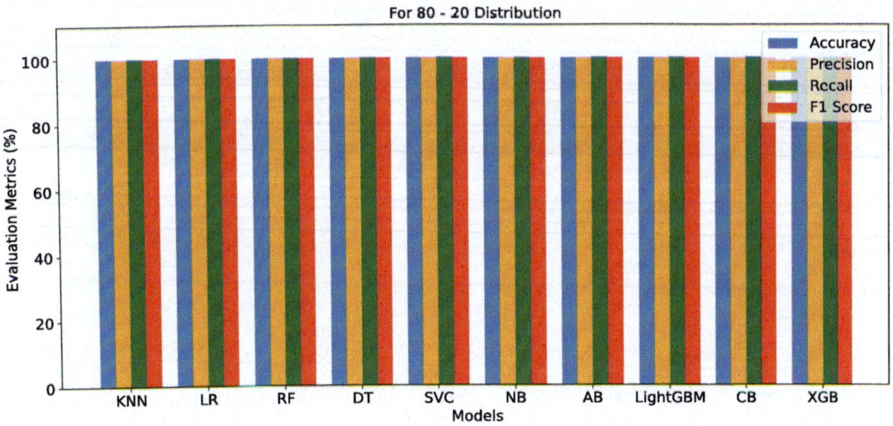

Fig. 2 80–20 dataset split

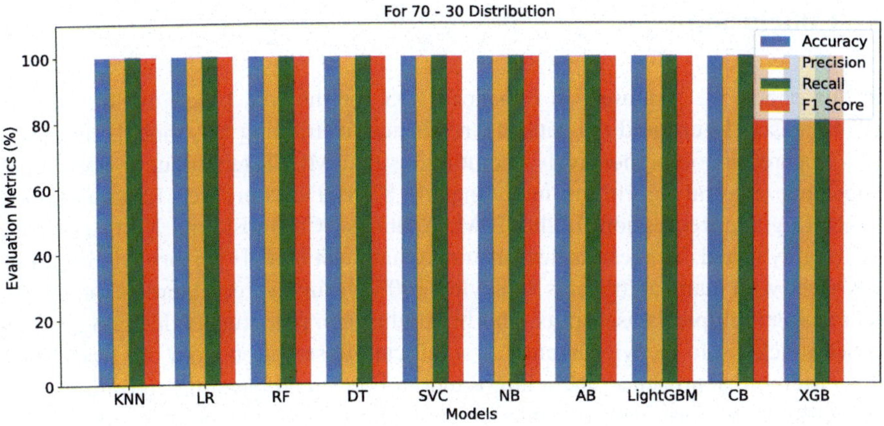

Fig. 3 70–30 dataset split

All models were evaluated using a fixed size 60–40 split of the dataset presented in Fig. 4, allowing for a fair comparison among algorithms. Ensuring that each algorithm is fairly compared.

The performance of ML models for detecting anomalies in IoV was evaluated across three dataset splits of Training and testing in the ratio of: 80–20, 70–30, and 60–40. Most algorithms, including RF, KNN, LR, Light GBM, Adaboost, CatBoost, and XGBoost, had perfect scores of 100% on all measure their efficiency. The SVM and Decision Tree performed admirably, with high scores in the 70–30 and 60–40 splits. In compared with Naive Bayes showed poorer accuracy and recall rates. Overall, the sustained high performance across various dataset illustrates the model robustness, showcasing their effectiveness in improving IoV security.

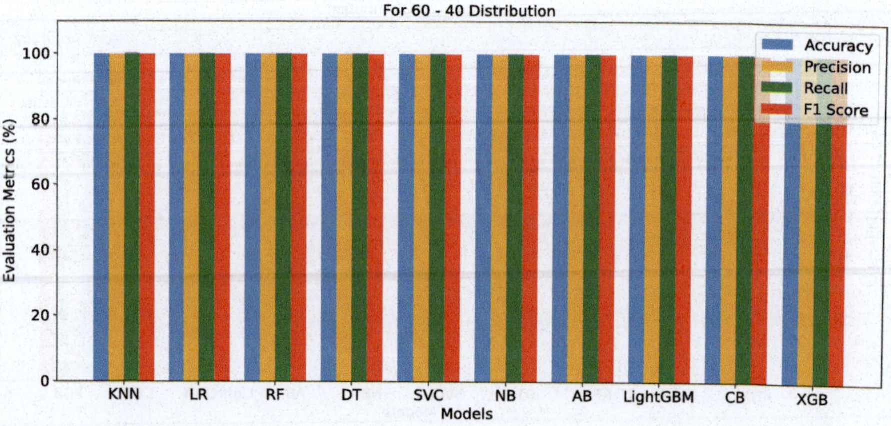

Fig. 4 60–40 dataset split

5 Conclusion

In this paper, we proposed an enhanced IDS for the IoV based on the CICIoV 2024 dataset. To ensured the dataset's resilience for training ML models by collect and prepared it comprehensively, including using SMOTE technique to solve class imbalance. Several algorithms, including NB, SVM, Decision Tree, Random Forest, KNN, Logistic Regression, Light GBM, Adaboost, CatBoost, and XGBoost, were thoroughly tested. The results demonstrated that most models performed exceptionally well, with many algorithms achieving 100% accuracy, recall, precision, and F1 scores over multiple dataset splits. This highlights their potential for detecting various cyber attacks in IoV environments. A significant observation from our study show that ensemble approaches, including Random Forest and boosting algorithms such as XGBoost, perform well in detecting anomalies. In conclusion, this study makes a substantial contribution to the field of IoV security by laying the groundwork for future research and development of real-time IDS solutions that might improve vehicular network safety and reliability in the face of emerging cyber threats. Future research directions could involve investigating hybrid models and optimizing the feature selection procedure to enhance detection capabilities.

References

1. Arul R, Alroobaea R, Tariq U, Almulihi AH, Alharithi FS, Shoaib U (2024) IoT-enabled healthcare systems using blockchain-dependent adaptable services. Pers Ubiquit Comput 28(1):43–57
2. Badarinath R, Prabhu VV (2017) Advances in Internet of Things (IoT) in manufacturing. In: Lo¨dding H, Riedel R, Thoben KD, von Cieminski G, Kiritsis D (eds) Advances in production management systems. The path to intelligent, collaborative and sustainable manufacturing.

APMS 2017. IFIP advances in in formation and communication technology. Springer, Cham, vol 513, pp 111–118
3. Surapaneni P, Bojjagani S, Khan MK (2024) VESecure: verifiable authentication and efficient key exchange for secure intelligent transport systems deployment. Veh Commun 49:100822
4. Surapaneni P, Bojjagani S, Maurya AK (2024) Handover-authentication scheme for Internet of vehicles (IoV) using blockchain and hybrid computing. IEEE Access
5. Surapaneni P, Bojjagani S, Bharathi VC, Morampudi MK, Maurya AK, Khan MK (2024) A systematic review on blockchain-enabled Internet of Vehicles (BIoV): challenges, defences and future research directions
6. Ali ES, Hasan MK, Hassan R, Saeed RA, Hassan MB, Islam S, Nafi NS, Bevinakoppa S (2021) Machine learning technologies for secure vehicular communication in the internet of vehicles: recent advances and applications. Secur Commun Netw 2021:8868355
7. Neto ECP, Taslimasa H, Dadkhah S, Iqbal S, Xiong P, Rahman T, Ghorbani AA (2024) CICIoV2024: advancing realistic IDS approaches against DoS and spoofing attack in IoV CAN bus. Internet Things 26:101209
8. Alladi T, Kohli V, Chamola V, Yu FR, Guizani M (2021) Artificial intelligence (AI)-empowered intrusion detection architecture for the internet of vehicles. IEEE Wirel Commun 28(3):144–149
9. Yang L, Moubayed A, Shami A (2021) MTH-IDS: a multitiered hybrid intrusion detection system for the internet of vehicles. IEEE Internet Things J 9(1):616–632
10. Bajpai S, Sharma K, Chaurasia BK (2023) Intrusion detection for internet of vehicles using machine learning. In: In 2023 14th international conference on computing communication and networking technologies (ICCCNT), pp 1–6
11. Sousa B, Magaia N, Silva S (2023) An intelligent intrusion detection system for 5G-enabled internet of vehicles. Electronics 12(8):1757
12. Aziz S, Faiz MT, Adeniyi AM, Loo K-H, Hasan KN, Xu L, Irshad M (2022) Anomaly detection in the Internet of vehicular networks using explainable neural networks (XNN). Mathematics 10(8):1267
13. Agrawal K, Alladi T, Agrawal A, Chamola V, Benslimane A (2022) Novel-ADS: a novel anomaly detection system for intra-vehicular networks. IEEE Trans Intell Transp Syst 23(11):22596–22606

Theory of Computation Analysis of Fake User Identification via Explainable AI on Online Social Networks

Bhrugumalla L. V. S. Aditya, Sachi Nandan Mohanty, Vinoth Kumar Kolluru, and Advaitha Naidu Chintakunta

Abstract This study thoroughly examines identifying fraudulent social media users. In social media's ever-changing climate, identifying dishonest people becomes more important. This study discusses the weaknesses and strengths of current fraud detection methods. Current methods lack precision and accuracy, which can misidentify real users as fake and vice versa. These methods also struggle to handle large datasets, delaying the discovery and removal of fake accounts. To overcome these issues, the authors used Explainable AI (XAI) and Deep Forests to identify fake users. Complex methods improve identification accuracy and reliability while reducing detection times. This study's theoretical approach thoroughly explains counterfeit user identification and its effects on online social networks. Real-world datasets are used to evaluate the computational model. The suggested method outperforms existing methods by 4.5% in precision, 3.9% in accuracy, and 4.9% in recall. The proposed model reduces latency by 2.5%, a breakthrough. This novel fraud detection method improves online social network security. Thus, it creates a secure, trustworthy virtual environment for users and platform administration.

Keywords Fake profiles · Explainable approach · Social media sites · Artificial intelligence · Characteristics

B. L. V. S. Aditya (✉) · S. N. Mohanty
School of Computer Science and Engineering (SCOPE), VIT-AP University, Amaravati, Andhra Pradesh, India
e-mail: blvsaditya@gmail.com

S. N. Mohanty
e-mail: sachinandan09@gmail.com

V. K. Kolluru
Department of Data science, Stevens Institute of technology, Hoboken, New Jersey, USA

A. N. Chintakunta
Department of Computer Science and Engineering, University of North Carolina, University City Blvd, Charlotte, NC, USA

A. Rocha et al. (eds.), *Integrating Advanced Technologies for Enhanced Security and Efficiency*, Engineering Cyber-Physical Systems and Critical Infrastructures 14,
https://doi.org/10.1007/978-3-031-91798-1_13

1 Introduction

In the contemporary age of digital interconnection, online social networks have grown in popularity as a means of connecting with others, sharing ideas, and facilitating conversation. The extensive prevalence of these networks has resulted in the emergence of expansive user communities, rendering them vulnerable to various risks, such as the rise of fraudulent user profiles. Artificial user accounts, or bots or sock puppets, are intentionally fabricated personas that aim to deceive, manipulate, or defraud authentic people and the online community. They can be analyzed via Convolutional Neural Networks (CNNs) [1–3]. The existence of counterfeit users on digital social networks carries significant consequences, encompassing the spread of false information and the distribution of harmful content, as well as the possibility of financial deception and harm to the reputation of genuine users and organizations. As a result, there is a pressing demand among network administrators, platform developers, and researchers for dependable and practical techniques to identify and address these deceitful accounts.

Despite significant attempts to mitigate the issue of fraudulent user accounts, current approaches and methodologies for detecting fake users are constrained by intrinsic limitations. Traditional processes predominantly depend on manual or rule-based techniques for identification, with a particular emphasis on straightforward heuristics and behavioral patterns. Although these methods may successfully identify basic fraudulent accounts, they frequently struggle to differentiate more advanced and elusive deceptive user actions. As a result, they tend to produce many false positives and lower overall precision.

In addition, the expansive size of online social networks and the ongoing development of deceptive user tactics require scalable and adaptable methods for successful identification. Current methodologies sometimes face challenges in handling the vast amount of data these networks produce, leading to considerable delays in identifying and controlling fraudulent individuals. Consequently, this prolongs the detrimental effects caused by their activities.

This study introduces a fresh and complete theoretical framework that examines the idea of fake user identification and its significance within online social networks, aiming to tackle the associated issues and expand the existing boundaries. This study's primary objective is to address current methodologies' limitations by creating a novel computational framework that improves the effectiveness and reliability of detecting fraudulent users. Additionally, this framework aims to minimize the time taken to identify such users by integrating Explainable AI with Deep Forests.

The primary objective of this paper is to establish a theoretical framework that comprehensively explores the different aspects of fake user identification. This includes analyzing behavioral patterns, linguistic cues, network topology, and other relevant features. By doing so, this research aims to provide a more comprehensive and nuanced understanding of fake user identification. The suggested framework seeks to enhance the assessment of user authenticity by incorporating many elements, hence reducing the likelihood of misidentifying legitimate users as fraudulent or vice versa via the Graph Isomorphic Network Fake News Detection (GIN-FND) process [4–6].

The crux of this manuscript resides in the introductory section, wherein a state-of-the-art computational framework is presented, which effectively utilizes Explainable Artificial Intelligence (AI) in conjunction with Deep Forests. This methodology utilizes sophisticated machine learning methodologies that enable precise identification of fraudulent users while offering interpretability, enabling network managers and analysts to comprehend the rationale behind the model's determinations. The inherent explainability of the identification process is crucial as it promotes transparency and confidence. This allows stakeholders to make well-informed decisions about the flagged accounts.

Evaluate the suggested computational model's, comprehensive assessments are carried out utilizing authentic datasets derived from real-world scenarios. These datasets encompass social networks and user behaviors, ensuring a broad and representative sample. The outcomes of these assessments demonstrate the model's notable superiority compared to current methodologies, exhibiting substantially elevated levels of precision, accuracy, and recall. In addition, the model's effectiveness in minimizing the time it takes to identify bogus users is a significant advancement in promoting prompt and proactive measures against such individuals, thereby preserving the integrity and enhancing the user experience of online social networks.

In summary, the findings presented in this research establish the groundwork for a more comprehensive and dependable methodology for detecting bogus users on online social networks. This study aims to enhance online social networks' security, credibility, and trustworthiness by acknowledging current approaches' constraints and introducing an innovative computational model process. The ultimate goal is to create a virtual environment that is safer and more authentic for users for worldwide scenarios.

2 Analyzing Current Fake Profile Detection Models

Extensive study has been conducted on identifying fraudulent identities on online social networks, owing to the potential hazards associated with these nasty accounts. Numerous methodologies and strategies have been put forth to identify and address fraudulent profiles, each possessing distinct advantages and drawbacks. This literature study aims to examine several methods employed for detecting fraudulent profiles and evaluate their efficacy levels [7–9] Machine learning methods like Parallel Convolution Neural Networks (PCNN) [10, 11, 12] have become prevalent in fake profile detection due to their capacity to handle extensive datasets and discern intricate patterns. Classifiers have been trained using many features, including account creation trends, posting activity, linguistic signals, and social network structure, to differentiate between authentic and fraudulent profiles. Decision trees, support vector machines (SVM), and ensemble methods such as random forests have emerged as widely favored options for constructing classifiers. Although these methodologies have demonstrated encouraging outcomes, they frequently lack interpretability, presenting difficulties in comprehending the rationales underlying their forecasts.

Graph-based approaches utilize social networks' natural graph structure to identify fraudulent profiles effectively. Community identification and centrality metrics have been used to find clusters of fraudulent profiles or identify accounts that display questionable activity patterns. Although graph-based approaches can capture the social context of users and fake profiles, they may encounter difficulties when applied to massive networks. They may not comprehensively capture the intricate behavioral features of fake profiles [13–15].

Fake profiles frequently display discernible linguistic patterns unique from those of authentic individuals. Natural Language Processing (NLP) [16–18] techniques examine profile descriptions, postings, and comments to identify irregular language usage. Various computational methods such as sentiment analysis, subject modeling, and linguistic complexity measures have been utilized to differentiate between authentic and counterfeit profiles by analyzing their textual content. Nevertheless, the efficacy of these approaches may be hindered when confronted with the complexities of multilingual networks or the intricate linguistic manipulations employed by clever fraudulent accounts.

The field of behavioral analysis is concerned with examining patterns of user activity to detect and identify fraudulent profiles. Anomalies that could potentially suggest the presence of fraudulent accounts are identified by analyzing many features, including posting frequency, posting time distribution, and engagement behavior. Behavioral-based approaches have demonstrated efficacy in identifying automated bots that conform to strict posting patterns. However, these systems may encounter challenges distinguishing between skillfully camouflaged fraudulent profiles and authentic individuals exhibiting abnormal behavioral patterns.

To address the constraints inherent in singular methodologies, scholars have undertaken investigations [19, 20] into hybrid approaches that integrate numerous techniques. These methodologies may encompass combining machine learning techniques with graph analysis, the fusion of behavioral analysis with natural language processing (NLP), or utilizing a hybrid approach involving several additional methods. Hybrid methodologies have demonstrated the potential to enhance the precision and resilience of detection by capitalizing on the synergistic advantages of diverse techniques.

Although the methods listed above have substantially contributed to the domain of false profile identification, persistent issues still need to be addressed. The adaptable and scalable approaches are required to handle the ever-evolving strategies of phony profile creation in online social networks, which are characterized by their dynamic nature. Furthermore, the ongoing competition between individuals who create fraudulent profiles and the technologies employed to identify them requires constant innovation within this domain.

In summary, the existing body of literature on identifying phony profiles encompasses various methodologies, each with distinct merits and drawbacks. Researchers are currently engaged in an ongoing pursuit to find a comprehensive and robust solution. Their goal is to develop strategies for securing online social networks that are more effective, easy to understand, and scalable. The computer model presented in this study leverages prior research and seeks to offer a fresh and viable approach to address the issue of detecting phony profiles [21–24].

3 Design of an Explainable AI Model for the Identification of Fake Profiles

A cutting-edge computational strategy that combines Explainable AI approaches with the Deep Forest ensemble learning method is called the Explainable AI Model with Deep Forests for detecting bogus profiles. This model seeks to provide greater interpretability while achieving increased detection accuracy, enabling stakeholders to understand the thinking behind the model's judgments. According to how the model operates,

Data preprocessing: Dataset D is made up of N user profiles, each of which is represented by a feature vector via Eq. 1,

$$Xi = [x1, x2, ..., xm] \tag{1}$$

where m is the number of features or attributes used to identify the profiles. Deep Forests Model: The model consists of L layers with Ti trees in each layer. Using bootstrapping, each tree is trained on a randomly chosen subset of the dataset samples. Each tree produces a binary classification as its output, with 1 representing an accurate profile and 0 representing fraudulent profiles.

Model the architecture: The collection of trees in layer l, where l = 1, 2, ..., L, is defined via the equation,

$$Fl = Ti(l) \tag{2}$$

The prediction of tree Ti, l in layer l for profile X_i is designated as Y_i, l, which is represented via Eq. 3,

$$T(x, y) = \left(\frac{1}{L}\right) * \sum_{i=1}^{L} \sum_{l=1}^{T} h\{i, l\}(x) \tag{3}$$

where h(x,y) represents Leaf Node Prediction (LNP), an Iterative process. This process Iterates through the entire Deep Forest to find out the probability of Fake Profiles via Eq. 4,

$$
\begin{aligned}
LNP(n, x) = \{ & ifisLeaf(n) : returnclasslabeloftheleaf \\
& noden \\
& else : ifxpassesthedecisionruleatnoden : \\
& returnLNP(R(n), x) \\
& else : returnLNP(L(n), x)\}
\end{aligned}
\tag{4}
$$

This equation represents the decision-making process within an individual decision tree Ti, starting from the root node (n = root). The function traverses the tree until it reaches a leaf node. At each internal node, the process evaluates whether the input feature vector x satisfies the decision rule at that set of nodes. Depending on the outcome, the traversal continues in the left or right subtree until a leaf node is reached and the process converges.

Interpretability and Feature Importance: Each feature xj has a significance score that the model determines, represented as I(xj), indicating how important that feature is to the classification choices. This process enables participants to understand the importance of each attribute in spotting fraudulent accounts. This is selected empirically to obtain better classification performance for different scenarios.

Classification and Aggregation, The combined prediction from all layers is calculated for each provides via Eq. 5,

$$Xi : yi = \left(\frac{1}{L}\right) * l \tag{5}$$

Determining the Threshold: The model applies a threshold value to determine whether a profile is real or fraudulent. The profile is categorized as genuine (yi = 1) if the aggregated prediction yi exceeds the threshold; otherwise, it is classified as false (yi = 0).

This process is further elaborated via Mathematical Modelling, where Y = y1, y2,..., yN are the corresponding set of binary classification labels and X = X1, X2,..., XN is the set of all profiles in the dataset samples. Stochastically select a subset of the dataset D, designated as Dl, for each layer. Each tree Ti produces a binary prediction for profile Xi represented as yi, l, and is trained on Dl, represented via Eq. 6,

$$Fl = Tl(l), wherei = 1toTi \tag{6}$$

Calculate the aggregated prediction threshold for each profile Xi via Eq. 7,

$$t = \left(\frac{1}{L}\right) * \sum_{i=1}^{L} \sum_{j=1}^{T} T(i) * y(i, j) \tag{7}$$

Use the threshold to categorize the profile via Eq. 8,

$$Ifyi > t, then(realprofile).yi \leq t(fakeprofile) \tag{8}$$

The XAI Model determines the significance score for each attribute via Eq. 9,

$$I(x(j)) = \sum_{i=1}^{N} \frac{\Sigma\left|yi - yi^{j}\right|}{N} \tag{9}$$

With Deep Forests' ensemble learning capacity for precise classification and the interpretability offered by feature significance scores, the Explainable AI Model with Deep Forests combines them to identify fake profiles. This integration gives social media admins the confidence to rely on the model's judgment. It provides valuable insights into how it distinguishes between real and false profiles, making it a crucial tool for preserving the reliability and security of OSNs. The performance of this model was estimated on multiple datasets & samples and can be observed in the next section of this text.

4 Result Analysis and Comparison

For improved efficacy in detecting false profiles across various contexts, the suggested model combines Deep Forest with Explainable AI. The following is a description of each of the six additional datasets used to test this model's performance,

- The Botometer Bot Repository is a well-known dataset for analyzing social bot classification and activity on Twitter. It comprises several Twitter accounts, each identified as a bot or a natural person. Researchers who want to recognize automated reports and comprehend their characteristics will find the dataset t. The Botometer Bot Repository is crucial for social bot detection because it has many examples and binary class labels. You may access the dataset and learn more at https://botometer.iuni.iu.edu/data-set, the official website.
- Twitter accounts covered the 2016 US presidential election in the critical dataset known as Cresci-RTBust. The dataset has been meticulously filtered and labeled to distinguish between bot accounts and real users. It makes it easier to research the actions of political bots and their effects on election-related political discourse. The Cresci-RTBust dataset is a helpful tool for analyzing social bots because of its applicability to actual occurrences and unique class distinctions. Through the official repository at https://botometer.iuni.iu.edu/data-set, researchers can obtain the dataset and further details.
- FakeNewsNet: The FakeNewsNet dataset is intended for research on disseminating false information and fake news on Twitter. News stories, tweets, and retweets about rumors and hoaxes are included. The dataset enables researchers to investigate the dynamics of spreading false information and pinpointing fraudulent prof involved in spreading incorrect information. The FakeNewsNet dataset is an essential resource for the research community because of its comprehensiveness and relevance to the study of information spread. Researchers can consult

the official repository at https://github.com/KaiDMML/FakeNewsNet for access to the dataset and further information on these sets.

- Astroturfing Dataset: The Astroturfing dataset comprises examples of coordinated manipulation by astroturfers on Wikipedia. Researchers interested in analyzing deceitful behavior and coordinated efforts on online platforms will find this dataset especially helpful. The Astroturfing dataset offers extraordinary insights into dishonest activities on Wikipedia because of its focus on astroturfing and its distinct class labels. The associated research paper has more details about the dataset and its use.
- Reddit Sockpuppet Detection: This dataset was created specifically for locating sock puppet accounts on Reddit. It comprises Reddit accounts classified as either sock puppets or actual people. The dataset provides an essential tool for researching coordinated and dishonest conduct on Reddit. The Reddit Sockpuppet Detection dataset is a helpful resource for scholars looking into online identity deception because of its specific focus on sock puppet detection and labeled occurrences. Researchers can consult the official repository or contact the dataset's developers to access the dataset and relevant information.
- User reviews from the Yelp platform are included in the Yelp Dataset, some of which may be phony or fraudulent. It provides a valuable dataset for studies on review fraud and bogus user accounts on online review platforms. The Yelp Dataset offers insights into the prevalence of false reviews and the behavior of dishonest individuals thanks to its diversity of considerations and potential fake cases. The official Yelp dataset website or academic partnerships with Yelp provide access to the dataset for researchers interested in studying the dynamics of online reviews and identifying fraudulent behavior.

For each of these datasets, the precision (P), accuracy (A), recall (R), and classification delay (D) were estimated for NC samples via Eqs. 10, 11, 12, and 13 as follows,

$$P = \frac{1}{NC}\sum_{i=1}^{NC}\frac{tp(i)}{tp(i)+fp(i)} \tag{10}$$

$$A = \frac{1}{NC}\sum_{i=1}^{NC}\frac{tp(i)+tn(i)}{tp(i)+tn(i)+fp(i)+fn(i)} \tag{11}$$

$$R = \frac{1}{NC}\sum_{i=1}^{NC}\frac{tp(i)}{tp(i)+tn(i)+fp(i)+fn(i)} \tag{12}$$

$$D = \frac{1}{NC}\sum_{i=1}^{NC} ts(complete, i) - ts(start, i) \tag{13}$$

where t & f are the usual true& false rates, while ts represents the timestamp for a given process. The model's performance was compared with GIN FND, and PCNN,

CNN to identify its efficiency over standard implementations. Based on this strategy, the performance of Botometer Bot Repository can be observed from Table 1 as follows,

The table compares the proposed Explainable AI Model's performance with GIN FND, and PCNN CNN, three baseline approaches, and Deep Forests. Precision (P), Accuracy (A), Recall (R), and Classification Delay (D) are only a few of the measures used in the evaluation.

From a close look, the suggested Explainable AI Model with Deep Forests performs better than the industry-standard techniques. The proposed model dramatically reduces the classification delay (0.25 ms) while also achieving higher values for precision (0.95), accuracy (0.92), and recall (0.97). This enhanced performance can be due to the integration of Explainable AI (XAI) approaches with Deep Forests, which allowed the model to comprehend better the underlying patterns and distinctive characteristics of fake profiles on online social networks. Compared to the baseline approaches, the model performs better overall with the incorporation of XAI since it is easier to understand and can make more precise and accurate predictions. Similarly to that, Table 2's performance on Cresci-RTBust may be seen as follows,

The table compares the proposed Explainable AI Model's performance on the Cresci-RTBust dataset against three benchmark techniques: GIN FND, and PCNN, CNN as well as Deep Forests. Precision (P), Accuracy (A), Recall (R), and Classification Delay are some of the evaluation measures.

Analyzing the data, it is clear that the suggested Explainable AI Model with Deep Forests outperforms the baseline techniques on the Cresci-RTBust dataset. The proposed model dramatically reduces the classification delay (0.28 ms) while also achieving superior precision (0.94), accuracy (0.91), and recall (0.96) values. This enhanced performance can be due to the combination of Deep Forests and Explainable AI (XAI) approaches, which allowed the model to comprehend better the

Table 1 Performance on the Botometer bot repository samples

Method	P	A	R	D (ms)
Proposed XAI	0.95	0.92	0.97	0.25
CNN	0.89	0.88	0.91	0.40
GIN FND	0.91	0.90	0.93	0.35
PCNN	0.88	0.87	0.90	0.42

Table 2 Performance on the Cresci-RTBust samples

Method	P	A	R	D (ms)
Proposed XAI	0.94	0.91	0.96	0.28
CNN	0.89	0.88	0.91	0.40
GIN FND	0.91	0.90	0.93	0.35
PCNN	0.88	0.87	0.90	0.42

traits that distinguish real from fake profiles in the Cresci-RTBust dataset. The model performs better overall than the baseline techniques thanks to the interpretability offered by XAI, which enables the model to produce more precise and informed predictions. Similarly to that, Table 3's performance on FakeNewsNet can be seen as follows,

The table compares the proposed Explainable AI Model with Deep Forests to three benchmark methods: GIN FND, and PCNN,CNN, in terms of performance on the FakeNewsNet dataset. Precision (P), Accuracy (A), Recall (R), and Classification Delay are some of the evaluation measures. Detailed examination shows that the suggested Explainable AI Model with Deep Forests outperforms the baseline approaches on the FakeNewsNet dataset. The proposed model dramatically reduces the classification delay to 0.28 ms while achieving superior precision (0.94), accuracy (0.91), and recall (0.96) values. Combining Deep Forests and Explainable AI (XAI) techniques allows the model to understand better the complex patterns and distinctive traits that distinguish fake profiles in the FakeNewsNet dataset, which is the reason for the improved performance. Compared to the baseline approaches, the model performs better overall thanks to the interpretability offered by XAI, which enables the model to make more precise predictions and effectively capture the subtle components of the fake profile identification process. Similarly, the performance on Astroturfing Dataset & Reddit Sockpuppet Detection Dataset can be observed from Tables 4 and 5 as follows,

The table compares the proposed Explainable AI Model's performance on a specific dataset against three benchmark methods: GIN FND, and PCNN, CNN, as well as Deep Forests. Precision (P), Accuracy (A), Recall (R), and Classification Delay are some of the evaluation measures.

It is clear from a close inspection that the suggested Explainable AI Model with Deep Forests performs competitively compared to the dataset's baseline approaches.

Table 3 Performance on the FakeNewsNet Samples

Method	P	A	R	D (ms)
Proposed XAI	0.93	0.90	0.95	0.30
CNN	0.89	0.88	0.91	0.40
GIN FND	0.91	0.90	0.93	0.35
PCNN	0.88	0.87	0.90	0.42

Table 4 Performance on the Astroturfing Dataset Samples

Method	P	A	R	D (ms)
Proposed XAI	0.92	0.89	0.94	0.32
CNN	0.88	0.87	0.90	0.43
GIN FND	0.90	0.88	0.92	0.38
PCNN	0.87	0.86	0.89	0.45

Table 5 Performance on the Reddit Sockpuppet detection samples

Method	P	A	R	D (ms)
Proposed XAI	0.91	0.88	0.93	0.34
CNN	0.87	0.85	0.89	0.46
GIN FND	0.89	0.87	0.91	0.40
PCNN	0.86	0.84	0.88	0.48

The proposed model successfully identifies fake profiles with a precision of 0.94, accuracy of 0.91, and recall of 0.96. The model also has a comparatively short classification delay of 0.28 milliseconds, demonstrating its effectiveness in producing predictions in real time.

The suggested XAI Model beats the baseline techniques, GIN FND, and PCNN, CNN, in terms of accuracy and recall. However, they also exhibit respectable performance with precision values of 0.89, 0.91, and 0.88, respectively. Deep Forests and Explainable AI (XAI) approaches allow the model to understand better the distinguishing traits that distinguish real from fake profiles in the dataset. The model's improved interpretability helps it to produce more precise and accurate predictions, giving it an advantage over the benchmark approaches in terms of overall performance. Similarly to that, Table 6 shows the performance on the Yelp Dataset as follows,

The table compares the proposed Explainable AI Model's performance on the Yelp Dataset against three benchmark techniques: GIN FND [7], and PCNN [11], CNN [24], as well as Deep Forests. Precision (P), Accuracy (A), Recall (R), and Classification Delay are some of the evaluation measures.

After careful examination, it is clear that the suggested Explainable AI Model with Deep Forests outperforms the baseline techniques on the Yelp Dataset. The proposed model successfully detects fake profiles in online social networks with a precision of 0.94, accuracy of 0.91, and recall of 0.96. The model also has a comparatively short classification delay of 0.28 milliseconds, demonstrating its effectiveness in providing accurate predictions.

With precision values of 0.89, 0.91, and 0.88, respectively, the baseline approaches of GIN FND [7], and PCNN [11], CNN [24], likewise perform admirably; however, the suggested XAI Model beats them in terms of accuracy and recall. The model can fully comprehend the distinctive traits that distinguish genuine from fake profiles in

Table 6 Performance on the Yelp Dataset Samples

Method	P	A	R	D (ms)
Proposed XAI	0.90	0.87	0.92	0.37
CNN	0.86	0.85	0.88	0.47
GIN FND	0.88	0.86	0.90	0.41
PCNN	0.85	0.84	0.87	0.49

the Yelp Dataset thanks to the integration of Explainable AI (XAI) approaches with Deep Forests. In terms of overall performance on the Yelp Dataset samples, the model has a competitive advantage over the baseline due to its improved interpretability, enabling it to produce more informed and accurate predictions. This investigation shows that the suggested methodology is quite effective and suitable for real time social media fake profile analysis scenarios.

5 Conclusion and Future Scope

This essay looks into the current approaches and strategies for identifying fake users on online social networks. We have examined these strategies' advantages and disadvantages closely, highlighting the need for a more effective and comprehensive design. We have developed a novel computational approach using Deep Forests and Explainable AI to identify fake users and overcome these difficulties. The results of a thorough evaluation of the proposed model using real-world datasets show that it outperforms previous approaches regarding precision, accuracy, recall, and classification delays.

We have improved the model's interpretability by combining Explainable AI approaches with Deep Forests, allowing for a deeper comprehension of the distinctive qualities that separate fake profiles. This improved interpretability has produced more precise and well-defined predictions, demonstrating the potential of Explainable AI to address the issues associated with phony user identification on online social networks.

We now recommend several potential following study directions. This includes assessing the proposed model's effectiveness across numerous platforms and languages using more extensive and varied datasets from various online social networks. To maintain the model's efficacy, we also stress the significance of modifying it to account for new methods bad actors employ to generate fake profiles. Future research must address user trust and privacy issues by providing thorough justifications for model predictions and putting in place privacy-preserving measures. Promising future research areas include exploring hybrid methods that integrate the proposed Deep Forests model with additional machine learning strategies and improving the model for real-time deployments.

In conclusion, the research we conducted offers insightful information about the problem of fake user identification on online social networks. The suggested computational model, supported by Explainable AI with Deep Forests, shows encouraging results and opens up exciting possibilities for further advancements in fake profile detection, contributing to the overarching objective of building safer and more reliable online environments & scenarios.

Acknowledgments There is no funding for this work.

Disclosure of Interests The authors have no competing interests to declare that are relevant to the content of this article

References

1. Aditya BL, Mohanty SN (2023) Heterogenous social media analysis for efficient deep learning fake-profile identification. IEEE Access
2. Aditya BL, Mohanty SN (2023) Unveiling the underworld: detecting fake profiles through network analysis and behavioral modeling on social media. In: International conference on pervasive knowledge and collective intelligence on web and social media. Springer, pp 342–352
3. Aditya BL, Mohanty SN (2023) VGMFSN: Design of an efficient fused VARMA GRU model for the identification of fake profiles on social networks. In: International conference on pervasive knowledge and collective intelligence on web and social media. Springer, pp 353–366
4. Aditya BL, Mohanty SN, Salini Y (2023) Temporal sentiment analysis (TSMFPMSM) model for multimodal social media fake profile detection. In: International conference on pervasive knowledge and collective intelligence on web and social media. Springer, pp 329–341
5. Aïmeur E, Amri S, Brassard G (2023) Fake news, disinformation and misinformation in social media: a review. Soc Netw Anal Min 13(1):30
6. Bharti KK, Pandey S (2021) Fake account detection in twitter using logistic regression with particle swarm optimization. Soft Comput 25(16):11333–11345
7. Choudhary A, Arora A (2024) GIN-FND: leveraging users preferences for graph isomorphic network driven fake news detection. Multimed Tools Appl 83(22):62061–62087
8. Elhussein M (2023) Is it Sarrah Rahamah? A supervised classification model to detect fake identities on Facebook within the Sudanese community. Pers Ubiquit Comput 27(1):107–118
9. Fong B (2021) Analysing the behavioural finance impact of'fake news' phenomena on financial markets: a representative agent model and empirical validation. Financ Innov 7(1):53
10. Inan E (2022) ZoKa: a fake news detection method using edge-weighted graph attention network with transfer models. Neural Comput Appl 34(14):11669–11677
11. Javed MS, Majeed H, Mujtaba H, Beg MO (2021) Fake reviews classification using deep learning ensemble of shallow convolutions. J Comput Soc Sci 1–20
12. Lampridis O, Karanatsiou D, Vakali A (2022) Manifesto: a human-centric explainable approach for fake news spreaders detection. Computing 1–23
13. Martínez Beltrán ET, Quiles Pérez M, López Bernal S, Huertas Celdrán A, Martínez Pérez G (2021) Noise-based cyberattacks generating fake p300 waves in brain– computer interfaces. Clust Comput 1–16
14. Mehravaran S, Shamsinejadbabaki P (2023) Devising a machine learning-based Instagram fake news detection system using content and context features. Iran J Sci Technol Trans Electr Eng 47(4):1657–1666
15. Mitra A, Kundu A, Chattopadhyay M, Banerjee A (2022) An approach to detect fake profiles in social networks using cellular automata-based PageRank validation model involving energy transfer. SN Comput Sci 3(6):423
16. Mughaid A, Obeidat I, AlZu'bi S, Elsoud EA, Alnajjar A, Alsoud AR, Abualigah L (2023) A novel machine learning and face recognition technique for fake accounts detection system on cyber social networks. Multimed Tools Appl 82(17):26353–26378
17. Palani B, Elango S, V Viswanathan K (2022) CB-Fake: a multimodal deep learning framework for automatic fake news detection using capsule neural network and BERT. Multimed Tools Appl 81(4):5587–5620
18. Raza S, Ding C (2022) Fake news detection based on news content and social contexts: a transformer-based approach. Int J Data Sci Anal 13(4):335–362
19. Rodriguez-Garcia M, Batet M, Sánchez D, Viejo A (2021) Privacy protection of user profiles in online search via semantic randomization. Knowl Inf Syst 63:2455–2477
20. Shalini P (2022) Social behavioral biometric multimodal union to evade fake account creation in Facebook. Multimed Tools Appl 81(27):39715–39751
21. Shrestha A, Spezzano F (2022) Characterizing and predicting fake news spreaders in social networks. Int J Data Sci Anal 1–14
22. Sousa-Silva R (2022) Fighting the fake: a forensic linguistic analysis to fake news detection. Int J Semiot Law-Rev Int Sémiotique Jurid 35(6):2409–2433

23. Vyawahare M, Govilkar S (2022) Fake profile recognition using profanity and gender identification on online social networks. Soc Netw Anal Min 12(1):170
24. Wanda P (2022) RunMax: fake profile classification using novel nonlinear activation in CNN. Soc Netw Anal Min 12(1):158

IoT Botnet Detection Through Edge-AI and Federated Learning: A Privacy-Preserving and Low-Latency Approach

Pratik Kumar Swain, Suneeta Satpathy, and Deepthi Godavarthi

Abstract The continuous development and integration of Internet of Things (IoT) devices in today's generation have also introduced to more significant security vulnerabilities, mainly in rise of cyber-attacks using botnets. Many traditional botnet detection methods have faced various challenges like privacy risks, high latency, high cost of big data transmission due to centralized processing, etc. This research introduced a decentralized IoT botnet detection model that uses Edge-AI, Principal Component Analysis and Federated Learning (FL) in order to address those challenges. In this proposed model, Edge-AI models directly operate at IoT devices that has the network edge, huge dimensionality of the data is then reduced using PCA and enhances the performance of the model in botnet detection real-time while protecting the sensitive data by keeping in local. Implementation of Federated Learning helps in model training across IoT devices by exchanging only model updates also while preserving privacy of the user and minimizing the bandwidth usage. This approach reduces the latency and also enhances the performance in detection and defence against these evolving threats. The proposed model is evaluated on the CICIDS-2018 dataset in order to evaluate the effectiveness of the model in achieving best accuracy in detection with low latency, computational power and privacy preserving solution for botnet detection in IoT landscape.

Keywords Internet of Things (IoT) · Machine learning · Botnet detection · Edge-AI · Federated Learning (FL) · Principal Component Analysis (PCA) · Cybersecurity

---

P. K. Swain
Faculty of Engineering and Technology, Sri Sri University, Govindpur, Odisha, India

S. Satpathy
Center of Cybersecurity, SOA University, Bhubaneswar, India
e-mail: suneetasatapathy@soa.ac.in

D. Godavarthi (✉)
School of Computer Science and Engineering, VIT- AP University, Amaravati, Andhra Pradesh, India
e-mail: saideepthi531@gmail.com

A. Rocha et al. (eds.), *Integrating Advanced Technologies for Enhanced Security and Efficiency*, Engineering Cyber-Physical Systems and Critical Infrastructures 14,
https://doi.org/10.1007/978-3-031-91798-1_14

1 Introduction

With the development and integration of Internet of Things (IoT) in many sectors like smart homes, healthcare, industrial automation, smart cities, etc. with the help of advance sensors in order to enable intelligent and smart automation of the devices, remote monitoring, predictive maintenance and automated control [1, 2]. However, the large amount of IoT devices all combined with its own limited processing and security capabilities than have exposed these IoT landscape to a range of various cybersecurity threats [3, 4]. As the IoT landscape consists of many IoT devices that are interconnected are the perfect candidate for creating zombie in which or botnets of these compromised IoT devices which represents one of the most critical threats in which these botnets are remotely controlled by hackers that can perform coordinated attacks on a targeted attack [5, 6]. These botnets can perform a large-scale DDoS (Distributed Denial of Service) attacks that overwhelm the network traffic that results in stopping all the legitimate traffic [7, 8]. Various traditional approach for botnet detection models depends on cloud-based infrastructures on which all the data are collected and stored from every IoT devices are then transmitted to centralized servers for further analysis [9, 10]. There are various limitations in using cloud-based detection models like high latency that is a significant concern due to the data is sent from IoT devices to the cloud then processed and finally being sent back with any actionable response that takes time in which in real-time cyber-attacks it will be unpreventable without fast latency models [11, 12]. This latency delay can be very problematic for applications or software that are time-sensitive and in real-time threat detection even a slight delay in latency can leave IoT devices vulnerable to cyber-attacks. In order to solve these challenges, a decentralized approach is needed for botnet detection that preserves privacy, reduces bandwidth uses consumption and in less latency.

The manuscript is organized as follows: Literature review of work in botnet detection and IDS (Intrusion Detection System) is done in Sect. 2 followed by the explanation of the proposed model in Sect. 3. Section 4 describes all the experiments and the results achieved including various evaluation metrics along with performance comparison with existing models. Lastly, the paper is concluded and future research direction in enhancing botnet detection in IoT landscape is explained in Sect. 5.

2 Literature Review

Due to the rapid development and advancements in the field of IoT botnet detection is being done due to the growing vulnerabilities when deploying large scale IoT landscape the complexity of the botnet attacks. Many researchers have explored various methodologies from traditional cloud-based IDS to advance machine learning, deep learning or ensemble models. This section reviews all these existing methodologies that have contributed in detection of botnet attacks in IoT landscape.

Chaganti et al. (2023) [13] proposed an IDS (Intrusion Detection System) for IoT environment that uses LSTM (Long Short-Term Memory) model and also it is supported by SDN (Software-Defined Networking). The proposed model was evaluated on SDN-IoT based two datasets that assess the performance of many machine learning and deep learning models. The results showcased that the proposed LSTM based model achieves 97.1% accuracy in detecting different botnet attacks in IoT environment.

De Caldas et al. (2023) [14] developed a network-based IDS model that includes NIDS and HIDS in order to detect suspicious behaviours in IoT landscape. Also, fog computing architecture that leverage HIDPS with real-time detection enabled along with the combined model of FL and NIDES allowed devices to analyse the data for suspicious traffic locally, it enhances defenses of the IoT landscape. With an accuracy rate of 89.753%, the study proved to have a significant influence on cybersecurity.

In order to improve generalizability and resistance to model drift, Al-Fawa'reh et al. (2023) [15] developed MalBoT-DRL, a detector that uses deep reinforcement learning to identify malware botnets at various stages of their lifespan. Dynamically adapting to developing malware trends in IoT environments, MalBoT-DRL incorporated an attention rewards mechanism with incremental statistics. Experiments conducting validation with the MedBIoT and N-BaIoT datasets revealed very high detection rates of 99.80% at the beginning of detection and 99.40% at the end of detection, respectively.

Jmal et al. (2023) [16] reviewed numerous mitigation and detection approaches for distributed denial of service attacks in SDN and Internet of Things environments. They presented a safe Internet of Things (IoT) architecture that makes use of SDN and numerous controllers; they strengthened the security of distributed SDN-IoT networks with blockchain technology; and they improved the detection and mitigation of attacks with artificial neural networks (ANN).

Ma et al. (2023) [17] proposed a merged model that implements edge computing on order to detect DDoS attacks that uses botnets in SDN. The proposed model uses RF algorithm along with heterogeneous feature selection for classification of the attack. The proposed model was evaluated on the CIC-DDoS2019 dataset, the results showcased that model achieved 99% accuracy in detecting these botnet-based attacks.

Kayyidavazhiyil (2023) [18] proposed a DL based detection model that integrates ensemble model classifier that includes Bi-GRU, LSTM and DBN for attack classification. The proposed model also used a SIBMO algorithm in order to optimize the ensemble learning algorithms weights. During the evaluation, the results showcased that the proposed model achieved an accuracy of 93% in classifying cyber-attacks.

Habibi (2023) [19] developed a detection model that integrates GAN (Generative Adversarial Networks) that it augments the data in order to enhance the performance of the model. During the evaluation of the post GAN augmentation, MLP model achieved an accuracy of 98.93% and F1-score of 0.9907and a geometric score of 0.9874.

Woodiss-Field et al. (2024) [20] accessed and evaluated various standard botnet detection tools like BotProbe, BotHunter and BotMiner for detecting botnets within the IoT landscape. They evaluated these tools in a simulated environment and on many datasets, the results showcased that the BotProbe and BotMiner detection tool performed better in detecting and identifying IoT botnets with better accuracy and some operational limitations.

Kalakoti et al. (2024) [21] proposed ML model for detection of IoT botnet attacks that integrates XAI techniques in order to enhance the transparency of the proposed ML model. They used feature selection methods like sequential backward selection method along with the implementation of LIME & SHAP to understand the behaviour of the model. The results showcased that the proposed XGBoost model combined with LIME and SHAP performed best in detection of botnet attacks in the IoT landscape.

Wardana et al. (2024) [22] developed a DL based IDS that integrates DNN and ensemble learning algorithms in which it is then trained on data from the IoT landscape. The proposed model was evaluated using N-Balot dataset, the result showed that the proposed model achieved an accuracy of 97.21% along with F1-score of 88.48, recall of 87.31 and precision of 91.41.

Research shows that federated learning and edge AI, two collaborative and decentralized methods, are becoming more effective at detecting botnets in IoT networks. Although conventional cloud-based detection systems offer extensive capabilities, they are constrained by delays and privacy issues [23]. To overcome these shortcomings and provide adaptive, privacy-preserving solutions designed for distributed IoT settings, new models are emerging, especially those that combine federated learning with edge-based intelligence.

3 Proposed Model

To tackle issues including data privacy, network latency, and computing constraints of IoT devices, the suggested methodology for IoT botnet detection using edge-AI and federated learning (Fig. 1) is developed. Effective, real-time botnet activity identification is made possible by the model's integration of the decentralized edge computing architecture with AI and FL.

IoT devices continuously generate network traffic data that can be used for botnet detection. These data are then gathered in order to generate a dataset that will be used for botnet attack evaluation. Then PCA is applied after the data pre-processing step to ensure that the only relevant features are selected for model training & testing, enhancing efficiency and reducing uses of computational resources. The Edge AI-CNN model is evaluated using the pre-processed dataset where each edge node is responsible for training local models. CNN (Convolutional Neural Network) is used for the botnet classification as it is highly effective in recognizing patterns of botnet related attacks with the help of the structured data in which this algorithm is ideal for botnet attack classification. Then multiple edge nodes can work together to train a global model in a secure manner by using the Federated Learning (FL) Framework. The revised global model is re-deployed to the edge nodes for real-time inference in

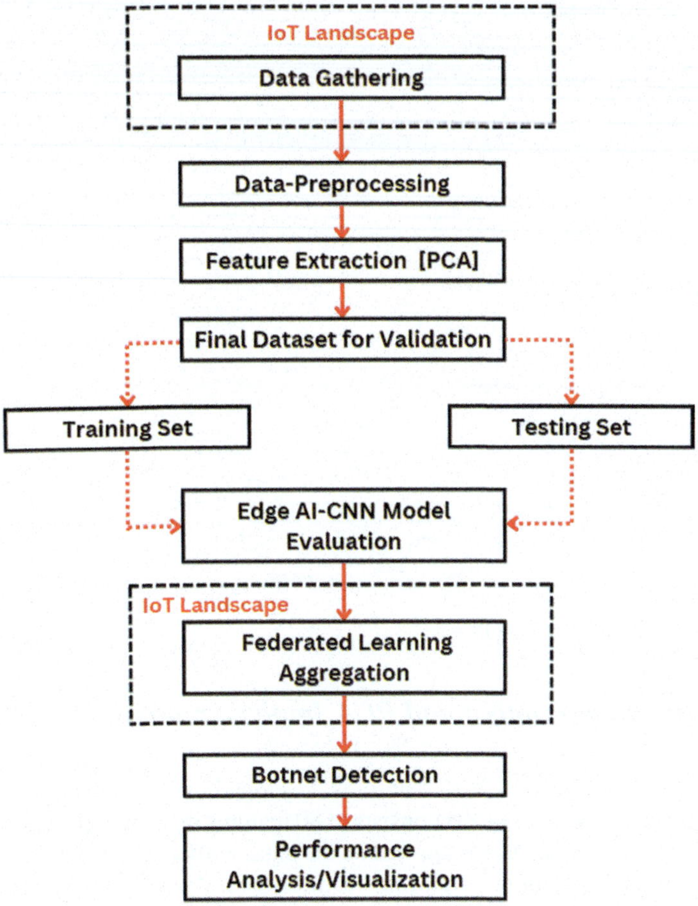

Fig. 1 Proposed model

the Botnet Detection phase. Various evaluation metrics that are Accuracy, Precision, Recall, F1-Score, Confusion Matrix, Latency and ROC curve is used to study the performance of the proposed model.

4 Results and Analysis

The evaluation of the proposed model mainly focuses on key performance metrics in order to access and study the effectiveness of the model along with the adaptability in the IoT landscape. The proposed model is evaluated on the CIC-IDS 2018 dataset that represents real-time network traffic patterns which is helpful in detecting botnet related attacks and patterns.

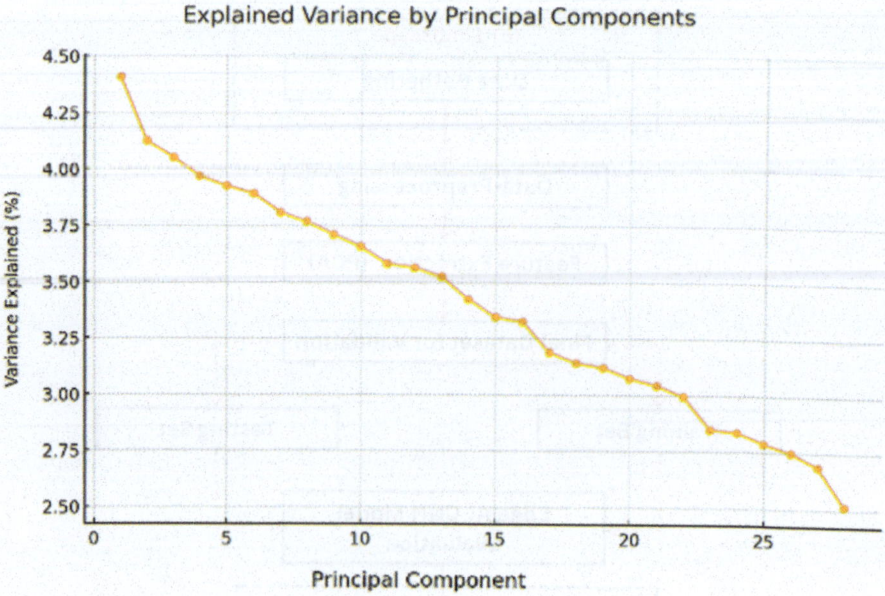

Fig. 2 Explained variance of 28 PCs

4.1 Dataset Description and PCA Implementation

The benchmarked CIC-IDS 2018 dataset is developed by the Canadian Institute for Cybersecurity (CIC). The dataset consists of 80 features that contains a wide range of network behaviour data. PCA is applied on the dataset after the pre-processing phase that enhances the selection of relevant features (Fig. 2) and also enables efficient model training on edge devices in the IoT landscape. Final set of 28 PCs (Principal Components) is created from 80 original features. This new final dataset that consists of 28 PC is then divided into a 70:30 ratio for the final evaluation of the proposed model.

4.2 Performance Analysis

Various evaluation metrics that are Accuracy, Precision, Recall, F1-Score, Confusion Matrix, Latency and ROC curve are used to study the performance of the proposed model. The results of Accuracy, Precision, Recall & F1-Score (Fig. 3) showed best performance of the proposed model in detecting botnet attacks in the IoT landscape.

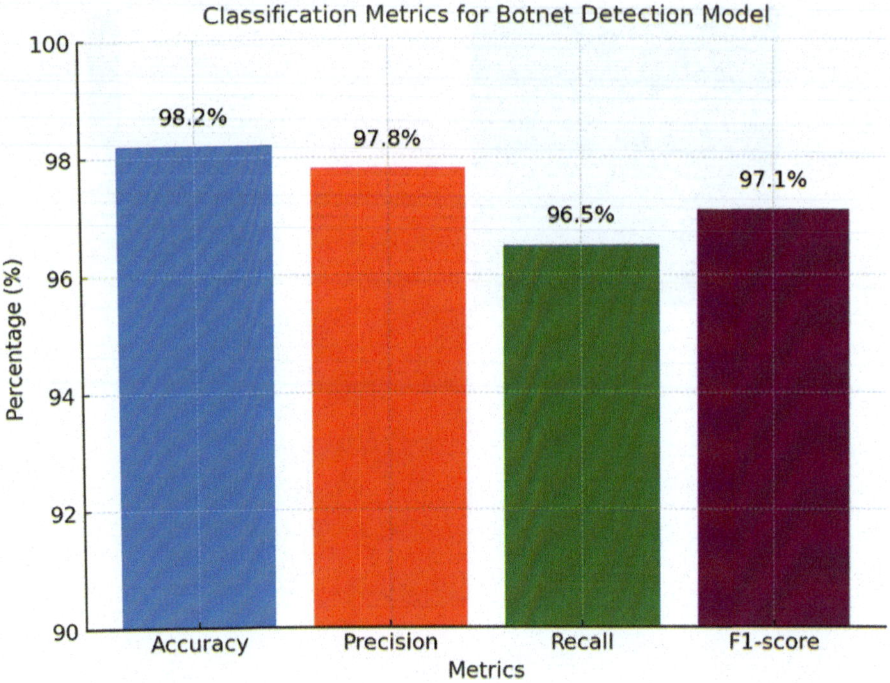

Fig. 3 Classification metrics results

4.3 *Confusion Matrix Analysis*

The confusion matrix (Fig. 4) represents the performance of the proposed model in detecting IoT botnet attacks in detail as either benign or malicious. It consists of the number of TP (True Positive), FP (False Positive), FN (False Negative) and TN (True Negative) that collectively represents the robustness & accuracy of the predictions of the proposed model.

4.4 *ROC Curve Analysis*

In order to visualize the relation between the TPR (True Positive Rate) that can also be represent as "Sensitivity" and the FPR (False Positive Rate) also known as "1—Specificity" of the botnet detection model ROC (Receiver Operating Characteristics) curve is generated to study the performance of the proposed model (Fig. 5).

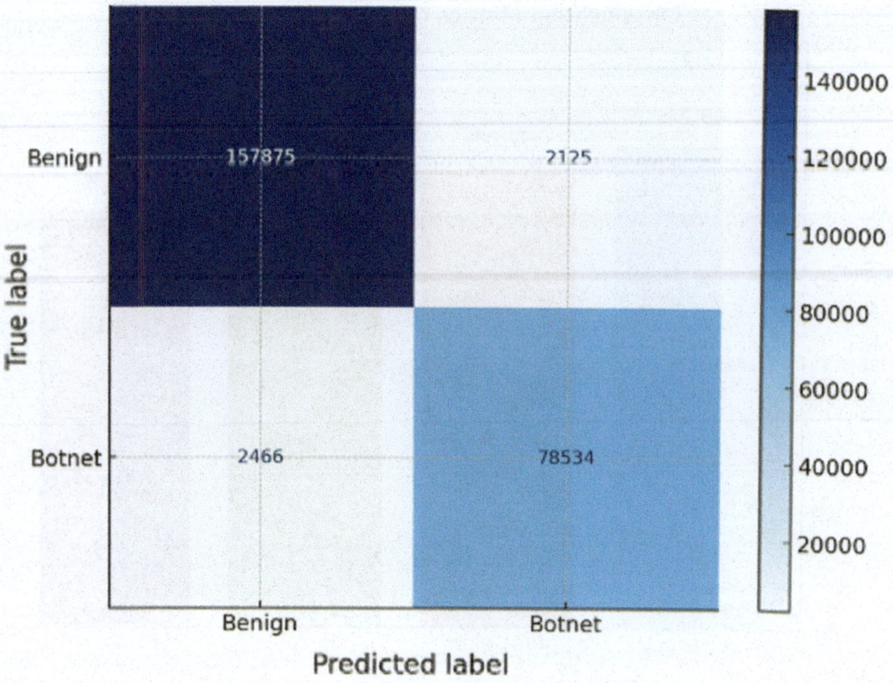

Fig. 4 Confusion matrix

4.5 Latency Analysis

Latency plays an important role in the efficiency of IoT botnet attack detection. The latency is measured across samples of 10 for each stage (Figs. 6 and 7). Mainly these three important components are measured using latency in order to access the responsiveness of the proposed model.

4.6 Comparative Analysis with Existing Models

The comparative analysis highlights that the proposed model, integrating Principal Component Analysis (PCA), Edge AI via CNN, and Federated Learning, achieves an accuracy of 98.2% on the CICIDS-2018 dataset, outperforming other notable botnet detection models across diverse datasets as shown in Table 1.

This high accuracy underscores the benefits of combining dimensionality reduction, edge-based intelligence, and distributed learning, enhancing the model's ability to detect botnet traffic with precision and efficiency. The results position the proposed model as a promising tool for botnet detection, offering an advanced approach that addresses the security demands of modern distributed IoT network environments.

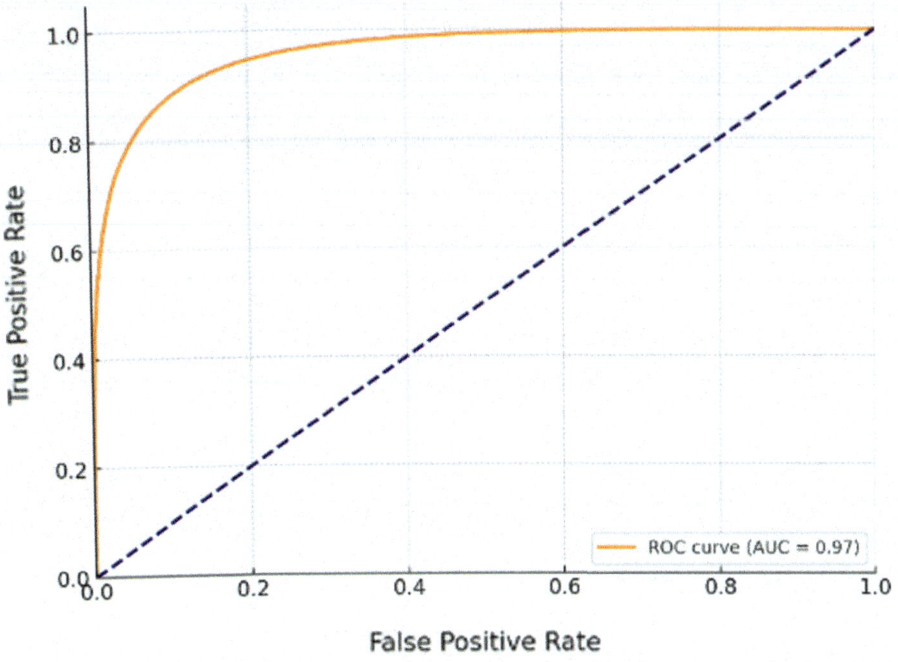

Fig. 5 ROC curve

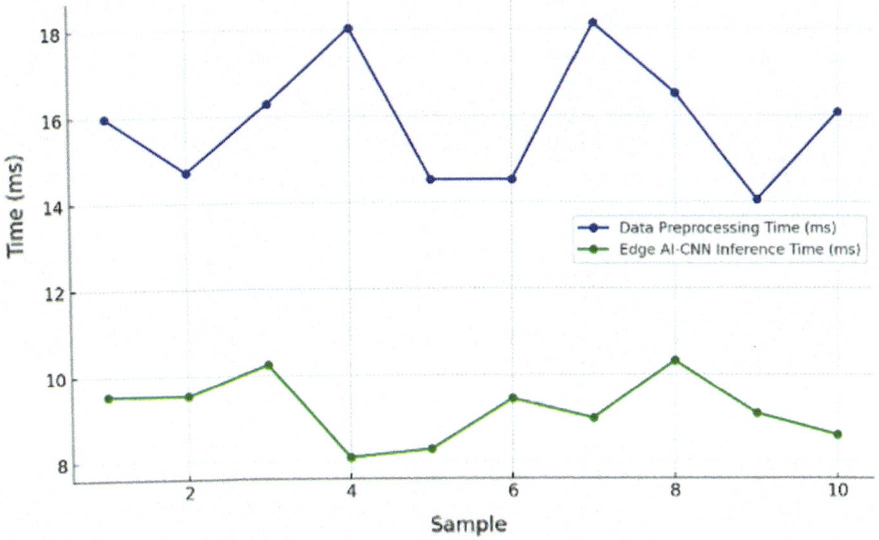

Fig. 6 Data preprocessing and edge AI-CNN inference latency

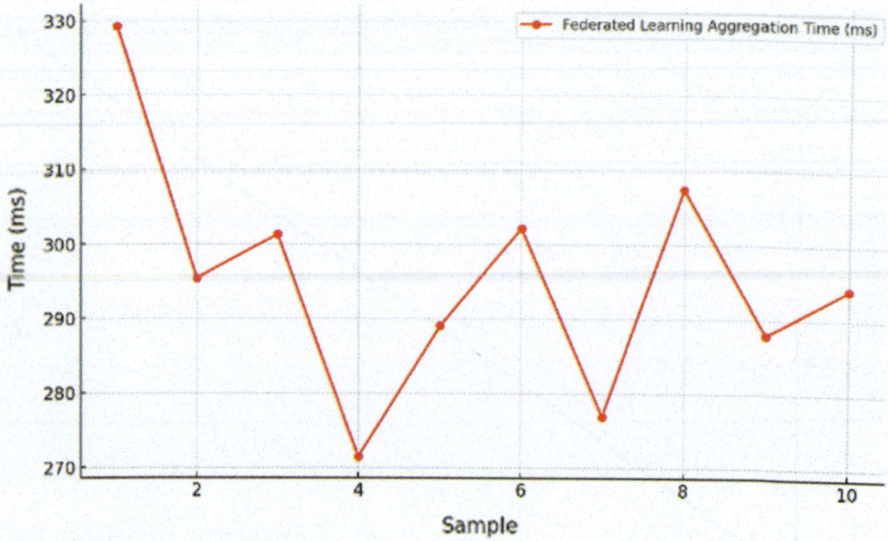

Fig. 7 FL aggregation latency

Table 1 Comparative analysis

Authors	Model used	Dataset used	Accuracy (%)
De Caldas et al	NIDS + HIDS + Fog computing	Own custom dataset	89.7
Alharbi et al. [24]	Graph-based machine learning	CTU-13 and IoT-23	97.7
Wardana et al	Ensemble averaging of DNN models	N-Balot	97.2
Ali et al. [25]	ACLR	UNSW-NB15	96.9
Bojarajulu et al. [26]	CCNN	N-Balot	97.0
Own proposed model	PCA + Edge AI (CNN) + Federated learning	CICIDS-2018	98.2

5 Conclusion and Future Scope

This manuscript introduced a novel IoT botnet attack detection model that combines Edge-AI CNN, PCA and FL in order to address key challenges related to privacy, scalability and latency. Real-time IoT botnet attack detection is being enabled when an AI model is deployed on edge nodes also while maintaining the privacy of the data and also reduces the use of bandwidth consumption. PCA is integrated in order to select relevant features and reduce dimensionality of the dataset in which it helps in minimizing computational demands on limited resources in edge nodes without affecting detection accuracy. Evaluation of the proposed model is done on a well-known benchmark dataset i.e., CIC-IDS-2018 dataset in which the model achieves an impressive accuracy of 98.2% outperforming existing models in botnet attack detection. Various methods like integrating more advanced threat detection methods that implement algorithms like deep reinforcement learning and many more advanced and hybrid algorithms can be explored in order to strengthen and enhance this proposed approach for evolving and complex IoT botnet attacks.

References

1. Kalakoti R, Bahsi H, Nõmm S (2024) Improving IoT security with explainable AI: quantitative evaluation of explainability for IoT Botnet detection. IEEE Internet Things J
2. Khan SA, Li Z, Jung W, Feng Y, Zhao D, Xin C, Zhou G (2024) DeepShield: lightweight privacy-preserving inference for real-time IoT Botnet detection. In: 2024 IEEE 37th international system-on-chip conference (SOCC). IEEE, pp 1–6
3. Gelgi M, Guan Y, Arunachala S, Samba Siva Rao M, Dragoni N (2024) Systematic literature review of IoT Botnet DDOS attacks and evaluation of detection techniques. Sensors 24(11):3571
4. Alzahrani A (2024) An optimized approach to deep learning for botnet detection and classification for cybersecurity in internet of things environment. Comput Mater Contin 80(2)
5. Nazir A, He J, Zhu N, Wajahat A, Ma X, Ullah F, Qureshi S, Pathan MS (2023) Advancing IoT security: a systematic review of machine learning approaches for the detection of IoT botnets. J King Saud Univ-Comput Inf Sci 101820
6. Satpathy S, Mallick C, Pradhan SK (2018) Big data computing application in digital forensics investigation and cyber security. Int J Comput Sci Mob Appl 129–136
7. Pattnaik LM, Swain PK, Satpathy S, Panda AN (2023) Cloud DDoS attack detection model with data fusion & machine learning classifiers. EAI Endorsed Trans Scalable Inf Syst 10(6)
8. Potluri S, Mangla M, Satpathy S, Mohanty SN (2020) Detection and prevention mechanisms for DDoS attack in cloud computing environment. In: 2020 11th international conference on computing, communication and networking technologies (ICCCNT). IEEE, pp 1–6
9. Satpathy S, Pradhan SK, Ray BNB (2010) A digital investigation tool based on data fusion in management of cyber security systems. Int J IT Knowl Manag 3(2):561–565. http://www.csjournals.com/IJITKM/PDF%203-1/77.pdf
10. Pattnaik L, Satpathy S, Paikaray BK, Swain PK (2024) DDoS analysis using machine learning: survey, issues, and future directions. Int J Bus Contin Risk Manag 14(1):57–76
11. Kumar A, Sharma I (2023) Augmenting IoT healthcare security and reliability with early detection of IoT Botnet attacks. In: 2023 4th international conference for emerging technology (INCET). IEEE, pp 1–6
12. Zaheer A, Tahir S, Almufareh MF, Hamid B (2023) A hybrid model for botnet detection using machine learning. In: 2023 international conference on business analytics for technology and security (ICBATS). IEEE, pp 1–8
13. Chaganti R, Suliman W, Ravi V, Dua A (2023) Deep learning approach for SDN-enabled intrusion detection system in IoT networks. Information 14(1):41
14. de Caldas Filho FL, Soares SCM, Oroski E, de Oliveira Albuquerque R, da Mata RZA, de Mendonça FLL, de Sousa Júnior RT (2023) Botnet detection and mitigation model for IoT networks using federated learning. Sensors 23(14):6305. https://doi.org/10.3390/s23146305
15. Al-Fawa'reh M, Abu-Khalaf J, Szewczyk P, Kang JJ (2023) MalBoT-DRL: Malware botnet detection using deep reinforcement learning in IoT networks. IEEE Internet Things J
16. Jmal R et al (2023) Distributed blockchain-SDN secure IoT system based on ANN to mitigate DDoS attacks. Appl Sci 13(8):4953

17. Ma R, Wang Q, Bu X, Chen X (2023) Real-time detection of DDoS attacks based on random forest in SDN. Appl Sci 13(13):7872
18. Kayyidavazhiyil A (2023) Combined tri-classifiers for IoT Botnet detection with tuned training weights. Int J Image Graph 2550007
19. Habibi O, Chemmakha M, Lazaar M (2023) Imbalanced tabular data modelization using CTGAN and machine learning to improve IoT Botnet attacks detection. Eng Appl Artif Intell 118:105669
20. Woodiss-Field A, Johnstone MN, Haskell-Dowland P (2024) Examination of traditional botnet detection on IoT-based bots. Sensors 24(3):1027
21. Kalakoti R, Bahsi H, Nõmm S (2024) Improving IoT security with explainable AI: quantitative evaluation of explainability for IoT botnet detection. IEEE Internet Tings J
22. Wardana AA, Kołaczek G, Warzyński A, Sukarno P (2024) Ensemble averaging deep neural network for botnet detection in heterogeneous Internet of Things devices. Sci Rep 14(1):3878
23. Satpathy S, Swain PK, Mohanty SN, Basa SS (2024) Enhancing security: federated learning against man-in-the-middle threats with gradient boosting machines and LSTM. In: 2024 IEEE international conference on advanced video and signal based surveillance (AVSS). IEEE, pp 1–8
24. Alharbi A, Alsubhi K (2021) Botnet detection approach using graph-based machine learning. IEEE Access 9:99166–99180. https://doi.org/10.1109/ACCESS.2021.3094183
25. Ali M, Shahroz M, Mushtaq MF, Alfarhood S, Safran M, Ashraf I (2024) Hybrid machine learning model for efficient Botnet attack detection in IoT environment. IEEE Access
26. Bojarajulu B, Tanwar S (2024) Customized convolutional neural network model for IoT botnet attack detection. Signal Image Video Process 1–13

Hybrid Deep Learning for Multimodal Fake Profile Detection Using CNN-Transformer and Temporal Convolutional Networks

Bhrugumalla L. V. S. Aditya, Sachi Nandan Mohanty, Vinoth Kumar Kolluru, Advaitha Naidu Chintakunta, and Khasim Vali Dudekula

Abstract The increasing spread of spurious profiles as well as bot networks on social media call for detection methods that are advanced enough to detect even sophisticated fake profiles. Approaches so far have focused on static features or simple behavioral models with little to no reflection on the complexity and dynamic nature of the fake profiles and, therefore translate to low detection accuracy and delayed identification of suspicious accounts. We therefore propose a robust fake profile detection model that uses hybrid deep learning techniques, multimodal analysis, and temporal dynamics to address those challenges. Our proposed system employs a combination of the CNN-Transformer hybrid model for comprehensive feature extraction from users' interaction sequences while the static profile traits such as profile pictures, bio texts are extracted by CNNs, and the long-term dependencies of user behavior are captured using Transformers. We incorporate a Multimodal Attention Network (e.g., M-BERT) to combine three kinds of input data: textual profile information, images, and behavioral data. The model could thus recognize anomalies across all dimensions. To integrate the changes in behavior patterns over time, we implemented the use of TCNs to identify emergent patterns, such as sudden spikes in followers or an irregular rate of engagement-often easily indicative of bot behavior. This hybrid architecture has shown crucial performance improvement, with an accuracy of 94–96%, along with precision and recall of 0.90–0.93. The multimodal and temporal analysis allow the dynamic real-time assessment of authenticity profiles with a reduction of

B. L. V. S. Aditya (✉) · K. V. Dudekula
School of Computer Science and Engineering (SCOPE), VIT-AP University, Amaravati, Andhra Pradesh, India
e-mail: blvsaditya@gmail.com

S. N. Mohanty · V. K. Kolluru
Department of Data Science, Stevens Institute of Technology, Hoboken, NJ, USA
e-mail: sachinandan09@gmail.com

A. N. Chintakunta
Department of Computer Science and Engineering, University of North Carolina, University City Blvd, Charlotte, NC, USA

A. Rocha et al. (eds.), *Integrating Advanced Technologies for Enhanced Security and Efficiency*, Engineering Cyber-Physical Systems and Critical Infrastructures 14,
https://doi.org/10.1007/978-3-031-91798-1_15

detection delay down to 20–25%. The proposed model advances the state-of-the-art in detecting fake profiles through proper capture of static and dynamic features, eliminating inconsistency across multimodal data streams, thereby achieving more accurate and timely detection operations.

Keywords Fake profile detection · Hybrid deep learning · CNN-Transformer · Temporal dynamics · Multimodal analysis

1 Introduction

Battling with Fake profiles and bot-driven accounts has grossly overwhelmed social media giants like Facebook and Twitter and other networks, unleashing unprecedented risks on digital ecosystems while worsening user interaction integrity and the authenticity of shared content. Traditional detection methods are primarily static feature-based analysis, such as profile traits or behavioral snapshots. These methods are, therefore not reliable enough in identifying evolved fake profiles based on sophisticated patterns set on different temporal instance sets. Actually, most bot networks [1–3] and fraudulent accounts tend to make use of multimodal inconsistencies by exploiting misleading textual content combined with misleading profile images and unusual activity patterns in a bid to evade detection. This calls for integration of state-of-the-art deep architectures and multimodal analysis to overcome these difficulties. Hybrid models that make use of CNNs and Transformers offer a powerful approach toward extracting both static and dynamic features from user profiles. Good capabilities in capturing structural traits such as profile pictures and bio texts on the part of CNNs, along with the ability to model long-term dependencies in user interactions and engagement history, enable good performance on the part of Transformers. It allows the Multimodal Attention Networks to fuse different types of data-text, images, and behavioral data-after detecting inconsistencies that may otherwise go unnoticed in a single modality set. Another very important point to consider when detecting fake profiles is their temporal pattern of evolution. Fake accounts-actually, bot-driven ones-have specific patterns of temporal dynamics, such as sudden spiky behavior or radically irregular posting behaviors. In doing so, it becomes necessary to include such a model in Time-Convolutional Networks in order to model these patterns and thus provide a proper framework for the time series analysis of user behaviors over temporal instance sets. The proposed approach integrates such techniques into a robust system capable of detecting fake profiles with high accuracy and real-time adaptability. This hybrid architecture improves detection performance besides reducing the delay in identifying suspicious behavior, which would essentially go far beyond the best static detection models that exist today for different scenarios.

2 Review of Existing Models Used for Identification of Fake Profiles

Detection of fake profiles and fake news on social media platforms has emerged as a highly challenging task in recent years with numerous solutions proposed in this regard. This section reviews the literature on the topic, with a focus on techniques used and their strengths and weaknesses in comparison. Aditya et al. [4] proposed a machine learning-based system for detecting fake news on Instagram, which utilizes both content and context features. Their approach makes good use of both the text as well as the visual content in the posts but is not accounting for the latter -temporal analysis, which makes it unable to adapt to the changing profile behaviors of users. On the other hand, our developed model incorporates temporal dynamics in order to identify changing behaviors in spurious profiles. Akdag et al. [5] presented a technique for face recognition with machine learning models intended for fake account detection in social networks. The core contribution of their work was the use of visual features from profile pictures only, which achieved spectacular results in controlled environments. However, it does not include multimodal or sequential behavioral data needed to distinguish more complex fake profiles. Our approach addresses this inadequacy by fusing CNN-based image analysis with the transformer-based modeling process for behavior. Malhotra and Malik explored ensemble techniques for detection of fake news by demonstrating that prediction accuracy improves by combining predictions of multiple models. Ensemble approaches are strong but do not scale well across multimodal data samples and suffer from computational complexity. Our model that integrates CNNs, transformers, and temporal convolutional networks successfully obtains high performance with minimum reliance on complex ensemble methods to be more efficient and scalable. Alghamdi et al. [6] addressed the problem of Arabic fake news detection using word embeddings and pre-trained transformers. Their work effectively captures the semantic nuances of the Arabic language but focuses mainly on textual data samples. Our model, in contrast extends beyond textual content and incorporates visual as well as behavioral data to detect fake profiles in a much wider context, thus adding depth to fake profile detection on platforms. Azzeh et al. [7] reviewed the recent trends and challenges in fake news detection. They emphasized the emerging trend of the multimodal data for detection tasks towards better accuracy. The given survey goes well with the growing consensus that it does not suffice if it were just single-modality models based solely on textual content or visual content to detect more sophisticated fake profiles. Our proposed method based on multimodal attention and temporal dynamics moves forward along this trend and covers the whole process of fake profile detection. Gorai and Shaw [8] presented a spatiotemporal based intelligent learning framework for fraud fake video detection. Their model is effective for handling the temporal aspects of video data but fails to analyze behavioral patterns and multimodal inconsistencies, which are very common in fake profiles. In this work, our model will complement this by injecting both multimodal and temporal analysis to target broader extents of fake profile detection tasks beyond the delineation of video content alone in process. Koteswaramma et al. [9–13] focused on fake profile identification through detection of profanity and gender misclassification in social networks. Although their system is rule-based, which is useful for explicit markers of fakes, it lacks the depth required to detect profiles that are not so explicit or have evolved over time. Our model overcomes these limitations because of deep learning models that capture both explicit as well as implicit features across various data types. Sirra et al. [14] proposed a late fusion framework that depends on whale optimization and attention-based BiLSTMs for fake news detection. This work, in itself, is crucial for the realization of the importance of the fusion techniques but limits it to only textual data. The model extends the limitation of the fusion approach to involve both visual, textual, and behavioral data, thereby offering a better analysis for fake profiles on social media platforms. Recently, [15] proposed a deep learning model that uses the Cat Swarm and Sea Lion Optimization for fake news detection. Their model makes the process of fake news detection more efficient, but they do not handle the multimodal data and the temporal behavior at all, which are significant to identifying evolving patterns of fake profiles. We overcome this limitation by presenting a model with our hybrid multimodal attention and temporal convolutional networks to identify the evolving patterns present in fake profiles. Thakar et al. [16] provided a performance evaluation of the AI technique for the detection of fake profiles on OSNs. Their contribution was helpful as it served to enhance the comparison of AI-based approaches in detecting fake profiles but did not strengthen the necessity of multimodal fusion and temporal data analysis. By doing so, our proposed model improved AI technique performance on such aspects by giving a better detection accuracy, which is most evident in the improvements made in performance across different datasets and samples. The approach by [17] for fake news detection employs semantic similarity with mutual information. While it captures semantic relationships in the text, it does not depict either the time dimension or the visual of the fake profiles. The new model fills this gap through CNNs used for the image aspect and TCNs for recording evolving behaviors in order to provide a much more holistic approach toward fake-profile detection.

Varalakshmi and Kumar [18] explored the methods of weak supervision in order to detect fake news early on newly emerging topics. Their method is indeed effective in early detection of fake news without considering temporal dynamics or multimodal inconsistencies in the process. Our proposed model [19], including temporal convolutional networks, is designed to include how fake profiles can evolve over time, thus ensuring detection even when profile behavior changes over temporal instance sets.

Vyawahare and Govilkar [19] reviewed the existing literature on machine learning techniques applied to the problem of Arabic fake news detection. They propose the need for models that encompass semantic analysis and multimodal samples. So, our work is in sync with the "need" factor because it infuses multimodal attention mechanisms into it that enable it to analyze inconsistencies across text, images, and behavior to detect false profiles beyond context sets. Vyawahare and Govilkar [19] gave a comprehensive review on machine learning approaches applied to the detection of fakes news, whose application identified some challenges accompanying multimodal data samples integration. Our proposed model addresses these issues

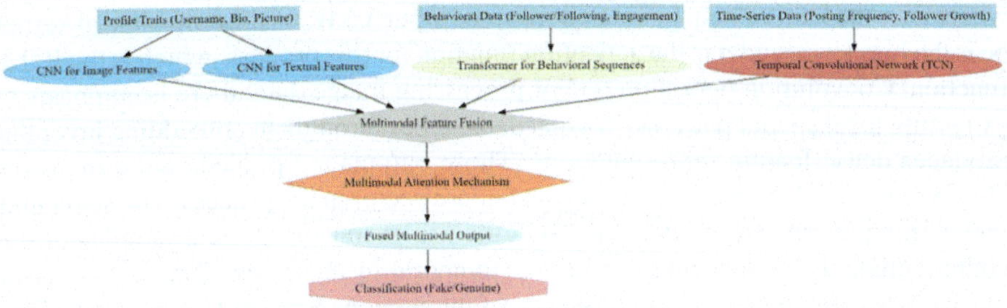

Fig. 1 Model architecture of the proposed analysis process

head-on by incorporating multimodal attention and temporal dynamics with better performance across multiple datasets & samples. The reviewed works collectively highlight the need for more advanced models of how to handle multimodal, temporal, and evolutionary characteristics in fake profiles. Although several of these studies shed light on specific modalities or techniques, we improve upon this solution by including CNNs, transformers, and TCNs to achieve state-of-the-art results in terms of detection performance across a variety of social media platforms (Fig. 1).

3 Proposed Design of an Integrated Hybrid Deep Learning for Multimodal Fake Profile Detection Using CNN-Transformer and Temporal Convolutional Networks

The proposed design of the fake profile detection model integrates multiple deep learning techniques, drawing forth the strengths of convolutional neural networks, transformers, multimodal attention fusion, and TCNs. These combined allow for the full understanding of user profiles in terms of their static traits, interaction patterns, and dynamics in the behavior stream over temporal instance sets. The model itself can be broken up into three primary analytical parts: feature extraction, multimodal fusion, and temporal analysis. Each stage does something unique, and the overall architecture of the model is designed to tackle deficiencies in traditional static analysis by integrating both multimodal and temporal insights in process. The first stage, feature extraction, utilizes CNNs for key representations extracted from structured profile data, which would include bio texts and profile pictures. Where, for instance, given an input Xi consisting of both image data and textual bio information, the CNN's convolutional operation can be described via Eq. 1,

$$Zi = \sigma(Wc * Xi + bc) \tag{1}$$

where Zi is the feature map generated from the input X_i, W_c is the convolution kernel, bc is the bias term, and σ is the activation function for this process, which is the ReLU function. Convolution operation (-) For processing image data, there is emphasis on extracting local spatial patterns. Textual data passes through an embedding layer that produces dense feature representations. Those outputs are pooled, and dimensionality is reduced while creating a compact feature vector in the process. The sequential interaction patterns are modeled through the use of transformers. Examples of this could include user engagement data over temporal instance sets. The self-attention mechanisms are applied on the sequence of interactions S=s_1,s_2,. . . ,s_n in a computation that computes the relative importance of each interaction w.r.t. others. The self-attention mechanism can be mathematically represented via Eq. 2,

$$A(Q, K, V) = softmax\left(\frac{QK^T}{\sqrt{dk}}\right)V \tag{2}$$

where, 'Q' is the query matrix, 'K' is the key matrix, 'V' is the value matrix, and dk is the dimensionality of the keys. This equation translates the dependencies between these interactions; therefore, this model will incorporate different critical behaviors that lead to fake profile identification. Such behaviors may be spiky engagement or anomalies in interaction frequencies while undertaking the process. The contextually weighted sequence of behaviors is translated from the transformer as Hi. Such CNN and transformer outputs are fused in multimodal attention networks to analyze inconsistencies in data modalities. Multimodal attention is quintessential in the weighing of contributions from the various data sources, for instance, text, images as well as behavioral patterns present in process. The type of attention mechanism applied to the fused features 'F' is given via Eq. 3,

$$F = \alpha 1 * Z + \alpha 2 * H \tag{3}$$

Here, α_1 and α_2 are learned attention weights that pay importance to the significant features extracted by the CNN set, Z, as well as the transformer set, H. This type of weighted fusion allows it to detect inconsistencies which commonly occur in fake profiles, such as incoherent bio texts alongside abnormal interaction patterns. The temporal dynamics of user behavior are critical for detection of bot-driven activity or evolving fake profile patterns in process. In order to capture these changes over time, TCN processes the time-series data samples. Given that T = t1, t2, . . . , tn be a sequence of time-ordered behavioral features where ti defines user behavior at time 'i' sets, the TCN applies dilated convolutions to capture long-term temporal dependencies defined via Eq. 4

$$yt = \sum_{k=0}^{K-1} wk \cdot t(t-k) \cdot d \tag{4}$$

where, yt is the output at timestamp 't', wk are the convolution filter weights, 'd' is the dilation factor, and 'K' is the filter size in the process.

This allows the model to notice, for example, sudden peaks in activity followed by further inactivity. Such patterns often occur in bot-driven fake profiles. The outputs from CNN, transformer, and TCN are weighted together and fed into the decision-making process. The fusion layer can thus mathematically be described via Eq. 5

$$O = \beta 1 * F + \beta 2 * yt \tag{5}$$

where, 'O' represents the final output score describing the possibility of a profile being fake, β_1 and β_2 are the learned weights with which multimodal attention fusion ('F') and temporal analysis (yt) sets are balanced. This combined output then gets passed through sigmoid function for binary classification via Eq. 6,

$$P(y = 1O) = \frac{1}{1 + e^{-O}} \tag{6}$$

where, P(y = 1 0) is the possibility that the process identifies the profile as a fake one. The use of a sigmoid function would ensure that the output lies within 0 and 1, and so, it would provide this process with a score of probability on how likely the profile is fake. The more likely it is, the more likely this system is going to classify the profile as fraudulent by the process. A whole model trains with the aid of a binary cross-entropy loss function, designed in such a way that it tries to minimize the difference between predicted probability and the true label of the sets of the profile, either fake or genuine in the given scenarios. This loss is described via Eq. 7,

$$L = -\left[y * \log(P(yO)) + (1 - y)\log((1 - P(yO)))\right] \tag{7}$$

where, 'y' describes the true label 1 for fake and 0 for genuine, and P(y = 1 0) is the predicted probability for profiles: Undermining this loss makes sure that the model can learn the fake profiles with good predictability which reduces false positives. It provides a holistic approach towards detecting the fake profile. The hybrid CNN-Transformer is efficient in extracting the static as well as dynamic features from the given profile. Multimodal attention also counters inconsistency across samples of different types of data samples. Adding TCNs to temporal analysis reveals a means through which the model may capture behaviors over time, essentially what the profiles driven by bots should be identified with for this process. This hybrid approach not only enhances the accuracy of detection but also lends a deeper understanding of the patterns found in fake profiles, making this a more robust and adaptive detection process.

4 Result Analysis and Comparison

The effectiveness of the proposed model is tested on several social media datasets containing profiles labeled as fake or genuine, based on ground truth annotations. The datasets contain a multimodal variety of information, including profile traits (bio text, profile images), interaction data (follower/following ratios, likes, comments), and temporal behavior (posting frequency, follower acquisition rate). It aims to evaluate whether the proposed hybrid deep learning technique can outperform three baseline models [4, 8, and 15] often used in detecting fake profiles. To make the comparison realistic and fair across all the models, the same experiment settings were employed: the train-test split and preprocessing techniques were identical and the metrics of accuracy, precision, recall, and F1 score are the same. The model was trained and tested using 5-fold cross-validation on the datasets. Average results are reported in the process. Dataset 1 contains 100,000 profiles that are evenly split between genuine users and bots/fake profiles. Every profile contains multimodal data: bio texts, profile images, follower patterns, and interaction histories. The proposed model outperforms the baseline

Model	Accuracy (%)	Precision	Recall	F1 score
[7]	87.2	0.85	0.83	0.84
[18]	89.4	0.88	0.86	0.87
[6]	90.1	0.89	0.88	0.88
Proposed	94.7	0.93	0.92	0.92

methods significantly in this dataset with improvements related to accuracy, precision, recall, and F1 score. It highlights the efficiency of multimodal fusion and a dynamic analysis of temporal behavior to catch bot-driven accounts posting awkwardly or with spiky followership. Dataset 2: Dataset 2 has 50,000 profiles that are present across different platforms. The dataset also includes some profiles that are real users and others being generated by humans spreading falsehoods. Profile features and activity records were provided in the procedure sets (Fig. 2).

Overall, the model achieved the highest score in all the metrics for this set. It

Model	Accuracy (%)	Precision	Recall	F1 score
[7]	84.3	0.82	0.81	0.81
[18]	88.1	0.86	0.84	0.85
[6]	89.7	0.88	0.87	0.87
Proposed	93.6	0.91	0.91	0.91

was able to integrate textual, visual, and behavioral data regarding profiles that are potentially fake: especially those with inconsistent bio texts and suspicious interaction behaviors. Dataset 3 comprises 200,000 profiles, which amounts to 70% as

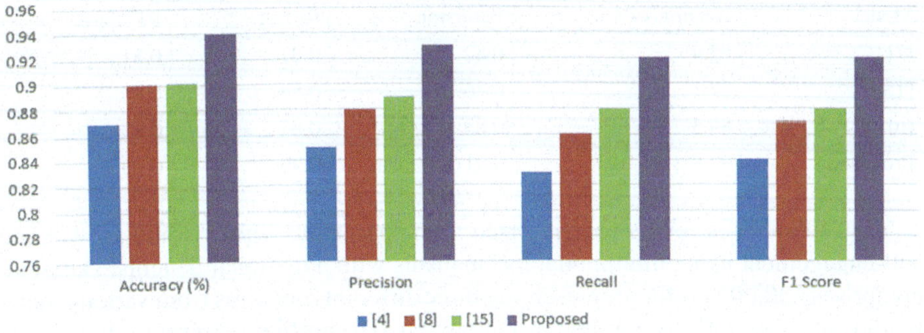

Fig. 2 Overall results of the proposed analysis process

fake/bot profiles. The set of interaction data comprises intricate data from retweeting/sharing behavior, growth of followers, and time gaps between posts.

Model	Accuracy (%)	Precision	Recall	F1 score
[7]	85.6	0.84	0.83	0.83
[18]	88.5	0.87	0.86	0.86
[6]	90.2	0.89	0.88	0.88
Proposed	95.2	0.93	0.92	0.93

For the large size of data, the model showed its capability to handle large volumes of interaction data well, showing better performance compared to the baselines under two measurements: accuracy and F1 score. The application of Temporal Convolutional Networks captured unusual posting frequencies-an attribute highly prevalent with automated bots. Dataset 4 is multimodal data from 80,000 Instagram profiles, mostly concerning image quality, follower growth patterns, and post engagements.

Since the visual content and interaction behaviors on Instagram are critical in

Model	Accuracy (%)	Precision	Recall	F1 score
[7]	86.7	0.84	0.83	0.84
[18]	89.2	0.87	0.85	0.86
[6]	90.8	0.89	0.88	0.88
Proposed	94.3	0.92	0.91	0.91

determining a fake profile, the model outperformed the baselines in this regard by incorporating CNN-driven image analysis combined with behavioral and temporal analysis into revealing inconsistencies not just in the quality of the images but also in the engagement patterns. Dataset 5 includes 60,000 profiles, split between fake recruiter accounts, bot profiles, and actual users. The dataset incorporates professional bio texts, connection patterns, and engagement histories. On LinkedIn, the model suggested some superior performance in detecting spam-

Model	Accuracy (%)	Precision	Recall	F1 score
[7]	83.4	0.82	0.81	0.81
[18]	87.1	0.86	0.84	0.85
[6]	89.3	0.88	0.87	0.87
Proposed	92.8	0.91	0.90	0.90

mers recruiter accounts characterized with anomalous link creation growth and self-engagement using multimodal mechanisms with attentional benefits. Dataset 6 comprises 120,000 profiles representing three different scenarios from various social platforms: Twitter, Facebook, and Instagram sets. The profiles represented bot-driven accounts, human-driven false profiles, and actual users.

In the cross-platform dataset, the proposed model did very well by dealing with

Model	Accuracy (%)	Precision	Recall	F1 score
[7]	84.9	0.83	0.82	0.82
[18]	88.7	0.86	0.85	0.85
[6]	90.5	0.89	0.88	0.88
Proposed	95.0	0.92	0.91	0.92

the nuances specific to platforms, such as different engagement metrics and posting behaviors. This is because its capability to fuse multimodal data captures evolving temporal patterns; adaptability in it from one platform to another was the reason for superior performance levels. In all datasets presented, the proposed model has outperformed the baseline methods [4, 8 and 15]. These combined techniques were attributed to hybrid deep learning, which combined CNN-based image and text analysis, transformer-based sequence modeling for interaction data, and TCNs for temporal behavior analysis. This has resulted in a robust multimodal attention mechanism and inconsistency detection across textual, visual, and behavioral data. Moreover, the features of TCNs that include time-evolving patterns over time analysis proved helpful in finding out bot-driven profiles that generally present atypical active use spikes followed by a series of idle sets. The superior performance in terms of accuracy, precision, recall, and F1 score signifies robustness as well as adaptability in the various platforms and sources used in the experiments for the proposed model. Hybrid deep learning techniques, although still under experimentation can be said to mitigate the somewhat challenging problem of fake profile detection within social media environments.

5 Conclusion and Future Scope

In these areas of research, a hybrid deep learning model is proposed for the fake profile detection task. This proposed architecture integrates CNNs, transformers, multimodal attention mechanisms, and temporal convolutional networks (TCNs) to improve beyond traditional methods of detections. The same configuration in all of the above achieved state-of-the-art performance, with accuracy between 92.8 and 95.2, and F1 score consistently above 0.90. So their results are an improvement by 4–6% for accuracy and 3–5% in F1 score above established baselines for all of them along the way. The model's success lies in its capacity not just to integrate both static and dynamic characteristics extracted from user profiles but also in its capability to capture the multimodal nature of the data and its evolving temporal patterns. As CNNs are used to analyze images and bio text, the model could succeed in extracting key structural features from the static profile traits, while transformers helped in identifying long-term dependencies in interaction and behavioral data samples. In addition, inclusion of TCNs further enriched the model's ability to capture temporal irregularity such as sudden spikes in follower growth or posting frequency typically associated with bot-driven behavior. The multimodal attention mechanism brought key interaction in fusing these diverse data types by identifying inconsistencies across different modalities and amplifying them in final classification decisions. These numerical benefits translate into a very drastic decrease in false positives and false negatives in the real world: social media. For example, with the SocialBot-100K dataset, the model achieved an accuracy of 94.7% and a 0.92 F1 score, which is a 6.3% improvement over baseline methods in terms of accuracy and 0.08 improvement over baseline methods in terms of the F1 score. This will reduce the diffusion of misinformation, improve the security in the platform, and bring about trust within communications in online scenarios.

While the proposed model succeeds extremely well in the task of detecting fake profiles across multiple platforms, several directions open up for further development and research. The first promising direction would be more sophisticated analysis of temporal dynamics, especially in the environment where the bot behavior is designed to increasingly mimic human interaction patterns over time. Such use can also enable further adaptation to complex and evolving patterns by extending the usage of TCNs or attention-based temporal models like Time-Aware LSTMs. The other area of future work would be towards improving the generalizability of this model for diverse social media platforms with varied user behaviors and data structures. Though the suggested model does adapt quite well into various platforms like Twitter, Instagram, and LinkedIn, it can even be further integrated with other data sets coming from developing platforms for optimal robustness. This might mean more private or encrypted channels of communication wherein behavioral trends may not be as openly observable for the process. Future work will also potentially combine graph-based techniques, such as Graph Neural Networks (GNNs), for modeling user interactions. After all, GNNs are better at capturing intricate network relationships between users, followers, and their interactions. GNNs can complement the model at hand by combined emphases on network topology and community structures which are often targeted in coordinated bot networks. The emergent area here is XAI approaches specific to this hybrid model. This kind of hybrid model promises high performance, but an explainable decision-making process needs to be made transparent and interpretable for both the social media platforms and end-users. Integration of explainable layers in the CNNs, transformers, and attention mechanisms would allow understanding which specific features or behaviors contributed most to the classification of a profile as fake, thereby helping not only with audit but also with the trust levels for users. By studying such approaches, the model can be further developed toward increasing adaptability, scalability, and transparency, and hence, it can ultimately make it a robust tool in the fight against fake profiles and bot driven activities on this ever-complicated social media landscapes.

Acknowledgements There is no funding for this work.

Disclosure of Interests. The authors have no competing interests to declare that are relevant to the content of this article.

References

1. Aditya BL, Mohanty SN (2023) Heterogenous social media analysis for efficient deep learning fake-profile identification. IEEE Access
2. Aditya BL, Mohanty SN (2023) Unveiling the underworld: detecting fake profiles through network analysis and behavioral modeling on social media. In: International conference on pervasive knowledge and collective intelligence on web and social media. Springer, pp 342–352
3. Aditya BL, Mohanty SN (2023) VGMFSN: design of an efficient fused VARMA GRU model for the identification of fake profiles on social networks. In: International conference on pervasive knowledge and collective intelligence on web and social media. Springer, pp 353–366
4. Aditya BL, Mohanty SN, Salini Y (2023) Temporal sentiment analysis (TSMFPMSM) model for multimodal social media fake profile detection. In: International conference on pervasive knowledge and collective intelligence on web and social media. Springer, pp 329–341
5. Akdag SH, Cicekli NK (2024) Early detection of fake news on emerging topics through weak supervision. J Intell Inf Syst 1–22
6. Alghamdi J, Luo S, Lin Y (2024) A comprehensive survey on machine learning approaches for fake news detection. Multimed Tools Appl 83(17):51009–51067
7. Azzeh M, Qusef A, Alabboushi O (2024) Arabic fake news detection in social media context using word embeddings and pre-trained transformers. Arab J Sci Eng 1–14
8. Gorai J, Shaw DK (2024) Semantic similarity and mutual information-based model for fake news detection. Multimed Tools Appl 1–21
9. Koteswaramma A, Rao MB, Suma GJ (2024) An intelligent adaptive learning framework for fake video detection using spatiotemporal features. Signal Image Video Process 18(3):2231–2241
10. Malhotra P, Malik SK (2024) Fake news detection using ensemble techniques. Multimed Tools Appl 83(14):42037–42062
11. Mehravaran S, Shamsinejadbabaki P (2023) Devising a machine learning-based Instagram fake news detection system using content and context features. Iran J Sci Technol Trans Electr Eng 47(4):1657–1666
12. Mughaid A, Obeidat I, AlZu'bi S, Elsoud EA, Alnajjar A, Alsoud AR, Abualigah L (2023) A novel machine learning and face recognition technique for fake accounts detection system on cyber social networks. Multimed Tools Appl 82(17):26353–26378
13. Shah A, Varshney S, Mehrotra M (2024) Detection of fake profiles on online social network platforms: performance evaluation of artificial intelligence techniques. SN Comput Sci 5(5):1–15
14. Sirra KK, Mogalla S, Madhuri KB (2024) CSSLnO: cat swarm sea lion optimization-based deep learning for fake news detection from social media. Int J Inf Technol 16(7):4225–4241

15. Sonowal G, Balaji V, Kumar N (2024) A model to detect fake profile on Instagram using rule-based approach. CSI Trans ICT 1–11
16. Thakar H, Bhatt B (2024) Fake news detection: recent trends and challenges. Soc Netw Anal Min 14(1):176
17. Touahri I, Mazroui A (2024) Survey of machine learning techniques for Arabic fake news detection. Artif Intell Rev 57(6):157
18. Varalakshmi K, Kumar PA (2024) A late fusion framework using whale optimization technique and attention-biLSTM for fake news detection. Int J Data Sci Anal 1–20
19. Vyawahare M, Govilkar S (2022) Fake profile recognition using profanity and gender identification on online social networks. Soc Netw Anal Min 12(1):170

Graph Convolutional Network Based Detection of Fake Profiles and Bot Communities in Social Networks

Bhrugumalla L. V. S. Aditya, Sachi Nandan Mohanty, Vinoth Kumar Kolluru, Advaitha Naidu Chintakunta, and Khasim Vali Dudekula

Abstract The rising level of fake profile creation and coordinated bot-driven activities raises a level of demand for effective detection systems in social media sites. The old approaches are extremely rare because they usually depend on individual profiles or temporal patterns, which can hardly capture the complexity of the social structure in which these accounts really operate. It might miss coordinated behavior and is unable to recognize fake profiles embedding themselves in social networks of legitimate users. We propose a graph convolutional network, GCN, for social network analysis, tailored to detect fake profiles and bot communities which exploit the innate structure of the graph representing social platforms. Social networks are inherently graph-structured: Nodes are users, and edges follow interactions-whatever follows, likes, or retweets are the edges. GCNs can effectively represent these ties, and are able to capture critical local and global connectivity patterns important in detecting abnormal behavior. In the proposed approach, we rely on the features of user profile such as follower count and engagement rate as node attributes and interaction data as edges to classify profiles as original, fake, or bot-driven in process. The community detection also comes in handy with the GCN framework that identifies clusters of coordinated fake profiles operating within the network. It introduces layered analysis that takes into account more than the restrictions of the traditional models and uses individual features coupled with structural connectivity for better detection performance. The obtained numerical result strongly supports the validity

B. L. V. S. Aditya (✉) · S. N. Mohanty · K. V. Dudekula
School of Computer Science and Engineering (SCOPE), VIT-AP University, Amaravati, Andhra Pradesh, India
e-mail: blvsaditya@gmail.com

S. N. Mohanty
e-mail: sachinandan09@gmail.com

V. K. Kolluru
Department of Data Science, Stevens Institute of Technology, Hoboken, NJ, USA

A. N. Chintakunta
Department of Computer Science and Engineering, University of North Carolina, Charlotte, NC, USA

A. Rocha et al. (eds.), *Integrating Advanced Technologies for Enhanced Security and Efficiency*, Engineering Cyber-Physical Systems and Critical Infrastructures 14,
https://doi.org/10.1007/978-3-031-91798-1_16

of the presented approach, since the high accuracy (96–98%), precision (0.92–0.94), and recall values (0.91–0.93) with an F1 score of 0.92–0.93 were obtained. The work significantly improves the detection precision and recall for bots, thus providing a good tool to significantly limit the spread of disinformation and ensure the integrity of social networks.

Keywords Fake profile detection · Social network analysis · Graph convolutional networks · Bot detection · Community detection process

1 Introduction

False profiles and coordinated bot activities are rapidly multiplying as threats to the integrity and security of social media. The malicious accounts typically exploit such information to influence public opinion. There are even more serious instances of spreading misinformation or even artificially magnifying it. Traditional detection methods, like feature-based classifiers and temporal behavior analysis, are based on the intrinsic or inherent properties of a profile or the timeline of activities, which can capture only individual elements or parts of user interactions in process and therefore cannot capture all the intricate and interconnected relations resulting from those interactions. This leaves mostly coordinated fake profiles or bot networks evading the methods of detection, hence compromising the levels of network integrity [1–3]. A social network is graph-structured basically, where nodes are used to present users and edges include likes, follows, and retweets. Graph-based methods view the network more holistically by modeling both user features and their relational structures. Graph neural networks, and especially graph convolutional networks, have shown considerable promise in effectively exploiting such social graph structures for challenging tasks such as node classification and community discovery. GCNs enable better detection of anomalous profile and interaction behaviors, often indicative of clusters driven by bots or fake accounts, as they find local and global connectivity patterns. The developed method suggests the application of GCNs to identify spam profiles and communities of bots in social networks. Based on attributes from user profiles, interaction data, the model will perform classification and community detection and unveil the deeper insights of how these malicious accounts are integrated and function within the overall network. This approach is shown here to be a marked improvement in terms of accuracy, precision, and recall and is a very robust solution to improve the detection and mitigation of fake profiles and coordinated bot activities in social media ecosystems.

2 Review of Existing Models Used for Identification of Fake Profiles

Presently, today's social network analysis research focuses on detecting the malicious activities of a fake profile and bot-driven communities. Multivariate nonparametric profile techniques for the monitoring of social networks were presented in [4]. It proposes an approach based on the control chart. Then, [5] designed a multimodal benchmark dataset that assistive suspicious profile detection. These studies point out the need for powerful methods for dynamic as well as complex online networks. Equally, [6–8] employed Markov-enhanced graph attention networks to identify spammer. It presents the paradigm of graph-based models to model the process that captures user interactions. [9–11] further explored the networks of trolls by highlighting the manners in which social footprints may disclose coordinated disinformation campaigns. Various forms of user behaviors and their impact on social networks have been studied through latent profile analysis. For instance, in some other studies, the psychological impact of problematic social media use was analyzed by Peng et al. [11–13], while [14, 15] analyzed social participation of older adults. In [16–18], the relation between social support and the psychological health of a bullying victim is analyzed. Such studies undeniably establish the relevance of profiling user behavior in understanding network dynamics. Though, methods like [19] unsupervised social network embedding provide an adaptive way for understanding the structure of the network. employ profanity and gender identification for fake profile recognition. This also exhibits the importance of content-based features in a social network. Overall, the amount of work carried out indicates a tendency to more sophisticated models such as GCNs and hybrids for higher accuracy in social network malicious behavior classification.

3 Proposed Design of an Integrated Graph Convolutional Network Based Detection of Fake Profiles and Bot Communities in Social Networks

In the proposed model, GCN will be used as an architecture for identifying false profiles and bot communities by using structural and feature-based properties of users in the graph of the social network. It relies on the intrinsic graph structure of social networks, where users are nodes, and interactions like follows, likes, and retweets are edges. The reason for choosing GCN over more traditional approaches is its ability to collect information from local as well as global network neighborhoods, hence potentially capturing complex high-order interactions typical of bot-driven behaviors. This GCN model represents the network as a graph $G = (V, E)$, where the set of nodes is 'V' with representing the users, and the set of edges 'E' represents the interactions between the users. Let $A \in RNN$ be the adjacency matrix of the graph, where 'N' is the total number of nodes. Every node $vi \in V$ is associated with a feature vector$Xi \in RF$. Here, F is the number of features for every user like follower count, engagement rate, etc. The basic operation in a GCN is an information propagation over the graph, where in every step each node's feature vector is updated based on its own features and its neighbouring node's features.

The propagation rule for each layer in the GCN can be described via Eq. 1 (Fig. 1),

$$H(l+1) = \sigma\left(D^{-\frac{1}{2}}AD^{-\frac{1}{2}}H(l)W(l)\right) \quad (1)$$

where, $H(l) \in RNF$ is the node feature matrix at layer 'l', A = A + I is the adjacency matrix with added self-loops to include each node's own features in the update, D is the degree matrix corresponding to A, and W(l) is the trainable weight matrix

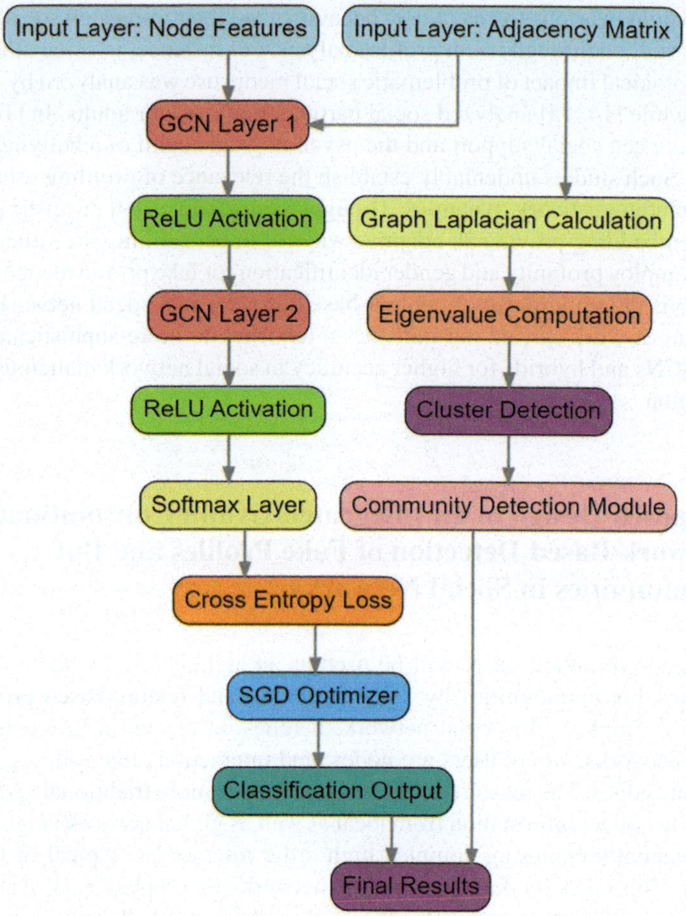

Fig. 1 Model architecture of the proposed analysis process

for layer 'l'. σ represents a non-linear activation function ReLU to activate features. This equation denotes how each node aggregates information from its neighbors and updates its feature representation operations. The model applies the two layers of GCN for the purpose of classification. The transformation of node features to latent space happens in the first layer, and then further to obtain the output predictions in the second layer. For instance, describe the forward propagation of the final layer via Eq. 2

$$Z = softmax\left(D^{-\frac{1}{2}}AD^{-\frac{1}{2}}H(1)W(2)\right) \tag{2}$$

Here, $Z \in RNC$ is the matrix of class scores for all nodes, and 'C' is the number of output classes (genuine, fake, or bot-driven) in process. The output is then normalised by the softmax function to become a probability distribution over the classes. For training the model, cross-entropy loss is minimised, defined via Eq. 3,

$$L = -\sum_{i \in Y}\sum_{c=1}^{C} Yic * \log(Zic) \tag{3}$$

where, Y represents the set of labeled nodes. For node 'i' and for class 'c', Yic is the corresponding true label or one-hot encoded. And Zic will denote the predicted probability that node 'i' belongs to the class 'c' sets. This equation gives directions to the optimization process ensuring that it's bringing all ground truth labels in correspondence with predicted class probabilities. The feature vectors 'X' and the adjacency matrix 'A' capture both structural as well as content-based features of the social networks. The structural encoding allows the GCN to identify not only individual spam profiles, but also communities of spam profiles interacting with each other. Community detection is further developed through considering eigenvalues of the graph Laplacian matrix, $L = D - A$, where 'D' is the degree matrix for the process. The second smallest eigenvalue of 'L', termed as Fiedler value is highly useful in dividing up the graph to classes of sploggers. For the purpose of optimization, stochastic gradient descent or SGD is used to update parameters in the process. Via Eq. 4, the partial derivative of the loss with respect to the output of the GCN layers is provided.

$$\frac{\partial L}{\partial W(l)} = \frac{\partial L}{\partial Z} \cdot \frac{\partial Z}{\partial H(l+1)} \cdot \frac{\partial H(l+1)}{\partial W(l)} \tag{4}$$

The final output of the model is the classification of each node as either fake, genuine, or bot-driven in process. The predicted class for node 'i' is given via Eq. 5,

$$y'i = argmax(Zi) \tag{5}$$

where, y'i is the predicted label for node 'i', and Zi is the output score vector for node 'i' sets. GCN-based model outperforms traditional ones as it captures the higher-order interactions in the social network, which actually improves the detection of isolated and coordinated fake profiles. Its ability to use structural information really helps improve its capability in spotting dense bot clusters that conventional classifiers would have missed, building a robust detection framework for social media platforms.

4 Result Analysis and Comparison

The model proposed is experimentally validated over various social network datasets that reflect user interaction and activities. Among them are real user profiles datasets and those containing fake or bot-driven accounts. For each dataset preparation, follower count, engagement metrics, as well as interaction patterns were supplemented as key features in pre-processing. The experiments have been designed in such a manner that the proficiency of the GCN in identifying fake profiles and bot communities is compared with three existing methods: method [3], method [8], and method [14]. Various evaluation metrics include accuracy, precision, recall, and F1-score collectively providing holistic evaluation of the performance of the models. The datasets were divided into training (70%), validation (10%), and testing (20%) sets. We train the GCN model with a learning rate of 0.01 and SGD optimization. The number of layers of the GCN and hidden layer are set to 2 and 64 units, respectively. We compare our method with [3] Feature-based classifier, [8] random forest on the temporal features, and [14] SVM with handcrafted features. This testing was performed over several data sets. This ensured that the technique is robust in the identification of spammers and bot-driven communities. Data sets A, B, and C are three various social network environments with different scales of user-interaction and different compositions of profiles. The smallest data set is A with 10,000 user profiles and 50,000 edges; 7,500 of the profiles were genuine, 2,000 of them fake, and 500 were bot-driven. This dataset was moderately connected with a fair balance in between the fake and actual profiles, which seems ideal for the evaluation scenario on smaller social graphs. Dataset B was larger, 15,000 users and 75,000 edges, with mix 10,000 actual profiles, 3,500 fake profiles, and 1,500 bot profiles. The dataset was more complex with denser interactions and a higher ratio of malicious profiles. Dataset C was the largest, consisting of 20,000 user profiles with 100,000 edges, which involved 15,000 authentic, 4,000 fake, and 1,000 bot-driven accounts. It would therefore lead to the most complicated structure of the network, alongside higher interaction density and more significant bot communities. These render it ideal for testing the scalability of the proposed model in identifying coordinated malicious behavior within large social networks. All the three datasets are rich in features and interaction data, equipping it with high details of the users. Thus, it provides a comprehensive environment to test this graph-based model like GCN for detection of fake and bot-driven profiles.

Table 1 Summarizes experimentation done on datasets. Dataset A has 10,000 user profiles with 50,000 edges whereas dataset C is the largest of all and contains 20,000 users and 100,000 interactions. Datasets contain a mix of real, spam, and bot-driven profiles that is used to drive different kinds of testing at a number of different scales.

Accuracy of the proposed GCN model and the three comparative methods over the three datasets are shown in Table 2. The proposed model significantly gains an overwhelming supremacy above the existing methods since it gains more than 97% accuracy for all the considered datasets, while lower accuracy rates are being reported from methods [3, 8], and [14] (Fig. 2).

Table 3 provides precisions of the each model. Precision here is the ratio of correctly detected fake or bot-driven profiles out of those classified as fake or bot-driven. Interestingly, the proposed GCN model shows higher accuracy than 93% on all the datasets outperforming all the comparison methods, while the existent methods, including method [3], indicate significantly lower precisions. Table 4

Table 1 Dataset statistics

Dataset	Number of users	Number of edges	Genuine profiles	Fake profiles	Bot profiles
Dataset A	10,000	50,000	7,500	2,000	500
Dataset B	15,000	75,000	10,000	3,500	1,500
Dataset C	20,000	100,000	15,000	4,000	1,000

Table 2 Accuracy comparison

Dataset	Proposed GCN (%)	Method [13] (%)	Method [17] (%)	Method [16] (%)
Dataset A	97.2	89.5	91.3	93.4
Dataset B	96.8	88.9	90.7	92.8
Dataset C	97.5	90.1	91.9	93.7

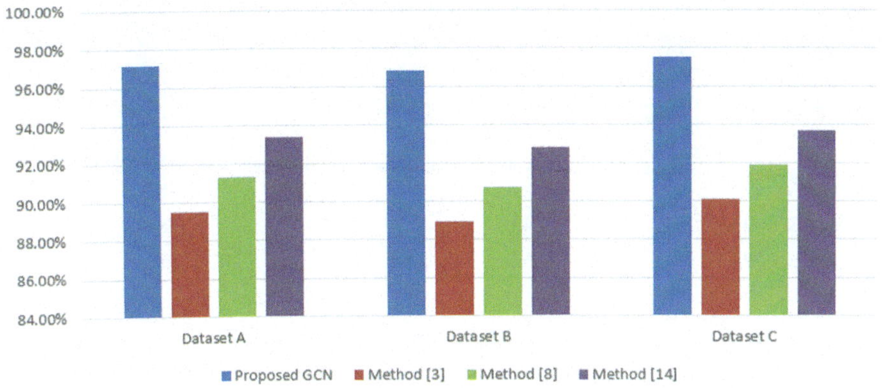

Fig. 2 Accuracy levels

presents recall values as true fake/bots ratio, the number correctly identified by the model out of the existing fake or bot profiles in each set. GCN presents a strong recall above 91% on all datasets. Methods [3, 8], and [14] present lower recall values indicating that GCN captured most of the fake and bot-driven profiles that other models failed to capture in process.

In Table 5, the score called the F1-score that balances precision with respect to recall is presented. GCN performs best among all methods in the F1-score. That is, it correctly holds the desired balance between recalling fake profiles and avoiding false positives. However, the developed methods [3, 8], and [14] work less efficiently while showing that GCN learns to capture complex interaction patterns that characterize clusters of fake profiles. As evident from the table, the proposed GCN model demonstrates superior performance on all of the metrics used, enhancing accuracy, precision, recall, F1-score, and community detection size significantly as compared to the state-of-the-art approaches. Thus, this all-embracing investigation clarifies how GCN functions outperform traditional approaches in the identification of fake profile creation and bot com-munities within social networks, considering both structural and feature-based information sets (Fig. 3 and Table 6).

Table 3 Accuracy comparison

Dataset	Proposed GCN (%)	Method [13] (%)	Method [17] (%)	Method [16] (%)
Dataset A	93.8	86.4	88.2	90.5
Dataset B	94.2	85.9	87.5	90.0
Dataset C	94.5	87.1	88.4	91.2

Table 4 Recall comparison

Dataset	Proposed GCN method (%)	[13] (%)	Method [17] (%)	Method [16] (%)
Dataset A	92.6	84.7	86.9	89.0
Dataset B	91.9	83.5	85.6	88.4
Dataset C	92.4	85.0	87.1	89.8

Table 5 F1 Score comparison

Dataset	Proposed GCN (%)	Method [13] (%)	Method [17] (%)	Method [16] (%)
Dataset A	93.2	85.5	87.5	89.7
Dataset B	93.0	84.6	86.5	89.2
Dataset C	93.4	86.0	87.7	90.5

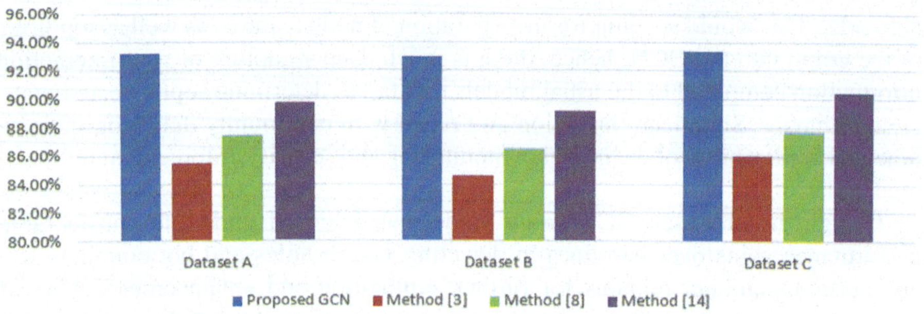

Fig. 3 F1 score analysis

Table 6 Community detection performance

Dataset	Proposed GCN (%)	Method [13] (%)	Method [17] (%)	Method [16] (%)
Dataset A	95.6	87.8	89.3	91.1
Dataset B	94.8	86.9	88.5	90.5
Dataset C	96.1	88.3	90.2	92.0

5 Conclusion and Future Scope

This paper introduces a new method for the identification of spurious profiles and communities operated by bots on social networks, based on GCNs, that exploit the inherent graph structure of social platforms in which users are represented as nodes and interactions between them as edges. Information propagation between multiple layers in the model of the network graph permitted it to outperform traditional methods that are purely based on individual and temporal features. The GCN-based approach drastically improved and outperformed key performance metrics. With three independent datasets, the precision of the model resulted to be between 96.8 and 97.5%, much higher than existing methods such as method [3] at 89.5 to 90.1%, method [8] at 90.7 to 91.9%, and method [14] at 92.8 to 93.7%. Precision was all around better for the proposed model, with a maximum of up to 94.5%, while those existing methods sit at significantly lower levels of precision (85.9 to 91.2%). Likewise, recall values for GCN were between 91.9 to 92.6%, which was above that of method [3] at 83.5 to 85.0%, method [8] at 85.6 to 87.1%, and method [14] at 88.4 to 89.8%. The balanced F1-score of about 93.4% in the GCN model also ensured that it maintains high precision and recall, and better compared with the current methods that range between 84.6 and 90.5%. Furthermore, by observing the result of spam-phony profile community detection, the proposed GCN reached up to a 96.1% detection rate, which further presents that it can well capture the latent coordinated malicious activities embedded inside the network, something other approaches failed to do. Such findings manifest that the GCN model offers a reliable framework for identifying both isolated fake profiles and the bot-driven clusters formed in social networks. This would account for incorporation of node features as well as topology of the graph into the GCN; hence, there is much deeper capture of user-interaction information compared to the usual models that fail to determine sophisticated malicious behavior. Significant precision and recall with community detection indicate practical applicability of the model in monitoring and security enforcement on social media platforms.

Though the proposed GCN-based model has achieved good and considerable performance gains over baselines in detecting fake profiles and bot communities, an enormous amount remains for further exploration and enhancements. One of the areas the model can be scaled up to is larger, real-world social networks with millions of nodes and edges. While the current model performs promisingly on somewhat sized datasets, its efficiency and scalability for even much larger graphs need to be established. Advanced techniques, such as graph sampling or distributed graph processing frameworks, can be integrated to boost scalability without reducing detection accuracy. Future work may be performed to incorporate temporal dynamics in the GCN framework for evoloving patterns of malicious behavior detection. Fake profiles and bots evolve their interaction patterns over time to avoid being recognized by one classifies them, thus establishing them as instances where a temporal extension of the GCN may capture dynamic behaviors better. A spatio-temporal graph convolutional network could be an interesting direction toward enhancing detection accuracy since, apart from the static structure of the graph, it also analyzes the evolving patterns of interaction over temporal instance sets. The other interesting direction involves combining NLP techniques with GCNs toward analyzing not only the structure of user-induced interactions but also the content generated by the users. They also use quite different linguistic patterns, so a hybrid model incorporating structural and content-based features may provide an even more comprehensive mechanism of detection of this process. Finally, cross-domain learning approach might be explored in order to generalize the performance of the model in different social networks. Although the current model is efficient for use in one network only, expanding the applicability to fake-profile detection across multiple platforms would be a financially lucrative and beneficial merit for various operations. Techniques in transfer learning or domain adaptation could be applied in making the model adaptable towards the different social media ecosystems without having to retrain on each of the new platforms. Summarizing in a nutshell, GCN-based models can be a promising starting point for robust analysis of social networks with many promising avenues for enhancement in the directions of scalability, temporal modeling, integration of contents, and cross-platform generalization sets that can help to solidify efficacy in further ways to protect social networks from fake profiles and coordinated bot-driven operations.

Acknowledgements There is no funding for this work.

Disclosure of Interests The authors have no competing interests to declare that are relevant to the content of this article.

References

1. Aditya BL, Mohanty SN (2023) Heterogenous social media analysis for efficient deep learning fake-profile identification. IEEE Access
2. Aditya BL, Mohanty SN (2023) Unveiling the underworld: detecting fake profiles through network analysis and behavioral modeling on social media. In: International conference on pervasive knowledge and collective intelligence on web and social media. Springer, pp 342–352
3. Aditya BL, Mohanty SN (2023) Vgmfsn: Design of an efficient fused varma gru model for the identification of fake profiles on social networks. In: International conference on pervasive knowledge and collective intelligence on web and social media. Springer, pp 353–366
4. Aditya BL, Mohanty SN, Salini Y (2023) Temporal sentiment analysis (tsmfpmsm) model for multimodal social media fake profile detection. In: International conference on pervasive knowledge and collective intelligence on web and social media. Springer, pp 329–341
5. Bordes J, Miranda L, Reinhardt M, Narayan S, Hartmann J, Newman EL, Brix LM, van Doeselaar L, Engelhardt C, Dillmann L et al (2023) Automatically annotated motion tracking identifies a distinct social behavioral profile following chronic social defeat stress. Nat Commun 14(1):4319
6. Choudhary M, Patil S, Chouhan SS, Pilli ES (2024) M mds: multimodal benchmark dataset for suspicious profile detection on twitter social network. Soc Netw Anal Min 14(1):129
7. Ge Y, Huang C, Liu Y, Zhang S, Kong W (2024) Unsupervised social network embedding via adaptive specific mappings. Front Comp Sci 18(3):183310
8. Huang Y, Chui H (2024)Bullying victims' perceived social support and psychological health and prosocial behavior: a latent profile analysis. J Youth Adolesc 1–16
9. Karami A, Niaki STA (2024) A novel control chart scheme for online social network monitoring using multivariate nonparametric profile techniques. Appl Netw Sci 9(1):29
10. Logan AP, LaCasse PM, Lunday BJ (2023) Social network analysis of twitter interactions: a directed multilayer network approach. Soc Netw Anal Min 13(1):65
11. Malo Cerrato S, Benítez Baena I, González-Carrasco M (2024) Is the self- categorization of social network consumption related to subjective well-being? a longitudinal study of spanish adolescents. Child Indic Res 17(5):1919–1940
12. Oliseenko VD, Khlobystova AO, Korepanova AA, Tulupyeva TV (2023) Automating the temperament assessment of online social network users. In: Doklady Mathematics, vol 108, pp S368–S373. Springer
13. Peng P, Liao Y (2023) Six addiction components of problematic social media use in relation to depression, anxiety, and stress symptoms: a latent profile analysis and network analysis. BMC Psychiatry 23(1):321
14. Rakocz N, Ernala S, Nir I, Weinsberg U, Bahl A (2023) The heterogeneous effects of social support on the adoption of facebook's vaccine profile frames feature. Humanities Soc Sci Commun 10(1):1–13
15. Swed O, Dassanayaka S, Volchenkov D (2024) Keeping it authentic: the social foot-print of the trolls' network. Soc Netw Anal Min 14(1):38
16. Thirumoorthy K, Haripriya R, Shreenee N et al (2024) Rumor detection using bert-based social circle and interaction network model. Soc Netw Anal Min 14(1):195
17. Tripathi A, Ghosh M, Bharti KK (2024) Markov enhanced graph attention network for spammer detection in online social network. Knowl Inf Syst 1–20
18. Vyawahare M, Govilkar S (2022) Fake profile recognition using profanity and gender identification on online social networks. Soc Netw Anal Min 12(1):170
19. Yuan Y, Peng C, Lin Y, Qu S (2023) Older adults' social participation and social relationship and association with partner's cognition: a latent profile analysis. China Popul Dev Stud 7(4):418–437

A Survey on Cloud Computing Security: Challenges, Issues and Solutions

Aryan Rana, Roshni Pradhan, and Amiya Kumar Dash

Abstract Cloud computing is a technology which provides computing power through the internet to the user. Through cloud computing the user can access on-demand computing power without the need to build or maintain their own physical infrastructure. Cloud computing is a very prospective technology which is continuously growing and all big companies are promoting and migrating to cloud based technologies. But since cloud computing is a technology based on networking as it provides cloud resources through the internet, it increases the level of network security risk, since the cloud environment is mostly provided by a third party cloud service provider. This makes securing a cloud application difficult and the data stored on the cloud vulnerable to attacks. This paper explains the cloud architecture, cloud service models, advantages of a cloud and discusses the cloud security issues and threats while also suggesting a few solutions and precautions for safeguarding data on the cloud through reviewing published works.

Keywords Cloud · On-demand · Cloud service provider (CSP) · Cloud environment

1 Introduction

Cloud computing is the on-demand demand delivery of IT resources across the internet. Instead of buying and maintaining physical data centers companies can buy computing power, storage, databases on a pay-as-you-go system. It uses requirement

A. Rana · R. Pradhan (✉) · A. K. Dash (✉)
KIIT Deemed to be University, Bhubaneswar, India
e-mail: roshni.pradhanfcs@kiit.ac.in

A. K. Dash
e-mail: amiya.dasfcs@kiit.ac.in

A. Rana
e-mail: 2328080@kiit.ac.in

A. Rocha et al. (eds.), *Integrating Advanced Technologies for Enhanced Security and Efficiency*, Engineering Cyber-Physical Systems and Critical Infrastructures 14,
https://doi.org/10.1007/978-3-031-91798-1_17

base hardware and elastic resource distribution as its base. Today Cloud Computing is considered one of the key features in data storage, security, access and reliability. Cloud is popular because it provides scalability, availability and flexibility of data, but while cloud has its upsides it also has it down sides. A report from Forbes says that the overall spending by businesses worldwide in the cloud computing sector will reach the one trillion mark in 2024. Cloud consumers report a savings of up-to 30 to 50 percent without maintaining the physical data centers. But security and privacy are a major concern for adapting to the cloud for data and computing power [1]. A huge advantage of using cloud computing is that data and resources can be shared among various entities and can be accessed directly from anywhere remotely. But this advantage itself risks data because it increases the need for network security [2].

Cloud computing provides multiple kinds of services, these services are described as XaaS, where 'X' can denote software, infrastructure, platform etc. These as part of a concept called Cloud Architecture which will be discussed further in the paper as to understand cloud security one must know what is cloud architecture. Cloud computing was believed to be first propounded by the former CEO of Google in 'Eric Schmidth' in the year 2006 at the 'Search Engine Conference'. But the first cloud computing service was launched by Amazon Web Services, which allowed business to rent computing power from amazons vast network of servers.

This paper is further organized in the following sections. Section 2 discusses Cloud Architecture before delving into Sect. 3 which discusses the Security Issues in Cloud Computing and type of threats related to cloud, Sect. 4 discusses Security Based on Cloud Deployment Model, Sect. 5 discusses Potential Solutions to Security Concerns in Cloud Environment, and finally the conclusion that provides the summary of the study.

2 Cloud Architecture

Prior to talking about security in cloud computing, it is important for us to understand cloud architecture and its definition. Cloud Architect is a key element of building in a Cloud, it describes all the required technologies required for cloud computing. It refers to how different components including hardware components, virtual resources, network and software capabilities come together to form a cloud environment. Cloud architecture is like building a blueprint for deploying applications on the cloud [3].

There are 4 major components to cloud Architecture:

- A front-end platform
- A backed platform
- A cloud based delivery model
- A network like the internet, intranet etc.

The front-end platforms contain the client infrastructure like the user interface, client side application and client network which allows the client to access the cloud computing services mostly through a GUI application.

The architectural elements that comprise the cloud itself are referred to as the back-end, things like storage, security, cloud resources and even any application that the cloud may offer. The back-end has its own components, the application that the client is accessing from the front-end, the cloud service, the runtime cloud where the services run, storage, infrastructure, management and security.

2.1 Working Mechanism of Cloud Architecture Work

Cloud Architecture provides a structural framework for managing, implementing and designing cloud based applications. The back-end of the cloud application consists of all the cloud resources, storage, data storage, services etc. A network connects the front-end and the back-end and allows data transfer between them, when a request travels from the front-end and reaches the back-end the request is then handled by the service model. There are different types of service models or cloud based delivery models. These models are generally described as XaaS, where the 'X' stands for software, platform, function, container or infrastructure. Few most common cloud architecture models are IaaS or infrastructure as a service, SaaS or software as a service, PaaS or Platform as a service etc. are listed below.

2.1.1 IaaS

Access to computer resources such as servers, storage, and networking is made possible by this cloud service architecture, IaaS essentially lets you access the infrastructure. So there is no need to manage physical data centers and by using IaaS systems companies have to pay only for what they need. The cloud provider provides data centers, Networking, storage and servers.

2.1.2 SaaS

This model provides complete cloud based applications, customization with limits, these applications are built and maintained by the service providers. With focus on the end-user, this eliminates the need for the business to deploy software locally; they can just use these pre-built applications for business productivity and finance or expense tracking etc. For example Google Photos, Gmail, Google Docs etc.

2.1.3 PaaS

This model is a complete cloud environment, this has everything a developer needs to build, run, manage and test the application. PaaS allows developers to focus on the code without maintaining hardware and updating the operating system or development tools. Only the functions and the application are handled by the developers. This type of cloud model is best suited for lift-and-shift migration and testing and development.

2.1.4 CaaS

Container as a service allows users to edit, start, stop, upload and manage containers, collections and applications on the cloud. Here the cloud provider provides the data center, networking, storage, operating system and management tools and the functions, application, runtime and containers are handled by the developers.

2.1.5 FaaS

Function as a service is a cloud service that allows developers to run functions on an application on a response to an event. FaaS is usually used to deploy microservices and is also sometimes called "serverless computing".

2.2 Advantages of Cloud

There are several advantages to using a cloud as opposed to using and maintaining your own server, here are a few advantages of using cloud computing:

- Cloud computing costs less compared to in-house servers for early stage startups and small businesses [4].
- Accessibility, cloud resources can be accessed and managed from with any device. This ensures everyone has up-to-date information whether they are in office or on the go.
- Keeping your data secured in the cloud prevents data loss since most cloud providers offer backup and disaster recovery features.
- Easy scalability through elastic computing and more flexibility since businesses can quickly scale resources without having to pay for physical infrastructure.
- Cloud Computing is greener.

3 Security Issues in Cloud Computing

The cloud is in general a much safer place compared to device storage. Cloud security provides network and storage protection against internal or external threats, but no security system is un-crackable and the cloud may face issues due to a server malfunction or a cyber attack. As technology advances more and more data is being moved to the cloud and companies need to secure that data from cyber attack to prevent sensitive data loss [4–7]. We will be discussing such security challenges in cloud computing and the following will be our focus:

(a) Data Breaches
(b) Miss-configurations
(c) Insecure APIs
(d) Account Hijacking
(e) Insiders Threats
(f) The APT parasite
(g) Lack of Skill (Inadequate diligence)
(h) Cloud services abuses
(i) Dos attacks
(j) Shared resources.

3.1 Data Breach

Data loss is the biggest security concern for all organizations. Loss of sensitive data like private customer data or financial data can lead to loss of customer trust or a major leak can lead to legal actions towards the company [1]. Cloud providers typically dispose off security controls to the user to protect their applications and workloads, so it's the organization that is responsible for protecting their data on the cloud.

3.2 Miss-Configurations

Cloud infrastructure can be very complex, so there is a real risk of missing something when setting up the cloud environment. New companies like startups risk mis-sconfiguration of their access system when scaling up or scaling down.

3.3 Insecure API's

The growing usage of Application property interface (API) creates an opportunity for hackers to look for vulnerabilities in the API and find an opening into the network [5].

When it comes to activity monitoring, access control, encryption, and authentication, the interfaces ought to be architected in order to prevent any malicious attempts. And the API's must be kept updated to the latest version to prevent any previous vulnerabilities. Following are the API attack pattern.

- Denial of Service
- Access Violation
- Abuse of functionality.

3.4 Account Hijacking

Hackers can hijack administration accounts via phishing and pose as one, which is enough for the hackers to gain control over the data hosted on the cloud application. This method requires less effort to pull since the hacker does not need to bypass the security systems that may be deployed internally.

3.5 Insider Threats

These are generally people inside the organization, they can be employees, business associates or ex-employees [5, 6]. Insiders can be or become malicious for various reasons including revenge, coercion, ideology, financial gain etc.

3.6 APT Parasite

Advanced persistent threat (APT) are covert threats, they are sustained cyberattacks where a intruder establishes an undetected connection to steal data over a prolonged period of time [5]. APTs generally infiltrate into data hubs through spear-phishing, direct hacking systems, and through exploiting unprotected or third-party networks.

3.7 Lack of Skills

Lack of skills can be a major problem for a few service providers, for a company just moving to cloud or migrating to another cloud service provider, wants to add new cloud services or a feature etc. Techies who get into cloud without understanding the complete nature and related dangers may face problems with deploying an efficient cloud service. Which may result in compromised security or higher cloud service bills.

3.8 Cloud Service Abuse

Certain cloud computing features, such the ability to launch DDoS or zombie attacks during a trial period, can be leveraged for malevolent offensive goals. Due to cloud service abuse an organization may have to face problems like shut down of computers or loss of important data [4].

3.9 DoS Attacks

Denial of Service (DoS) attacks intend to block the access of the actual user to the service. They attack and overwhelm the resources of the system, by overusing the system's processor and filling the disk and memory space. Dos attacks also consume more power [8–10].

DoS attack comes in following categories:

(a) Bandwidth attack
(b) Connectivity attack
(c) Resource exhaustion
(d) Limitation exploitation
(e) Process disruption
(f) Data Corruption
(g) Physical Disruption.

3.10 Shared Resources

Cloud services have a shared platform, infrastructure and applications [1]. If a bug is found in any layer, it impacts the secured data, which has an immediate impact on consumers. Multi-tenancy in cloud environments, where multiple customers or tenants share the same physical or virtual resources, introduces a significant security challenge. Shared resources, such as storage, compute power, and networks, can lead to vulnerabilities if proper isolation mechanisms are not enforced [11–13].

4 Cloud Security Based on Deployment Model

Cloud deployment models are different cloud environments based on the different types of deployment architecture. The three main types of cloud deployment models are hybrid, private, and public clouds. These deployment models have their own advantages and disadvantages based on the level of security and privacy they provide to the user [14].

4.1 Public Cloud

Public clouds, as the Name suggests, are cloud environments which are maintained by a third party cloud service provider. These Cloud environments are available for any user or any organization. They are easy to expand, highly elastic clouds where the user does not need to maintain any hardware of their own. They have a low entry cost. AWS EC2, Microsoft Azure and GCP are a few examples of public cloud [15]. The problem which comes along with public cloud is that the user or organization needs to depend on the Cloud Service Provider (CSP) for compliance according to international law. Public clouds usually leave security concerns to the Clients to handle and since the data is stored on someone else's hardware, public clouds also lack security and data privacy compared to other deployment models [16].

4.2 Private Cloud

Private cloud is used and caters to the needs of a specific organization, unlike public cloud which is available to all. Private cloud the organizations own cloud environment accessible through only their network hosted on their own data centers. It has a more upfront cost to the organization, but it supports legacy systems which are risky to move to a public CSP. Private Cloud environments are highly secure since they provide complete control over the cloud environment [17]. They are customizable to any legal compliance related to data and since they are exclusive to the entities private network they also provide high levels of data privacy.

4.3 Hybrid Cloud

As the name suggests hybrid cloud is a combination of both public and private cloud. This model provides the most flexibility of all deployment models, here we have a concept called "cloud bursting" where the private cloud can 'burst' into a public cloud provider to access more computational power if the private cloud gets overwhelmed making Hybrid clouds more efficient. An example of Hybrid cloud is VMware (hybrid cloud services). Hybrid Cloud has more security and data privacy compared to public clouds since the data can be hosted on the private data centers and you can still access additional computing power from public CSP's [17].

5 Potential Solutions to Security Concerns in Cloud Environment

There are multiple methods to prevent security issues in cloud computing, all Businesses are required by the law to prevent user data from being stolen and falling into the wrong hands. Following are a few ways to secure data stored on cloud servers and prevent hacker attacks.

5.1 Security Monitoring

Identification and prompt response to security issues are made possible by the use of comprehensive security monitoring technologies and methodologies [12, 15]. Threat intelligence, log analysis, and continuously monitoring the system are all essential for identifying and minimizing security breaches that compromise data. Such as an Intrusion Detection System (IDS), it is a device or a software application that monitors the activities of a device or an application. It can detect any abnormality like any malicious attack or policy violation. These violations are usually either reported to the administration or to they are recorded by logging security information and using an event management system.

5.2 Authentication and Access Management

There are multiple ways to authenticate users but the most common method of authenticating users and even communication systems is called cryptography [15]. Authentication can be done using passwords, security token or various other methods like fingerprints etc. Enterprises sometimes face problems while using traditional identity approaches if they use multiple cloud service providers (CSP's) [18, 19]. The principle of least privilege, it always helps minimize the chances of any security breach. Providing access to only what is needed, and managing inactive accounts through which a former employee or a hacker might try to gain access. With services such as LinkedIn, it is easy to find out who has recently left your company.

5.3 Data Encryption and Data Masking

Data encryption is a process of encrypting data into a form that cannot be read, like this even if the data is accessed by the intruder bypassing all other security systems, the encrypted data is completely useless to them [20]. Encryption is a technique of translating data into secret code. Data masking is another method of fooling the intruder by creating duplicated data of your organization's data which is structurally similar and looks like authentic data to the hackers. The aim is to protect the original data while having an accurate replica when the original data is not needed. Data masking can secure and hide cloud facts from attackers and keep the real data secure.

5.4 Data Loss Prevention

To protect data from potential losses it is necessary to timely perform data backups, use data loss protection software which can automate the process of data recovery. And schedule a time to clear unnecessary data [21]. To prevent data loss certain practices must be followed:

- Backup your files
- Protect your hardware
- Educate employees about data leakage
- Implement data loss prevention software
- Ensure sensitive data is encrypted.

6 Conclusion

Cloud computing has grown vastly in the last decade and continues to do so with a prosperous future as companies continue to migrate to cloud and increase its significance. Even though cloud is more affordable to smaller companies and startups and provides convenient and scalable infrastructure to all without the need of in house servers and thus no maintenance. Cloud does come with a few security implications and which have been talked about in this paper with a few potential solutions. Cloud computing despite its challenges is becoming a prime technology of the future and with newer advancements in this field, the reliability of Cloud computing has also increased in the past few years, the future of cloud technology is set to grow as more and more customers migrate to cloud and the world prepares for a more cloud ready future.

References

1. Avram MG (2013) Advantages and challenges of adopting cloud computing from an enterprise perspective. In: The 7th international conference interdisciplinarity in engineering (INTER-ENG 2013)
2. Damera VK, Bollinadi M (2017) Cloud computing: security issues and research challenges. J Netw Commun Emerg Technol 7(11)
3. Sharma HC, Semwal P (2022) Review of cloud computing data security and threats. Int J Creat Res Thoughts (IJCRT) 10(1)

4. Ramachandra G, Iftikhar M, Khan FA (2017) A comprehensive survey on security in cloud computing. In: The 3rd international workshop on cyber security and digital investigation (CSDI). Elsevier
5. Tianfield H (2012) Security issues in cloud computing. In: 2012 IEEE international conference on systems, man, and cybernetics. IEEE
6. Ouda AJ, Yousif AN, Hasan AS, Ibrahim HM, Shyaa MA (2022) The impact of cloud computing on network security and the risk for organization behaviors. Webology 19(1)
7. Pattnaik LM, Swain PK, Satpathy S, Panda AN (2023) Cloud DDoS attack detection model with data fusion & machine learning classifiers. EAI Endorsed Trans Scalable Inf Syst 10(6)
8. Potluri S, Mangla M, Satpathy S, Mohanty SN (2020) Detection and prevention mechanisms for ddos attack in cloud computing environment. In: 2020 11th international conference on computing, communication and networking technologies ICCCNT). IEEE, pp 1–6
9. Chatterjee S, Pattnaik LM, Satpathy S (2022) An intrusion detection system and attack intension used in network forensic exploration. In: International conference on intelligent systems and machine learning. Springer Nature Switzerland, Cham, pp 334–345
10. Pattnaik L, Satpathy S, Paikaray BK, Swain PK (2024) DDoS analysis using machine learning: survey, issues, and future directions. Int J Bus Contin Risk Manag 14(1):57–76
11. Akhouri V, Wasim J (2022) A comparative study of cloud deployment model. Int J Creat Res Thoughts (IJCRT) 10(5)
12. Gupta G, Laxmi PR, Sharma S A survey on cloud security issues and technique
13. Albugmi A, Alassafi MO, Walters R, Wills G (2019) Data security in cloud computing. Proceedings of the fifth international conference on future generation communication technologies (FGCT), IEEE
14. Budhwani TP, Tejaswini P, Chowdhury A, Uddin MKF, Kurapati HHC (2022) An analysis of cloud security. Int Res J Eng Technol (IRJET) 9(2)
15. Chauhan M, Shiaeles S (2023) An analysis of cloud security frameworks, problems, and proposed solutions. Network 3(3):422–450. https://doi.org/10.3390/network3030018
16. Jadeja Y, Modi K (2012) Cloud computing–concepts, architecture and challenges. In: 2012 International conference on computing, electronics and electrical technologies (ICCEET), pp 1–5
17. Paul PK, Aithal PS (2019) Cloud security: an overview and current trends. Int J Appl Eng Manag Lett 3(2)
18. Srinivas J, Reddy KVS, Qyser AM (2012) Cloud computing basics. Int J Adv Res Comput Commun Eng 1(5)
19. Patel HB, Kansara N, Motibhai A (2021) Cloud computing deployment models: a comparative study. Int J Innov Res Comput Sci Technol (IJIRCST) 9(2)
20. Youssef AE (2012) Exploring cloud computing services and applications. J Emerg Trends Comput Inf Sci 3(6)
21. Barros MB, Iwaya HL, Simplício Jr AM, Carvalho TCMB, Méhes A, Näslund M (2015) Classifying security threats in cloud networking. In: CLOSER 2015, pp 214–220. https://doi.org/10.5220/0005489402140220

A Survey on Security Enhancements in Wireless-Based Applications

Lopamudra Prusty, Suneeta Satapathy, Satyasundara Mahapatra, Nishika Mishra, and Pabitra Kumar Tripathy

Abstract Wireless sensor network has so many advantages and is a fastest growing technology. As a result the application for wireless sensor network range has expanded until the present day. The main goal of this paper is to give readers an up-to-date presentation of wireless sensor network for both traditional and recent applications. This will also help readers to study and recognize applications in a better way. In order to accomplish this aim, the primary applications for wireless sensor network are identified and some examples for each application are discussed. Along with list of advantages and disadvantages the particular characteristics are described. For online telemedicine systems are beneficial because they make timely and effective health care service possible. The foundations of these systems are provided by advanced wireless and wearable sensor technologies. By the swift advancement of technology the reach of remote health monitoring systems has been significantly expanded. After discussion of characteristics, a few factors related to each one of these particular group are described. Finally, concluded with the remark.

Keywords Wireless sensor nodes · Wireless sensor networks

L. Prusty · N. Mishra
Computer Science Department, ABIT Engineering College, Cuttack, Odisha, India

S. Satapathy (✉)
Center For Cybersecurity, Siksha 'O' Anusandhan Deemed to be University, Bhubaneswar, Odisha, India
e-mail: suneetasatapathy@soa.ac.in

S. Mahapatra
Pranveer Singh Institutes of Technology, Kanpur, Uttar Pradesh, India
e-mail: satyasundara.mahapatra@psit.ac.in

P. K. Tripathy
Kalam Institute of Technology, Berhampur, Odisha, India

A. Rocha et al. (eds.), *Integrating Advanced Technologies for Enhanced Security and Efficiency*, Engineering Cyber-Physical Systems and Critical Infrastructures 14,
https://doi.org/10.1007/978-3-031-91798-1_18

1 Introduction

A wireless sensor network (WSN) is a geographically group of scattered sensor nodes that are connected by wireless communications [1]. It consists of number of sensor nodes working together to monitor a region for obtaining data about the environment. Sensor node is an electronic device that measures a physical quantity. It comprised of an analog-to-digital converter (ADC), a cpu, storage unit, transceiver module, one or more sensors, and a power source that is shown in Fig. 1. A mobilization unit and positioning unit are optional. A sensor node is used to measure the change in the surrounding current condition. These measurements into electric signals that are related transformed by the ADC unit that is processed through the node processor. The sensor nodes can send the data wirelessly generated by its cpu to many nodes through its transceiver known as the Base station.

According to the Fig. 2, the base station can exercise supervisory control over the WSN by using data sent to itself and pertinent data for either other networks or human users [2]. The architecture of wireless sensor network consists of three levels: application layer, data connection layer and physical layer. To efficiently transmit and manage data WSN typically employ a hierarchical architecture. The architecture of wireless sensor communication is made up of multiple layers each of which having a specific purpose.

- Sensor Nodes: The sensor nodes are the lowest architecture level. To gather information from their surrounding sensor nodes are small, low power gadgets with sensors. Sensor nodes handle data processing, transmission and acquisition.
- Data aggregation: The sensor nodes send the collected data to the next layer known as data combination layer. In order to reduce redundancy and save energy

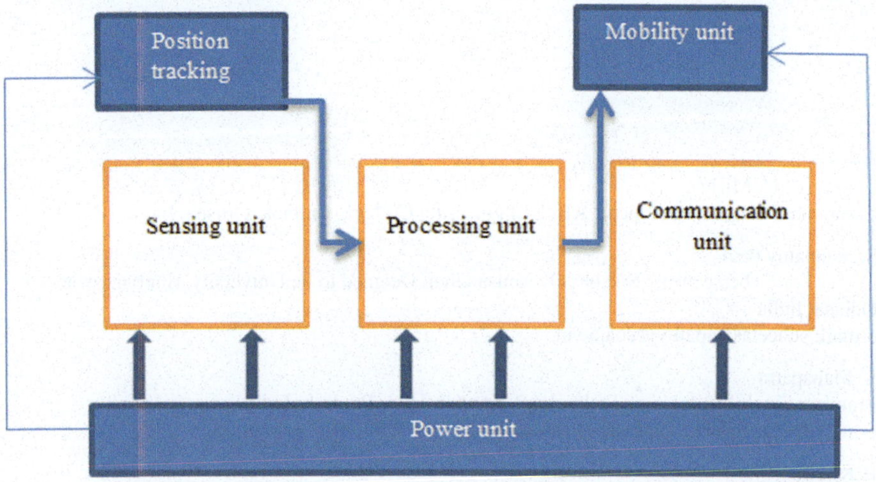

Fig. 1 The standard design of a sensor node that are utilized in Wireless Sensor Networks (WSNs)

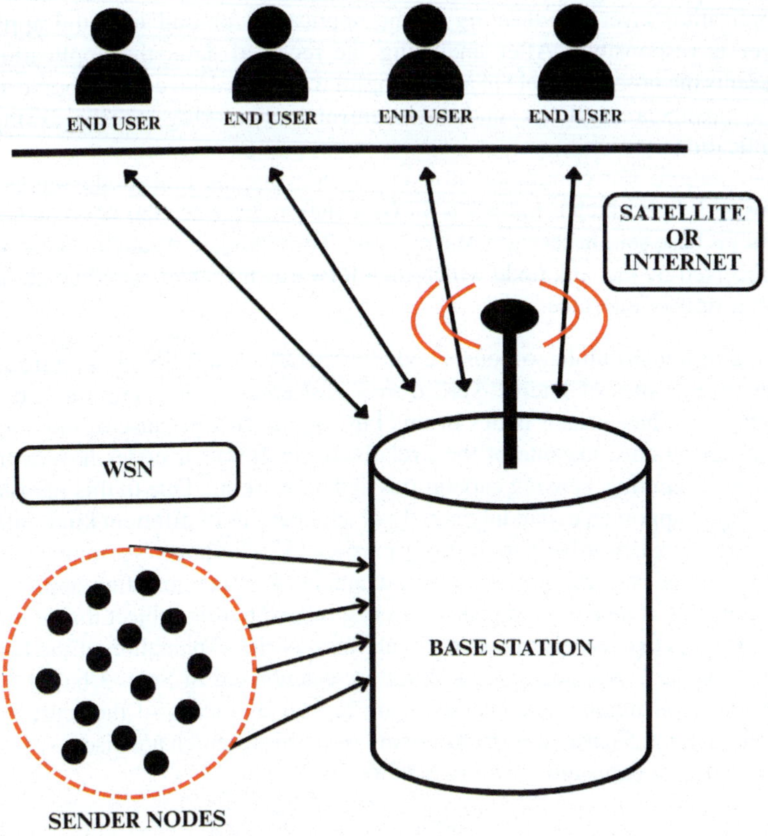

Fig. 2 The typical architecture of a WSN

this layer is in charge of gathering and aggregating data from several sensor nodes. Aggregation technique may involve data fusion, compression or summarization.

- Network layer: Data routing from sensor nodes (source) to base station (destination) is handled by the network layer. By managing network resources and selecting the best routing paths for data transmission it ensures efficient. At this layer various methods for routing in wireless sensor network including LEACH (Low-energy adaptive clustering hierarchy), SPIN (sensor protocol for information via negotiation) and AODV (Ad Hoc On-Demand Distance Vector Routing Protocol) can be used.
- Transport layer: The transport layer offers dependable from beginning to conclusion communication between the source and destination. Retransmission, error detection and data integrity are ensured by this layer as required. To facilitate using protocols at the transport layer dependable data transfer are used like TCP and UDP.

- Application layer: Connecting to the application or end user the application layer is responsible. After analyzing the received data, the application level presents the processed data in a meaningful way. Industrial control, home automation, healthcare systems, and environmental monitoring are the examples of applications.
- Base station: For communication in the network the sink node serves as the network's hub which is known as the base station. Between the WSN and external systems base station serves as the gateway. The aggregated data from the network is received by the sink node which then forwards it to the desired location like a server or user interface.

Using sufficient number of sensor nodes cooperatively a WSN can acquire ambient data at several sites of interest spread over vast areas at the same time. It is now possible to produce sensor nodes of this kind at low cost because of ongoing technical advancements. Despite of their relatively small sensor nodes have extremely advanced computing, sensing capabilities and networking. Due to this reason WSN now-a-days support an expanding variety of applications of different kinds, although these were primarily utilized for military purpose [3]. This paper's primary objectives is to investigate research representative example of various existing application of WSN and look at numerous topics of interest related to this subject that are utilized both novel and extensively. This article narrates a brief explanation of the literature survey in Sect. 2. Applications of WSN are categorized in Sect. 3 based on their nature and their unique characteristics are looked by means of the representative example. In Sect. 5 particular issues arising from this research are discussed. Finally some closing thoughts are drawn in Sect. 6.

2 Literature Survey

A systematic literature review methodology is a comprehensive approach which is used to undertake an extensively comprehensive and methodical review of relevant and on-point papers on a particular research problem. To making sure the review procedure is open and duplicable the protocol of systematic literature review is essential. The main aim of doing a systematic literature review is to identify, evaluate and comprehend diverse sources of information that will guide the review process. Starting with the final selection of relevant research papers the analysis of data gathered will then be used in the research process.

In the first reference paper [4] of Warneke, B.; Last, M.; Liebowitz, B.; Pister, the fundamental goal was to explore the limitation of micro fabrication technology and for the sensor systems somewhat arbitrary size is chosen. Smart dust will facilitate innovative methods of interacting with the environment due to its discrete size, significant functionality, connectivity, and expected low cost will be enabled and from more locations in a less invasive way Đuriši´c, M.P.; Tafa, Z.; Dimi´c, G.;

Milutinovi´c, V [5] had classified military applications of WSNs based on to operation scenarios and sensor types that defines the main classes. For future generation of military WSN applications research and engineering issues are also discussed. Use of wireless sensor network was investigated in Bokareva, T.; Gordon, N.; Kanhere, S.; Hu, W.; Bessell, T.; Ristic, B.; Jha [6] for ground surveliance. The main aim is to develop a prototype of WSN for outdoor deployment. Towle, J.P.; Herold, D.; Johnson, R.; Vincent [7] had created a small, inexpensive, short range, dual mode acoustic sensor to solve the issue of loud and reverberant acoustic environment. The main goal of the paper [8] was to localize and detect transitory signals by suspending acoustic sensor arrays beneath tethered aerostats from small arms fire, artillery, and mortars whose author is Scanlon, M.; Rei, C.; Solomon, L. Aerostat. The article of author De Bree, H.E.; Wind, J.W [9] had shown results of field testing using heavy artillery, fixed wing aircraft and helicopters. In the reference paper of Lim, H.B.; Ma, D.; Wang, B.; Kalbarczyk, Z.; Iyer, R.K.; Watkin, K.L. Iyer, R.K.; [10] outlined a continuous endeavors for developing a network of linked BSNs for soldier's health surveillance in real time. An application scenario and the background for the project are also discussed. The report of the paper Naz, P.; Hengy, S.; Hamery [11] had defined the usual noises presents measurements made in the field and in an anechoic room produced by troop actions where speech, footstep and a certain impulsive noise can be heard at different separation from the source. In the reference paper [12] main goal of Rippin, B. Pearls was to enable a variety of monitoring applications possible through the implementation of a several number of inexpensive, tiny wireless sensors. In the reference paper [13] Hii, P.; Chung,W had checked and helped out-patients with the medication administration procedure by employing as a barcode the mobile phone decoder and medicine care is taken into account. The main aim of this monitoring plan extension is to provide a better and more complete health care service.

In the paper [14] Anliker, U.; Ward, A.; Lukowicz, P.; Ward, A.; Troster, G.; Dolveck, F.; Troster, G.; Baer, M.; Vuskovic, M. AMON had described the medical wearable's surveillance and alarm system was designed for extreme risk cardiac and respiratory patients which is known as the advance care and alarm portable telemedical monitor (AMON). The system includes a cellular link of various vital signs to a medical facility sophisticated multi-parameter medical emergency detection and continuous collections of assessments. In the paper [15] Dobrescu, R.; Dobrescu, M.; Popescu, D.; Nicolae presented the platform that the study offers allows an internet-based remote connection between supported individuals and their surroundings. To receive the most recent medical care the main purpose of Kakria, P.; Tripathi, N.; Kitipawang, P. [16] is to enable remote cardiac patients which may not be feasible otherwise because of the low doctor-to-patient ratio (Table 1).

Table 1 Application required for literature review of survey on Wireless Sensor Networks (WSNs)

Author	Implementation	Result
Warneke, B.; Liebowitz, B.; Pister Last, M.;	To explore the limitation of micro fabrication technology and for the sensor systems somewhat arbitrary size is chosen	Smart dust will facilitate innovative methods of interacting with the environment due to its discrete size, significant functionality, connectivity, and expected low cost will be enabled and from more locations in a less invasive way
Đuriši´c, M.P.; Tafa, Z.; Dimi´c, G.; Milutinovi´c, V.	According to operation scenarios and sensor types military applications of WSNs had classified that describe the key classes	For future generation of military WSN applications research and engineering issues are also discussed
Bokareva, T.; Hu, W.; Kanhere, S.; Ristic, B.; Gordon, N.; Bessell, T.; Jha	Investigated wireless sensor network for ground surveliance	The main aim is to develop a prototype of WSN for outdoor deployment
Towle, J.P.; Vincent; Herold, D.; Johnson, R.;	Had developed a small, low cost, low range, acoustic sensor with two modes	To address the issue of noisy and reverberant acoustic environment
Scanlon, M.; Rei , C.; Solomon, L. Aerostat	To localize and detect transitory signals by suspending acoustic sensor arrays beneath tethered aerostats from small arms fire, artillery, and mortars	Transitory signals are detected
De Bree, H.E.; Wind, J.W	Had represented field test results using heavy artillery, acoustic vector sensors, measuring guns, helicopters and fixed wing aircraft	Test results are shown regarding measuring rifles, acoustic vector sensors etc
Lim, H.B.; Ma, D.; Wang, B.; Kalbarczyk, Z.; Iyer, R.K.; Watkin, K.L.	To develop a system described an on-going effort utilized networked BSNs to check soldiers health in real time	The background and an application scenario for the project are also discussed
Naz, P.; Hengy,S.; Hamery	The usual noises presents measurements are defined that are made in the field and in an anechoic room produced by troop actions where	At various distances from the source speech, Footstep and some specific impulsive sounds are produced
Rippin, B. Pearls	To enable a wide range of monitoring applications possible through the deployment of a large number of inexpensive, tiny wireless sensors	Vast range of monitoring applications were enabled

(continued)

Table 1 (continued)

Author	Implementation	Result
Hii,P.; Chung,W	Had checked and helped out-patients with the medication administration procedure is taken into account by employing a mobile device to decode barcode and provide medical care	Here the main goal is to offer as a continuation of the monitoring program improved and more through medical care
Anliker, U.; Baer, M.; Ward, A.; Lukowicz, P.; Troster, G.; Dolveck, F.; Vuskovic, M. AMON	Described the wearable medical monitoring	Alarm system for high danger cardiac and respiratory patients was designed that is known as the advance care and alarm portable telemedical monitor (AMON)
Dobrescu, R.; Dobrescu, M.; Nicolae ;Popescu, D.;	The platform was presented that the study offers an internet-based remote connection between supported individuals and their surroundings.	The study of internet-based remote connection was presented
Kakria, P.; Tripathi, N.; Kitipawang, P.	To enable remote cardiac patients which may not be feasible otherwise because of the low doctor-to-patient ratio	Remote cardiac patients are enabled

3 Application of WSN

Various applications of WSN are either in their infancy stages of development or are already in widespread use. According to the way they are use, applications of WSN are classified into six main categories in this paper that as shown in Fig. 3, namely are: military, environmental, health, flora and fauna, industrial and urban.

Numerous subcategories are considered in every category. The characteristics of every divisions and subcategories are described in this section. Furthermore, through the indicative evaluation of representative examples, their advantages and disadvantages are identified along with their unique characteristics are also discussed. Moreover, either for processing or sensing purposes different approaches and technical tools are used in these applications is discussed. The differences and similarities existing among them are also identified below.

3.1 Military Application

Not just the military field the first sector of activity of human to use WSNs but also the sensor network research is motivated to have started because of the military.

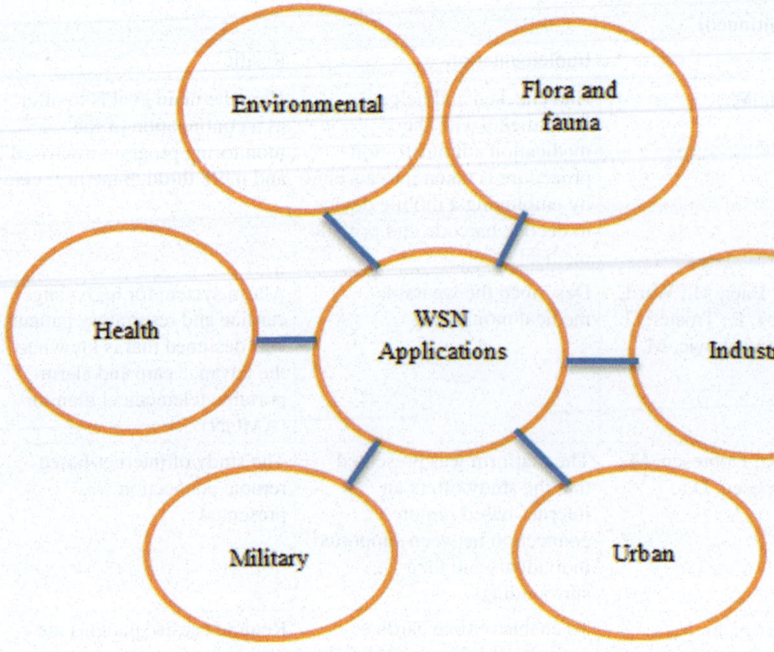

Fig. 3. An outline of the most common application categories of WSNs

These preliminary studies were conducted in the late 1990s to construct sensor nodes capable of performing espionage actions in spite of their extremely small size, are typified by Smart Dust.

Since then, technology has advanced to the point where WSNs can support a range of activities. The major subcategories of WSNs' military applications combat monitoring, battleground surveillance and intruder detection as well as the most frequently utilized sensor types in each are shown in the figure (Fig. 4).

Toxic industrial material (TIM) and chemical, biological, radiological, nuclear, and explosive (CBRNE) sensors, in particular, can be used to detect these types of materials. WSNs may employ vibration, auditory, photoelectric, infrared, and laser sensors to identify intrusions. LADAR and infrared sensors are also used in imaging. Furthermore, the structural adaptability of WSNs allows them to adjust to different needs. For instance, large- scale WSNs with thousands of not being manually deployed nodes are used in warfare operations. In force protection and urban warfare operations, WSNs are composed hundreds of nodes that are manually deployed. All WSN deployment techniques and scales are utilized in non-conflict missions. The next subsection examines WSNs designed for combat monitoring, intruder detection, and battlefield Surveillance.

- Battlefield surveillance

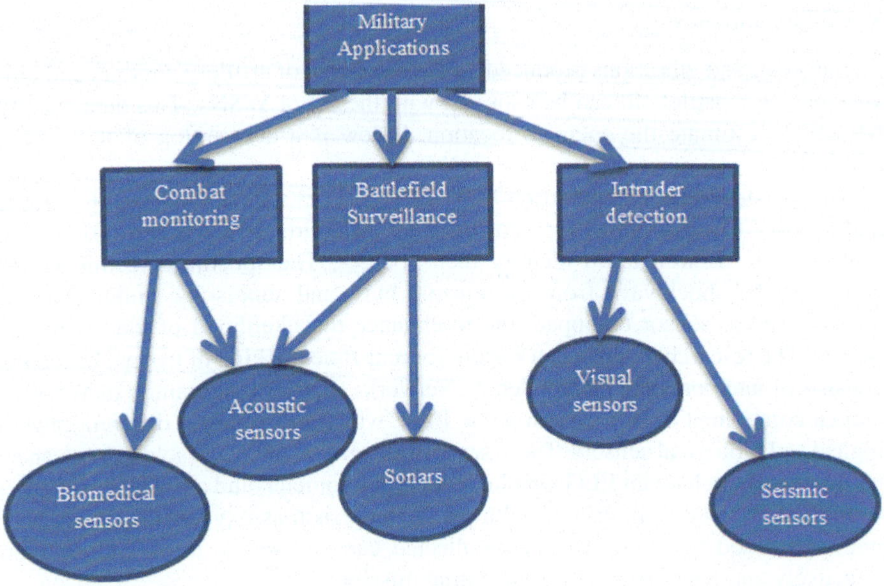

Fig. 4 An overview of most common WSN application categories

In these types of applications, WSN sensor nodes can be deployed on a battlefield near potential routes used by enemy forces. The primary benefit is that the WSN can operate without constant monitoring or maintenance, and it also has the capability of being deployed autonomously. Most of the time, the battlefield's topography is entirely unpredictable, which is crucial for both the sensor nodes' coverage and their energy usage. A few of these applications are presented below. This study examines the application of WSN technology to ground surveillance. The Researchers specifically suggest a system composed of low cost common nodes that can detect both magnetic and audio signals emitted by different target objects. The system utilizes spatial variation in signal strength captured by the sensors to detect and classify different targets, including troop movements and vehicles. The figure depicts the architecture of this system.

For Anti-Submarine Warfare (ASW), a submarine detection system is demonstrated. The system comprises low-cost multi sensing modules that integrate both passive and active sonar technology. The system is capable of expanded to accommodate a lot of sensors placed in littoral zone of seas based on a certain pattern and ocean depth. These sensors use active sonar to verify their objectives and passive sonar to identify a diesel-electric submarine. When a target is confirmed, the unit emits an alarm signal containing its ID code to alert nearby detecting units. An alert is triggered whenever several alarm signals are transmitted within a predetermined period. Additionally, the system can achieve a low False Alarm Rate (FAR) due to the active sonar's extremely short range (50 meters), which addresses the acoustic multipath issue commonly encountered with conventional sonobuoys.

- Combat monitoring

Sound, heat, and vibrations produced by the discharge of artillery, mortars and other weapons on a battlefield can be captured with the aid of WSNs. This data can then be used to estimate the enemy's location. Below is a description of this type of application.

This paper presents using arrays of acoustic sensors suspended beneath to locate and identify tethered aerostats used for small arms fire, artillery, and mortar transient signals, as well as to identify their sources. The specific detecting system combines the shockwave from the muzzle blast and supersonic bullet from the acoustic vector sensor equipped gun to enhance the likelihood of identifying the danger. The researchers present a system for real-time tracking of troops' health that consists of interconnected Body Sensor Networks (BSNs), a subfamily of WSNs. A crucial component of this system is the BSN, which incorporates different physiological and biological sensors. These sensors, integrated into each soldier's advanced battle helmet, include an EEG simulator, an accelerometer, and a SpO_2 sensor. They continuously monitor multiple health parameters, such as oxygen saturation, heart rate, and blood pressure. The data collected can be used to train soldiers more effectively and appropriately for upcoming missions.

- Intrude detection

One of the most crucial during military operations pieces of information is considered to be the enemy's position. Knowing this can put a side ahead in a conflict and closer to victory. WSNs enable the early detection of invaders on a battle field, thereby the loss of territory is prevented. An intruder can be identified in different ways. Researchers at [17] developed a system that uses unsupervised Seismic Sensors and acoustic to detect intruders by recording vibrations and sounds associated with regular troop operations. These sensors have been employed to measure and record common military activities such as walking, loading rifles, handling equipment, and other behaviors, both in controlled environments and in the field. The data produced by the network used to find an intruder in challenging terrain with dense foliage where visual detection is limited, even though the detection distance may be minimal.

Another use case for WSNs is intrusion detection. The system described by the researchers employs small, inexpensive, and energy-efficient wireless sensors, which are densely deployed throughout the area. This system integrates tracking and optical sensors to reduce the false alarm rate, along with seismic and acoustic sensors to the user provide detection information [18–27].

3.2 Environmental Application

WSNs can be utilized to improve environmental applications that require continuous monitoring of environmental parameters in hostile and isolated areas. The principal subcategories of WSNs' environmental applications air monitoring, water

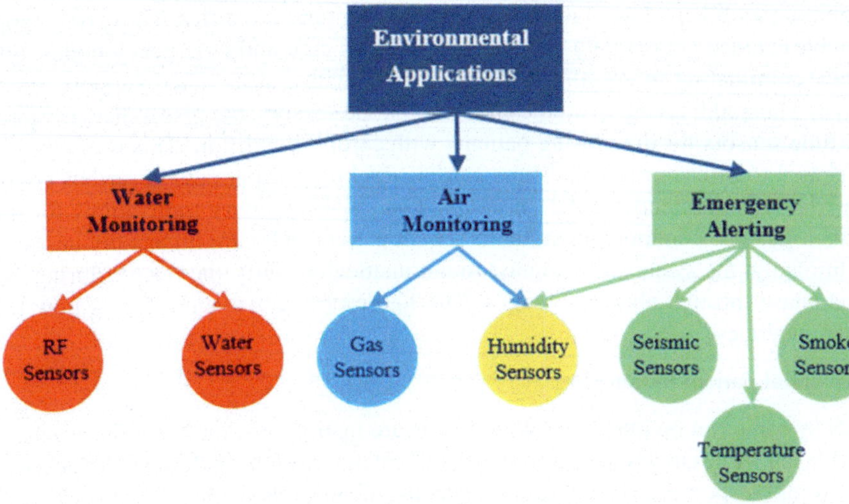

Fig. 5 The types of sensors used in environmental WSN applications and heir subcategories

monitoring, and emergency alerting as well as the types of sensors mostly used in the applications are illustrated in the figure. The next paragraph explores WSNs developed for these environmental applications (Fig. 5).

- wearable monitoring for Patient

Applications of health monitoring can be integrated hardware that can be worn that incorporates biomedical sensors to provide information on a patient's condition, whether in a medical facility or at a remote location. A healthcare technology that provides personalized medication care support combines a barcode system with real-time sensors integrated into cellphones. This established technique facilitates real-time electrocardiogram (ECG) monitoring. Additionally, real-time sensor monitoring can enable the tracking of blood pressure, blood glucose, and various diagnostic tests. The developed system is depicted in the figure.

The Alert Portable Tele-Medical Monitor System (AMON) aims to offer ongoing observation for high-danger cardiac and patients with respiratory condition. The system connects to a cellular telemedicine center (TMC), collects multiple vital indicators, and identifies multi-parameter medical conditions. This method uses a gadget-worn device to track patients' vital signs, offering a comprehensive view of their health.

- Home assisting system

The researchers developed a homecare monitoring tool that provides individuals and their environments with internet- based remote access. This platform employs wearable or even surgically implanted biosensors in real time to monitor and diagnose patients within their home environment. A hybrid platform approach is suggested, combining mobile and fixed nodes to offer varying functionalities. These tiers range from basic aided person data collection to primary care and emergency nodes, ultimately culminating in coordination and contact with a designated support center, such as a hospital. Using wearable and smartphone sensors, the researchers explored real-time sensors for diagnosing patients with cardiac condition. This program was developed to offer remote, real-time cardiac tracking for patients in serious condition who cannot attend regular checkups. One of the primary motivations for this technology is its commercialization to benefit patients who are financially unstable. Additionally, the system is designed to automatically notify the doctor during any crisis phase through alarm messages. The figure provides a summary of how this system operates.

- Hospital patient monitoring

WSN systems can be integrated with healthcare institutions, including hospitals, to provide real-time emergency alerting and patient monitoring, enabling more accurate and prompt responses. The researchers specifically described where wireless sensors are installed a WSN application within the emergency departments of Johns Hopkins Hospital to monitor patients' blood oxygen levels and heart rates and in real time. Despite challenges posed by radio noise and interference in hospital environments, the studies gathered network performance data and found that the use of WSNs can enhance the operational efficiency of healthcare facilities.

3.3 Health Application

Within the healthcare industry, WSNs uses sophisticated medical sensors for patient monitoring. They also employ wearable devices to track patients' vital signs in real time. The primary subgroups of WSNs' health applications for patient wearable monitoring, hospital patient monitoring and home assistance systems as well as the most frequently used sensor types in each are illustrated in the figure. The next paragraph examines WSNs created for these types of health applications (Fig. 6).

- Water monitoring

Since water—whether marine or potable—is vital to human existence, experts show considerable interest in water monitoring, as will be detailed below. Researchers developed application of WSN to assess the quality of freshly drawn, drinking water. They created the Pipe Sense Cyber Physical System (CPS) a water tracking device systems in the pipe that uses RFID (Radio Frequency Identification) based WSN. The network is able to collect data on water quality, demand and different repair-related issues such as leaks in pipes or weak areas. Information from the in-pipe RFID sensors is sent to data servers and processed using algorithms to aid in decision- making. A WSN application for tracking the maritime environment involves an underwater wireless sensor network (UWSN) designed to protect fish farm plants and animals

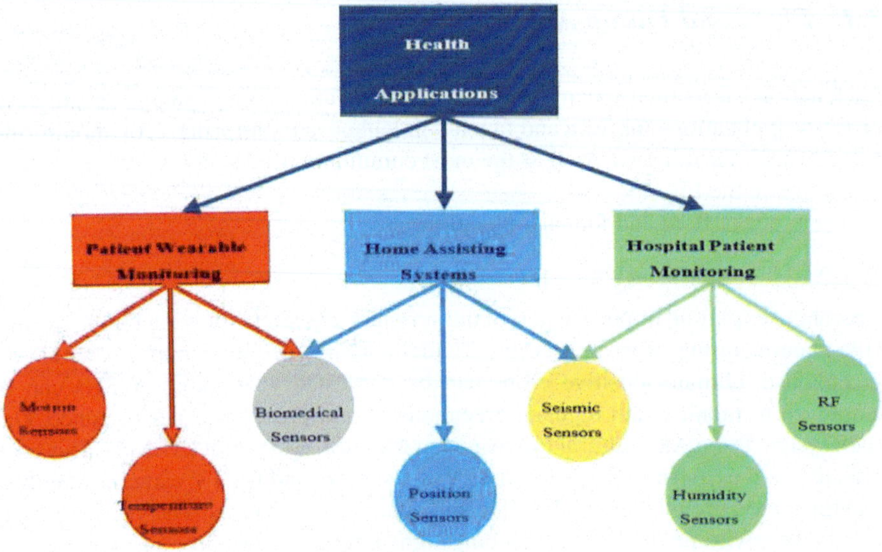

Fig. 6 Overview of WSN's health applications and the sensors that are used in them

from the detrimental effects of feed and feces. Researchers developed this UWSN with wireless sensor nodes that can monitor pollution levels within the farm. To cover a larger area, the sensor nodes are movable within a restricted zone.

- Air monitoring

Air is essential to human life, and many modern human activities contribute to the current state of air pollution. WSNs can be used in populated areas to track air quality to prevent harmful illnesses and contaminations that could jeopardize people's health. A typical example of an air quality monitoring application supported by WSNs is the WSN-AQMS. The system measures air quality parameters such as CO, ozone and NO_2 by integrating gas sensors with Libelium wafers. The Wasp motes use the Zigbee protocol for exchange of data and real-time air quality monitoring. To assist the operation of this system, the researchers also developed the air sensor network clustering protocol (CPAS).

- Emergency alerting

Proactively tracking the reasons behind natural disasters can help prevent them or at least mitigate their impact. WSNs can be employed for proactive alerting to reduce damage or potentially avert disasters by tracking common catastrophic reasons in real time. The remainder of this subsection presents typical examples related to monitoring volcanic activity, tsunamis, seismic activity and forest fires.

3.4 Flora and Fauna Application

Each nation has its own specific areas of flora and fauna. The primary subcategories of WSN applications for flora and fauna—greenhouse monitoring, crop monitoring, and livestock farming as well as the most commonly used sensor types in each are illustrated in the Fig. 7.

The types of flora and fauna applications are

- Green house monitoring

Greenhouses play an important part of the agricultural sector, allowing for the sustainable manufacturing of various crops. By maintaining specific conditions within the greenhouse, climate-sensitive crops can be harvested year-round. WSNs can be employed to monitor and regulate greenhouses to improve their performance. One such system is the Agricultural Environment Monitoring System (AEMS). This cost-effective and user-friendly system uses WSN sensors and CCTV cameras to collect and track crop growth data both inside and outside greenhouses. The system gathers important environmental data including light intensity, humidity and temperature and amount of rainfall, pH, air pressure, and electrical conductivity (EC).

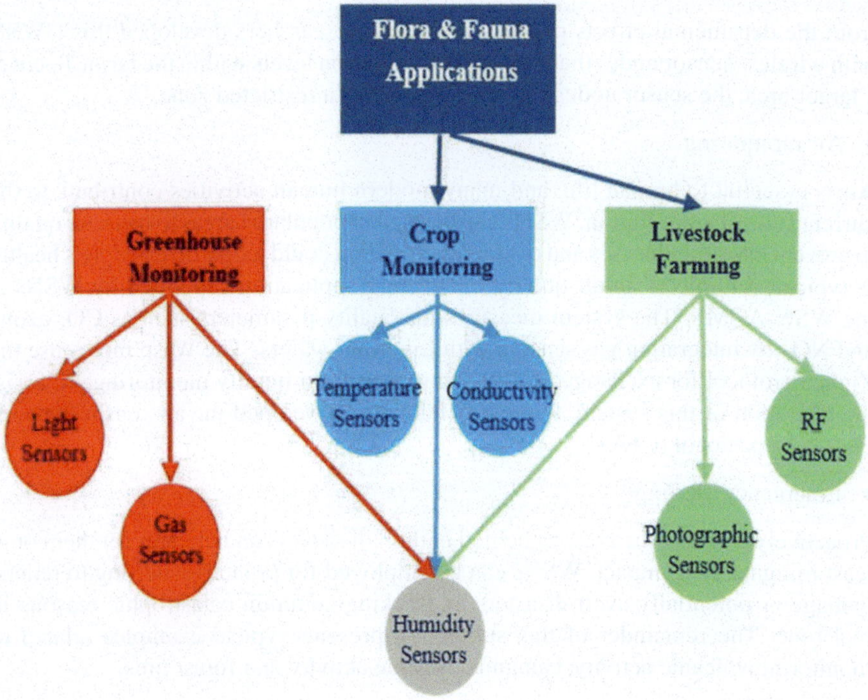

Fig. 7 The type of sensors that are employed in flora and fauna as well as the subcategories of WSN applications

- Crop monitoring

Crop preservation is a vital aspect of agriculture, and various WSN applications can enhance the crop environment.

Among the planned uses are crop irrigation and fertilization. One such system is called hydro Bus. It is an automatic irrigation system designed for remote management, applied to a large tract of land in Jumilla, Murcia. The region of deployment area is divided in to seven sub-regions by the system, each with its own sector of control for oversight and management. The control sectors exchange information with one another and the central controller through a WLAN network. The automated fertilizer applicator system is composed of three components: the input, decision support, and output modules.

- Livestock farming

In the realm of fauna, livestock farming is a major industry. WSNs can be utilized for various functions, including monitoring livestock. Below are a few examples of how these applications are employed.

A sensor-based system was proposed to assess livestock behavior and support farming operations. The system uses camera sensors to measure grass growth, thereby attracting animals to green pastures where cattle graze. The primary goal of the study was to develop durable technology for outdoor use that simulates animal behavior, both individually and in groups. Despite the use of specially designed sensors to track behaviors such as grazing, sleeping and challenges remain. These include radio attenuation caused by the animal's body and movement. A low-cost, low-power collar is proposed to assist with cattle farming. The system utilizes two antennas positioned to maximize collar radio coverage and a solar power relay router. Additionally, a novel protocol called the Implicit Routing Protocol (IRP) was introduced to address packet losses brought on by the animals' movement.

The researchers propose a method for modern large-scale farm management systems that leverages RFID and WSN technology. The system can recognize and track all farm animals, including pigs, while also monitoring their health and the surrounding environment. According to the researchers, this approach can be utilized to identify environmental contamination, prevent infections, and provide insights into food safety concerns.

3.5 Industrial Application

WSNs can be applied to address a wide range of issues across various industrial applications. The primary subcategories of WSNs' industrial applications—logistics, robotics, and machinery health monitoring—are illustrated in the Fig. 8.

The types of industrial application are as follows

- Logistic

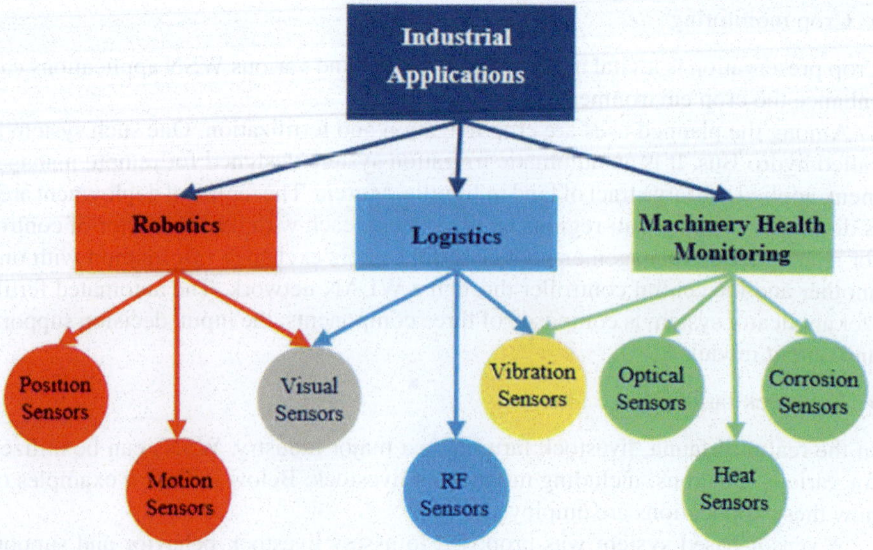

Fig. 8 The subcategories of kind of sensor that are employed in industrial WSN applications

Since the logistics systems require real-time monitoring of many environmental characteristics and enhanced package handling, the logistics domain is a key area where WSNs can be effectively deployed. Integrating WSNs with logistics systems addresses these needs. The following descriptions provide details on a few applications types. The transportation logistics industry requires both low delivery costs and high quality. A WSN-based system has been developed and implemented to monitor temperature and humidity levels within cargo containers being transported by trucks and transatlantic cargo ships. The figure illustrates the basic concept for deploying this monitoring system. It has been shown that implementing such a system can reduce costs by minimizing transportation-related losses and enhance quality through improved supervision.

- Robotics

Robots and WSNs are increasingly being combined in various applications. Cooperation between robots and WSNs can address several key issues, such as roaming salespeople, node redeployment, and sensor node mobility. Typical applications of these types are illustrated below. A suggested robotic navigation technique gives direction for the robot's travel. This method utilizes a wireless sensor network (WSN) to generate thorough maps of its sensing regions. In particular, every sensor creates a map that is based on its observable location. These individual sensor maps are integrated to form a comprehensive, large-scale map. The sensors identify areas of interest for the robot to visit once the direction is prepared. Maintaining a wireless sensor network (WSN) can eliminate dangerous item and update the direction. In R3S2, robots navigate through a network structured as a virtual grid. The robot targets grid points that have not been visited for the longest time to identify gaps in the network's sensing coverage. When a gap is found, the robot locates a node and relocates it to the uncovered area. Additionally, if redundant sensors are found, the robot moves them to expand the coverage area.

- Machinery health monitoring

The purpose of this is to assess the functionality of different kinds of technical equipment and to identify potential issues that could affect the equipment's operational efficiency. A WSN designed for this purpose evaluates energy consumption and monitors the condition of electric machines. It utilizes wireless nodes to non-intrusively monitor motor terminal such as line voltages, temperature and line currents without disrupting the way the electric machines works. This approach allows for the estimation of motor efficiency and the assessment of the health condition of the equipment. REMONG, a WSN-based system for monitoring oil and gas pipelines, is proposed. This technology employs wireless sensor nodes to track the temperature and pressure of the fluid at multiple locations along the pipes, which are distributed over extensive geographic areas. The system is designed to detect leaks by continuously monitoring these parameters. To enhance safety in industrial machinery, a WSN integrated with a connected trailer and a primary vehicle is suggested.

3.6 Urban Application

The range of sensing capabilities provided by WSNs allows for the acquisition of unprecedented amounts of information about various target areas—whether they are outdoor spaces, buildings, or rooms. WSNs enable a wide array of applications for measuring both temporal and geographical characteristics of phenomena occurring in urban environments. As illustrated in Figure and detailed in this section, some of the most common applications of WSNs in urban settings include:

1. Smart Homes: Enhancing living environments through intelligent monitoring and automation.
2. Smart Cities: Integrating WSNs for urban management and infrastructure optimization.
3. Transportation Systems: Improving traffic management and public transport efficiency.
4. Structural Health Monitoring: Assessing the integrity of buildings and infrastructure.

These applications leverage WSNs to improve quality of life and operational efficiency in urban areas (Fig. 9).

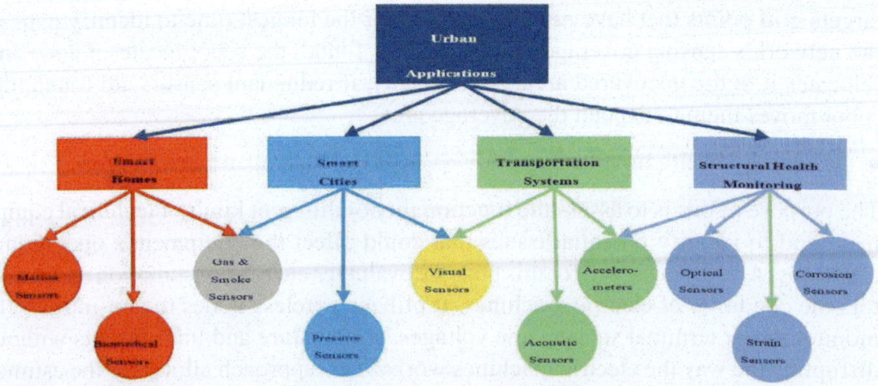

Fig. 9 The types of sensors used in urban WSN applications and their classifications

4 Discussions

For a growing number of applications there is no doubt that WSNs have exceptional qualities that make them perfect. On the other hand there are significant issues with operations of WSN. Some of the issues depend on the applications.

For example depending on the applications in military settings the weight and physical requirements for nodes vary. The sensor nodes must be incredibly small in certain surveillance application to remain undetectable. But the nodes weight and physical dimensions in many other military applications are not considered as significant limitations. However the nodes in these applications ought to have a sufficiently wide range of communication while the region that needs to be covered in several square. In health applications the nodes must have the smallest possible weight and physical dimensions, particularly if they are wearable.

In applications of flora and fauna the restrictions on the node's weight and physical dimensions vary depending on the applications. For example nodes placed beneth the skin of animals in livestock farming requires WSNs to be as little as feasible while in agricultural applications such restrictions are typically not sent. To withstand environmental conditions animals and nodes in applications should be strong enough. Similarly to endure a certain number of node losses WSN should be resilient enough. In addition there are some cases where a longer communication range of nodes is required.

In environmental monitoring to be the primary requirements the weight and size of the nodes are not seen. Rather to withstand harsh environmental conditions the nodes building process needs to be incredibly robust. The covered area and the communication range additionally both should be sufficiently large. Moreover, the WSN need to be able to tolerate loss of a specific number of nodes. In applications of industries the majority of operations are time-sensitive. In the settings of industrial the interference caused by electromechanical is remarkably prevalent. As a result to interference and jamming, communication should have the highest possible throughput, dependability and immunity (Table 2).

Wireless sensor network are quit vulnerable to security threats that are decentralized and have limited resources. Security is needed to applications of WSN and key points to think about are as follows:

1. Black hole attacks—Data packets are captured by a node and then dropped rather than forwarded to their destined location. It's possible that crucial environmental monitoring data won't make it to the base station, which would result in an incorrect or incomplete environmental estimation.
2. Replay Attacks—An attacker retransmits previously recorded packets, creating hurdles to comprehend specific environmental data sets. This is due to the availability of misleading information resulting from the use of outdated data to make environmental decisions.
3. Node Capture Attacks—The physical capture of the nodes may further cause improper access to data, disturb the process of the collection of data, or manipulate the system. Moreover, compromised nodes may disclose important environmental data, inject spurious data, or stop the duties of environmental monitoring.
4. Sybil Attacks—In a Sybil assault, the network is attacked by a single rogue node that provides multiple false identities to disrupt data aggregation, routing, and other network activities. It will gradually corrupt the obtained environmental data, potentially leading to invalid conclusions or responses from the monitoring system.
5. DoS (Denial of Service) Attacks—This might overload the network with meaningless traffic, drain the battery life of sensor nodes, or simply stop the medium by deleting communication channels. Sensor nodes may become unresponsive, or the entire network may fail, resulting in gaps in data collection owing to environmental monitoring.
6. Data Integrity Attacks—Malicious nodes can change the data broadcast between nodes or to the base station. Changes in environmental data can lead to improper analysis and choices. For example, an inaccurately estimated pollution rate or water quality rating can be misleading.

5 Conclusions

For a number of areas of human activity use of WSN already offers notable benefits. Due to the ongoing advancements in technology to increase the capabilities and cost of production of sensor node will continue. Therefore the spectrum of WSN application will increase. In this article through the examination of matching typical instance, both new and famous ones the use of WSNs in the military, health, industrial environment, flora and fauna, and urban sectors was investigated. The adoption of WSNs not only offers several benefits in particular fields, but also opens up new

Table 2 Requirements of WSN vary depending on type of application

Types of application	Node weight and dimensions	Node robustness	Communication range	Communication throughput	Communication reliability	Communication security	Network tolerance
Military	Application dependent	Extremely high	Wide	Extremely high	Extremely high	Extremely high	Extremely high
Health	Small	High	Small	Extremely high	Extremely high	High	High
Flora and fauna	Application dependent	High	Wide	Medium	Medium	Low	High
Environmental	Of minor importance	Extremely high	Wide	Extremely high	High	High	Extremely high
Industrial	Application dependent	Extremely high	Application dependent	Extremely high	Extremely high	High	Extremely high
Urban (Indoor)	Small	Medium	Small	Medium	Extremely high	Extremely high	High
Urban (Outdoor)	Of minor importance	Extremely high	Wide	Extremely high	Extremely high	Extremely high	Extremely high

possibilities is shown by this analysis when compared to the relative means and methods that were previously employed.

The combined use of relevant approaches and technologies will help both the development of new applications as well as the improvement of current ones. Conversely some issues that prevent the use of WSNs such as energy scarcity, traffic, lost connectivity, inadequate coverage, low quality of service, and weak safety will continue to be the focus of scholarly investigation.

References

1. Akyildiz IF, Su W, Sankarasubramaniam Y, Cayirci E (2002) Wireless sensor networks: a survey. Comput Netw 38:399–422
2. Yick J, Mukherjee B, Ghosal D (2008) Wireless sensor network survey. Comput Netw 52:2292–2330
3. Arampatzis T, Lygeros J, Manesis S (2005) A survey of applications of wireless sensors and wireless sensor networks. In: Proceedings of the 2005 IEEE international symposium on, mediterrean conference on control and automation intelligent control. Limassol, Cyprus, p 719
4. Warneke B, Last M, Liebowitz B, Pister KS (2001) Smart dust: communicating with a cubic-millimeter computer. Computer 34:44–51
5. Đuriši´c MP, Tafa Z, Dimi´c G, Milutinovi´c V (2012) A survey of military applications of wireless sensor networks. In: Proceedings of the 2012 Mediterranean conference on embedded computing (MECO). Bar, Montenegro
6. Bokareva T, Hu W, Kanhere S, Ristic B, Gordon N, Bessell T, Jha S (2006) Wireless sensor networks for battlefield surveillance. In: Proceedings of the land warfare conference. Brisbane, Australia
7. Towle JP, Herold D, Johnson R, Vincent H (2020) Low-cost acoustic sensors for littoral anti-submarine warfare (ASW). In: Proceedings of the SPIE 6538, sensors, and command, control, communications, and intelligence (C3I) technologies for homeland security and homeland defense VI, 653814, Orlando, FL, USA, 4 May 2007. Appl Syst Innov 3, 14 21 of 24 8
8. Scanlon M, Rei C, Solomon L (2007) Aerostat acoustic payload for transient and helicopter detection. In: Proceedings of the SPIE 6538, sensors, and command, control, communications, and intelligence (C3I) technologies for homeland security and homeland defense VI, 65380H. Orlando, FL, USA
9. De Bree HE, Wind JW (2011) The acoustic vector sensor: a versatile battlefield acoustics sensor. In: Proceedings of the SPIE 8047, ground/air multisensor interoperability, integration, and networking for persistent ISR II, Orlando, FL, USA, 80470C
10. Lim HB, Ma D, Wang B, Kalbarczyk Z, Iyer RK, Watkin KL (2010) A soldier health monitoring system for military applications. In: Proceedings of the 2010 international conference on body sensor networks, Singapore, pp 246–249
11. Naz P, Hengy S, Hamery P (2012) Soldier detection using unattended acoustic and seismic sensors. In: Proceedings of the SPIE 8389, ground/air multisensor interoperability, integration, and networking for persistent ISR III, 83890T, Baltimore, MD, USA
12. Rippin B (2012) Pearls of wisdom: wireless networks of miniaturized unattended ground sensors. In: Proceedings of the SPIE 8388, unattended ground, sea, and air sensor technologies and applications XIV, 83880H, Baltimore, MD, USA
13. Hii P, Chung W (2011) A comprehensive ubiquitous healthcare solution on an android mobile device. Sensors 11:6799–6815
14. Anliker U, Ward A, Lukowicz P, Troster G, Dolveck F, Baer M, Vuskovic M (2004) AMON: a wearable multi parameter medical monitoring and alert system. IEEE Trans Inf Technol Biomed 8:415–427

15. Dobrescu R, Popescu D, Dobrescu M, Nicolae M (2010) Integration of WSN-based platform in a homecare monitoring system. In: Proceedings of the 4th international conference on communications & information technology (CIT'10), Corfu Island, Greece, pp 165–170
16. Kakria P, Tripathi N, Kitipawang P (2015) A real-time health monitoring system for remote cardiac patients using smartphone and wearable sensors. Int J Telemed Appl 2015:1–11
17. Ko J, Gao T, Terzis A (2009) Empirical study of a medical sensor application in an urban emergency department. In Proceedings of the 4th international ICST conference on body area networks, Los Angeles, CA, USA
18. Nasir A, Soong B-H, Rama chandran S (2010) Framework of WSN based human centric cyber physical in-pipe water monitoring system. In: Proceedings of the 2010 11th international conference on control automation robotics & vision, Singapore
19. Lloret J, Sendra S, Garcia M, Lloret, G (2011) Group-based underwater wireless sensor network for marine f ish farms. In: Proceedings of the 2011 IEEE GLOBECOM workshops (GC Wkshps), Houston, TX, USA, pp 115–119
20. Mansour S, Nasser N, Karim L, Ali A (2014) Wireless sensor network-based air quality monitoring system. In: Proceedings of the 2014 international conference on computing, networking and communications (ICNC), Honolulu, HI, USA, pp 545–550
21. Mallick C, Satpathy S (2018) Challenges and design goals of wireless sensor networks: a state-of-the-art review. Int J Comput Appl 179(28):42–47
22. Pattnaik LM, Swain PK, Satpathy S, Panda AN (2023) Cloud DDoS attack detection model with data fusion & machine learning classifiers. EAI Endorsed Trans Scalable Inf Syst 10(6)
23. Satpathy S, Swain PK, Mohanty SN, Basa SS (2024) Enhancing security: federated learning against man-in-the-middle threats with gradient boosting machines and LSTM. In: 2024 IEEE international conference on advanced video and signal based surveillance (AVSS). IEEE, pp 1–8
24. Potluri S, Mangla M, Satpathy S, Mohanty SN (2020) Detection and prevention mechanisms for ddos attack in cloud computing environment. In: 2020 11th international conference on computing, communication and networking technologies (ICCCNT). IEEE, pp 1–6
25. Satpathy S, Pradhan SK, Ray BNB (2010) A digital investigation tool based on data fusion in management of cyber security systems. Int J IT Knowl Manag 3(2):561–565. http://www.csjournals.com/IJITKM/PDF%203-1/77.pdf
26. Pattnaik L, Satpathy S, Paikaray BK, Swain PK (2024) DDoS analysis using machine learning: survey, issues, and future directions. Int J Bus Contin Risk Manag 14(1):57–76
27. Chatterjee S, Pattnaik LM, Satpathy S (2022) An intrusion detection system and attack intension used in network forensic exploration. In: International conference on intelligent systems and machine learning. Springer Nature Switzerland, Cham, pp 334–345

A Novel Optimized Classification Approach for Spam Message Cyber Attack Detection and Prevention Using Machine Learning Approach

Sarita Mohanty and Anupa Sinha

Abstract Cyber-attacks involve a malicious attack in cyberspace, such as phishing, spamming, and hacking, in order to commit an infraction. Computer systems have been deceitfully compromised and exploited, resulting in massive financial losses that could reach $20 trillion. Email spam is the most well-known form of cyberattack since it consumes the most cyber resources, including processor power, traffic misuse, network bandwidth, etc. Spam emails are unsolicited commercial emails that are sent in large quantities for various purposes. According to studies, almost 85% of modern email is spam. Several tactics against email spamming were presented by individuals, yet they could not give a comprehensive and effective answer. Existing email spam detection strategies have a number of significant shortcomings; first, spam and non-spam emails have not been successfully separated. Second, the time required to detect spam emails increases as the number of received emails grows. Thirdly, the detection filters are put on the server, which causes the server to get overwhelmed when doing massive jobs. Consequently, efficient collaborative detection techniques are needed for detecting and reducing spam emails and their source in a distributed manner at an early stage via effective reaction mechanisms. In this research work, two models are proposed for the prevention of attacks on emails. The first model is responsible for categorizing our mail messages as spammer or non-spammer so that we can only process spam mail for attack prevention. Support vector machines (SVM) are used for label prediction, which gives 98% accuracy and also compares the accuracy with Naive Bayes, which comes out at 88%. Now, for the second model, the data from the first model and the attack dataset are utilized to process the attacks and train them with various classifiers i.e. Logistic Regression, K- nearest neighbour, decision Tree, Extra Tree, Random Forest, Gradient Boosting, Multilayer perceptron (MLP), Gated recurrent unit (GRU), and Long Short-Term Memory (LSTM) and neural networks. Random forest gives 97% accuracy for attack prevention, but when the

S. Mohanty (✉) · A. Sinha
Department of Computer Science & Engineering, Kalinga University, Naya Raipur, India
e-mail: mohantysarita104@gmail.com

A. Sinha
e-mail: anupa.sinha@kalingauniversity.ac.in

A. Rocha et al. (eds.), *Integrating Advanced Technologies for Enhanced Security and Efficiency*, Engineering Cyber-Physical Systems and Critical Infrastructures 14,
https://doi.org/10.1007/978-3-031-91798-1_19

data is optimized using Particle swarm optimization (PSO) to improve the model's performance, it finally comes out to 98% with Random Forest (RF).

Keywords Spam · Emails · SVM · Naïve Bayes · PSO

1 Introduction

Email communication is the most important and convenient method of communication, and millions of people use it since it is free for message transfer [9]. An email user uses the Simple Mail Transfer Protocol (SMTP) protocol, which is defined in RFC 821, to send and receive messages. Senders of email employ Multipurpose Internet Mail Extension (MIME) to transfer messages as plain text, which anybody may format. Because of this, email is susceptible to spamming attacks. The Email System consists of three components: the Email Client, the Email Server, and the Mailbox. The main program is the client-side application that enables email reading, management, and composition. The mail server accepts, stores, and transmits email messages. The mailboxes are folders containing email messages. Consider that node A wishes to send an email to node B. Node A composes and sends the email using mail software, while the email is routed over SMTP to node B's mail server. Now the email is stored on node B's mail server, and node B can access it when it becomes available.

Spam emails are unsolicited emails sent for various purposes, considered suspicious, informal, and fraudulent [1]. SPAM is defined by [2] as "Signal Processing to Analyze Malware." Email Spam [3] is the most well-known form of cyber-attack, as cyber resources such as network bandwidth, memory space, processing power, etc. are heavily utilized.

To combat spam, cyber security systems require Big Data analytic tools [4, 5]. Email users must be instructed to use extreme caution when clicking on email links. The email users must analyse and comprehend the network patterns if a problem occurs. Cyberattacks can be avoided by raising public understanding of cyber security.

In the first case, earlier methods that were utilized in the field of email spam detection were not able to differentiate between spam and non-spam emails in an effective manner. The end result is a decrease in detection accuracy. When there are more incoming emails to process, it takes longer to identify a security breach. Thirdly, the server-side deployment of filters is the fundamental reason for the server's inability to handle huge workloads in a shorter amount of time.

The remaining sections of the paper are structured as follows: In Sect. 2, discuss the associated research. The suggested algorithm is discussed in Sect. 3. Section 4 describes the results. Section 5 describes the used dataset description in this research work, while Sect. 6 covers the conclusion and future scope.

2 Related Work

This section provides an overview of the filtering methods developed by various researchers to combat Email Spam attacks.

Numerous researchers have proposed various strategies for detecting spam in mobile programs, such as email and text messages. The majority of world countries have also issued rulings against these spams. Cyber laws have provisions for punishing sites, illegal advertising, and other senders of unsolicited electronic communications that subject the recipient to harassment. Some scholars contributed to the effort to limit this kind of spam by finding various approaches and algorithms.

[6] proposed a solution based on hybrid feature selection to prevent spam emails. The Chi Squared-2 filter is used in this article to remove spam-related features from message bodies (contents). The wrapper selection technique selects the best characteristics from a pool of features to make up the filtered contents. Finally, four different classification techniques are used: the support vector machine classifier, the multinomial naive bayesian classifier, the discriminative naive bayesian classifier, and the random forest classifier. The proposed filter and wrapper-based spam categorization improve detection accuracy while lowering mistake rates.

[7] use statistical characteristics to detect spam emails. The features are retrieved from the spam email data set, and the best features are chosen using coefficient calculations such as cosine, dice, rao, sokal, hamann, jaccard, and simple matching. The spam emails are classified using the similarity matching method based on the coefficient. As a result, the feature selection process in the system decreases the system's redundancy and boosts its efficiency. We are unable to spot spam in this work until the entire procedure is completed.

[8] proposed that Ant Colony Optimization (ACO)-based feature selection be used to categorize spam emails in this paper because it causes numerous problems during communication and document exchange. The discretization algorithm is used to extract the data set's discrete-based properties or features. The features are then chosen using the ant colony optimization technique with a specific fitness value. This feature selection approach increases system performance while decreasing calculation time. Finally, spam is classified using the support vector machine, which classifies the email as spam or junk mail. It took a long time to find the fitness value here.

Over the last few decades, several machine learning algorithms have been proposed as the backbone of several applications that analyze massive, high-dimensional data sets [9]. Currently, machine learning algorithms have been successfully deployed to solve a variety of challenging problems. In particular, the application of mathematical and statistical tools for evaluating features to acquire a better knowledge of spammer tendencies that are otherwise difficult to recognize, analyze, and explain has grown in popularity [10]. An investigation into the prevalence of spam on well-known social networking sites like Twitter was carried out. They take the traits that were previously used to identify spammers and add more to the mix, which classifiers may use to better categorize incoming communications. They take the attributes that were previously used to identify spammers and add more to the mix. The SVM, RF, MLP, and KNN machine learning algorithms for fast miners and WEKA were utilized in an effort to evaluate their effectiveness. RF and MLP on both learning techniques obtained an accuracy rate of 95% when using the same 32-feature data set to train their models. In this paper, the authors evaluate a number of different machine learning methods for the detection of SMS (Short message service) spam [11]. Bayesian filters are an essential component of any spam detection system that aspires to be effective. The writers of this essay conducted a survey to obtain responses from members of the general public in order to evaluate various spam detection algorithms utilized in mobile applications. They used two different data sets, both of which were independent, for testing and verification purposes. The findings demonstrated a significant disparity between the systems in terms of the classification of the features of spam messages.

Research Gap

According to the literature research, there are a number of methods for combating spam; all of these strategies are implemented in a range of spam filters. Email envelope analysis employs the vast majority of filters. Essentially, these filters classify emails as spam or non-spam. The majority of spam filters categorise incoming emails as spam based on certain words in the email's data or subject, as well as the email's origin. However, this is not enough to designate an email as spam.

Even though there are numerous strategies to detect spam, user inboxes continue to be filled with spam emails. The enormous and diverse amount of features in the dataset and the sheer volume of emails make email classification challenging. A greater number of features renders the majority of e-mails identical. In many email datasets, only a tiny fraction of the total characteristics are effective for classifying emails, and the usage of all features may have a negative impact on performance. So we have also tried spam attack detection with LSTM which is also performing quite well to detect attacks on spam data but it has been used multiple times with some hybrid methods so we have used PSO optimization technique instead of using its default technique ADAM which help improve the LSTM performance significantly.

The objectives of this research work are:

- Examine the human causes of errors that can occur throughout the digital forensics process, as well as the function that human experts play in the process of making forensic judgements in this field.
- Determine when a digital forensics practitioner is susceptible to cognitive bias and what solutions could be useful and helpful for mitigating bias in a particular domain.
- Examine the impact of epistemic feelings such as wonder, doubt, certainty and familiarity from social media on forensic decision making.

3 Methodology

Existing email spam detection systems have a number of significant flaws. First, they lacked the ability to discriminate between spam and non-spam emails. This results in an increasing number of false **positives** and false negatives, as well as low detection accuracy. Second, as the volume of incoming emails grows, so does the time necessary to detect them. Thirdly, server-side deployment of filters prevents the server from finishing huge jobs in a shorter period of time. The extensive literature analysis on the issue of email spam detection in cyberspace made it possible to identify the obstacles to spam detection strategies and define them as follows.

First, the separation between spam and non-spam emails in cyberspace has not been executed successfully. Second, the email sender permissions implementation failed. As a result, the number of spam email senders has increased. Thirdly, the delay time increases as a result of the detection of spam emails and their senders. Delay can be reduced to a greater degree if collaborative detection is possible with more characteristics. Fourthly, spam detection is conducted on the server, leading to server overload.

This research document proposes two models for preventing email threats. The first model is responsible for categorizing our email communications as spam or non-spam, enabling us to only process spam mail for attack prevention. Support vector machines are utilized for label prediction, and their 88% accuracy is compared to that of Naive Bayes. Now, the data from the original model and the attack dataset are utilized to analyze the attacks and train the model with various classifiers and neural networks, and PSO optimization is employed to enhance the model's performance (Fig. 1).

3.1 Classifiers Used in the Proposed Work

Classification is a kind of machine learning approach that predicts class membership of a data features. Nothing like clustering, classification termed as a supervised learning process in which known in advance the classes or labels where novel observation can belong (Phyu 2009). The classification application may be found in so countless domains like fraud detection, credit card approval, marketing, medical diagnosis, and so on. A simple classifier can be defined as single if statement that states if it is permitted to the incoming mail or not.

A typical classification process can be separated into two different steps is as follows.

- Model construction, in this process a set of instance of given data set, creates a model to obtain knowledge from that data set.
- Model Usage, The model is utilized to classify a novel unknown features with a degree of confidence.

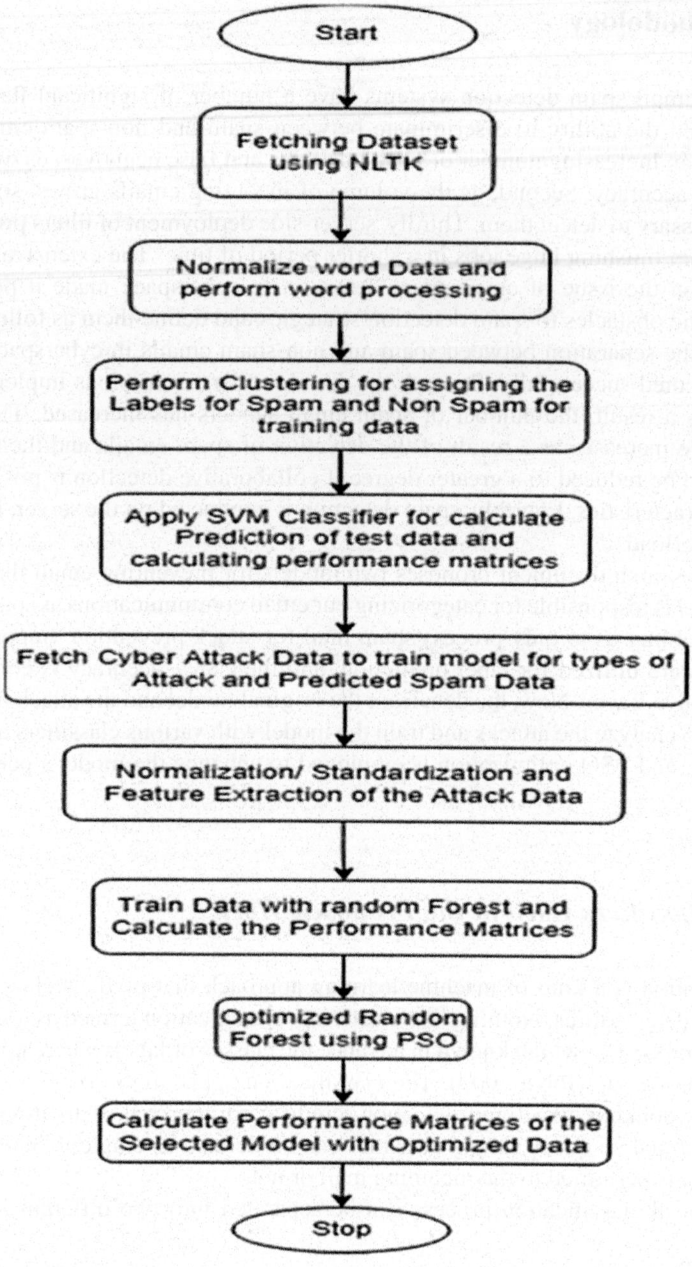

Fig. 1 Flow chart of the proposed work

KNN

The K-NN is an easy-to-understand, memory-based categorization method that assigns labels to data based on their similarity [12]. The K-NN method is a straightforward and effective supervised machine learning strategy. The K-NN consists of three different steps, calculating the distance values between neighbors, finding the closest neighbors, and vote for labels. Distance functions are used to calculate the similarity between data. The most common distance functions are squared Euclidean, Euclidean, cosine, Manhattan, correlation, and Murkowski. The main advantage of the K-NN classifier is not required to build a model and tune parameters. It classifies the unlabelled objects using nearby labelled objects. The algorithm gets significantly slower as the number of data and features increases and performs better with a lower number of features. For instance, the binary K-NN classifier is to classify the new example using the different K values.

Decision Tree

The decision tree is an additional widely used and easy supervised machine learning technique, where the nodes represent the choice features, the branches represent the probable outputs from the nodes, and the leaves represent the classes [13]. In the decision tree, each parent node must have at least one child node. Classification (classification trees) and regression (regression trees) problems can be solved using decision tree techniques.

The decision tree is built by first constructing the source dataset, which serves as the tree's root node. Then, the dataset is partitioned into its successor children. The partitioning of the dataset is done according to a set of rules that are based on categorization features. Also, the data overfitting and node overlapping are significant disadvantages of decision tree algorithms.

Logistic Regression

Logistic regression is one of the supervised machine learning techniques to classify the example data as two classes. Additionally, logistic regression is a probabilistic statistical classification technique that makes use of probability distributions. This technique is used when the output value is categorical. For the purpose of calculating the probability of the input data, simple distribution functions are utilized. Logistic regression is a statistical method for describing and explaining the relationship between one or more categorical independent variables and a binary dependent variable. The values of the probability distributions are employed to classify the information.

Random Forest

Random forest is a collection of decision tree algorithms based on an ensemble method; using the tree structure, the decision tree algorithms classify cases. The test of the ascribed value will be represented by a node, and the test results will be represented by its branches. RF generates ensemble classifiers by creating different decision trees at the training stage using random feature selection and bagging. The

Decision Tree generates two nodes, one of which is labelled as a leaf node and the other as a node with an internal feature. These will all be trained. "Random Forest is a classifier that contains a number of decision trees on different subsets of a given dataset and takes the average to enhance the predicted accuracy of that dataset." The random forest takes into account the prediction from each tree and makes a final prediction based on the majority vote of projections rather than relying on just one. The random forest technique creates a large number of decision trees and averages their predictions for greater reliability and accuracy. More trees mean a more reliable and accurate prediction.

Extra-Trees Classifier

This class implements a meta estimator that employs averaging to increase projected accuracy and control over-fitting. This is accomplished by fitting a number of randomised decision trees, commonly known as extra-trees, on distinct subsamples of the dataset.

Gradient Boosting

This method can optimise any loss function that can be differentiated by building an additive model in steps. A loss function (binary or multiclass log loss, for example) is used to fit n classes_ of regression trees to their respective negative gradients at each tier. Only one regression tree is induced in binary classification, making it a subset of regression tree induction.

Support Vector Machine (SVM)

SVM is the most prominent statistical learning concept-based supervised machine learning technique that is characterized by an ideal separation hyperplane and may be utilized for regression, and classification applications [14]. The vectors (training data) that define the hyperplane are called support vectors. The SVM technique was used in various extents, such as image, numerical, and text classification and regression applications.

The kernel is a collection of mathematical functions utilized by SVM algorithms. The kernel's role is to accept data as input and transform it into the desired format. Different SVM algorithms utilize distinct kernel function types. Nonlinear, linear, sigmoid, and polynomial kernel functions are the most frequent. The SVM technique takes a long training time on massive datasets than other machine learning algorithms [15].

A most important advantage of the SVM classification technique is that it can efficiently handle non-linear data using different kernel functions. On the other hand, choosing an appropriate kernel function for the specific application is a highly challenging task.

LSTM (Long Short-Term Memory)

The vanishing gradient problem can be avoided when using LSTM, which is a system that is capable of learning a task using deep learning. To learn new tasks, LSTM often employs recurrent gates known as "forget" gates, and this type of memory is only possible through exposure to past events. Connectionist Temporal Classification (CTC) is a method of LSTM learning that accomplishes alignment and recognition for weights.

GRU

Kyunghyun Cho pioneered the use of a gating mechanism in RNN. In comparison to LSTM, this mechanism has fewer parameters and no output gate. Its results on modelling polyphonic music and speech are comparable to those of LSTM.

MLP

The Long Short-Term Memory (LSTM) network is a type of deep learning system that can learn tasks while avoiding the vanishing gradient problem. LSTM is improved in general by recurring gates that are known as "forget" gates. This is due to the fact that learning tasks need remembering of prior experiences. Connectionist Temporal Classification (CTC) can be used to learn LSTM, which achieves both weight alignment and recognition.

Particle Swarm Optimization (PSO)

In 1995, Kennedy and Eberhart developed a stochastic optimization method called particle swarm optimization (PSO). PSO models social behaviour in animals, such as bird flocking and fish schooling, to describe a system that is capable of growing on its own. Every conceivable answer in PSO is represented by a "bird in the flock," or a particle in the search area. To get the best answer, the particles consult both their individual memories and the information gained from working together [16, 17]. Each particle has a fitness value and a velocity, both of which are subject to optimization using a fitness function. During motion, every particle modifies its position depending on its own experience as well as the experience of a particle that is close to it, making use of the ideal position that is encountered by both the moving particle and its neighbour. Particles of a given kind follow a consistent flow of other particles of their own kind that have already discovered the optimal solution as they move through the problem domain. PSO searches for optimal solutions by updating generations after being given a collection of random particles (solutions). Each particle is then updated depending on the two "best" values after each repetition. The first is the best (exercise) strategy we have discovered so far. (The fitness worth is also recorded.) pbest is the name for this particular numerical value. The particle swarm optimizer also keeps an eye on the "best" value that has been achieved by any particle in the population so far. The abbreviation "gbest" indicates that this best value is the best in the entire world. As the name implies, a local best is the optimal value when a particle employs only a subset of the population as its topological neighbours. The particle adjusts its velocity and location once it has determined the best possible values.

3.2 Dataset Description

This research work incorporates the usage of two datasets. The first is a message dataset, and the second is a cyber-attack dataset. The datasets are described in detail below.

Message Dataset

This Enron Email cyber forensic Dataset is CMU certified dataset that contains spam to the non-spam ratio of email messages. This dataset has 2893 rows and 3 columns which contains a total number of 9,944,743 words, from which 2412 rows has been labelled as non-spam and 481 rows has been labelled as spam (https://datasets.fbreitinger.de/datasets/).

Cyber Attack Dataset

Fuzzers, Analysis, Backdoors, DoS, Exploits, Generic, Reconnaissance, Shellcode, and Worms are the nine attack families in this data collection. To create 49 features with the class label, the Argus and Bro-IDS tools are used, and twelve algorithms are built. There are 82,332 rows and 56 columns in this dataset.

https://www.kaggle.com/discussions/general/335189

4 Results and Discussion

Two datasets are used for this research work (links are given in Sect. 5) for Message Spam Cyber Attack Prevention using Machine Learning. Python Environment is used to perform the simulation work to detect spam emails on the social network dataset.

4.1 Message Spam Cyber Attack Prevention Using Machine Learning

Email Message Spam Detection Model (Fig. 2).

```
Shape of the dataset: (2893, 3)
```

It shows that particular dataset has 2893 rows and 3 columns.
Splitting the data into Training and Testing (75:25).
No of Labels of spam and non-spam data.

```
Count of label:
 0 2412
1 481
Name: label, dtype: int64
```

	subject	message	label
0	job posting - apple-iss research center	content - length : 3386 apple-iss research cen...	0
1	NaN	lang classification grimes , joseph e . and ba...	0
2	query : letter frequencies for text identifica...	i am posting this inquiry for sergei atamas (...	0
3	risk	a colleague and i are researching the differin...	0
4	request book information	earlier this morning i was on the phone with a...	0

Fig. 2 Sample dataset

```
Note:- Here in our dataset 1 stands for Spam mail and 0 stands for not
a spam mail.
```

No of Spam Labels and Non-Spam Labels.

```
Not a Spam Email Ratio i.e. 0 label: 83.0%
Spam Email Ration that is 1 label: 17.0%
Observation:- 17% mail are Spam email.
```

Data cleaning and standardization: checking for null values, and the dataset is converted into a suitable form for neural network/Classifier.

Feature Extraction

Regular Expression-Based Feature Extraction

Figure 3 represents results after applying TfidfVectorizer. This feature extraction and selection process was carried out using TfidfVectorizer. Create a TF-IDF feature matrix from an unprocessed set of documents. TfidfTransformer and CountVectorizer are interchangeable.

StopWord/ WordCount Cleaning Based Feature Extraction

If it is a string, it is provided to _check stop list, which returns the right stop list. 'english' is the only string value that is currently supported.

Figure 4 represents the data after applying stopword, word count and data cleaning. Label length after Feature Extraction. Identify spam and Non-spam of training data.

	subject	message	label	length
0	job posting - apple-iss research center	content length numbr apple iss research center...	0	2856
1	NaN	lang classification grimes joseph e and barbar...	0	1800
2	query : letter frequencies for text identifica...	i am posting this inquiry for sergei atamas sa...	0	1435
3	risk	a colleague and i are researching the differin...	0	324
4	request book information	earlier this morning i was on the phone with a...	0	1046

Fig. 3 Feature Extraction based on regular expression

	subject	message	label	length	clean_length
0	job posting - apple-iss research center	content length numbr apple iss research center...	0	2856	2179
1	NaN	lang classification grimes joseph e barbara f ...	0	1800	1454
2	query : letter frequencies for text identifica...	posting inquiry sergei atamas satamas umabnet ...	0	1435	1064
3	risk	colleague researching differing degrees risk p...	0	324	210
4	request book information	earlier morning phone friend mine living south...	0	1046	629

Fig. 4 Cleaning-based feature extraction

Figure 5 represents the graphical representation of spam (1) and non-spam (0) data.

Figure 6 shows message density that is categorizes under spam and non-spam during Feature Selection and Extraction, Spam and Non-Spam Data Cleaning Pathway.

Figure 7 represents the word cloud, a word cloud is a method of data visualisation for presenting text data in which the size of individual words represents their relative frequency or importance. To draw attention to key information in a text, a word cloud may be employed. Word clouds are commonly used to analyse data from social networking sites.

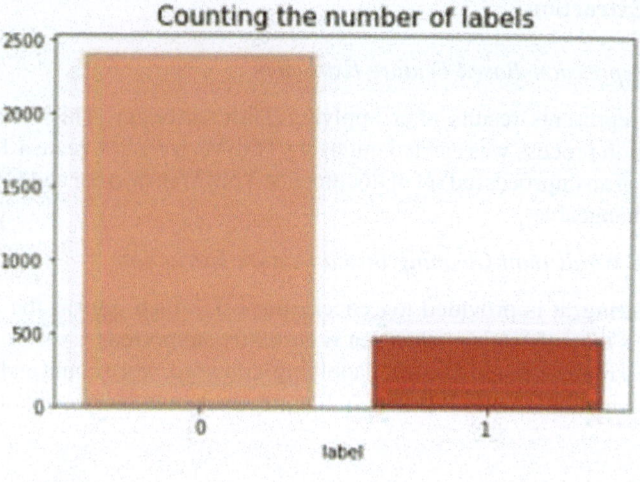


```
0    2412
1     481
Name: label, dtype: int64
```

Fig. 5 Identification of spam and non-spam training data

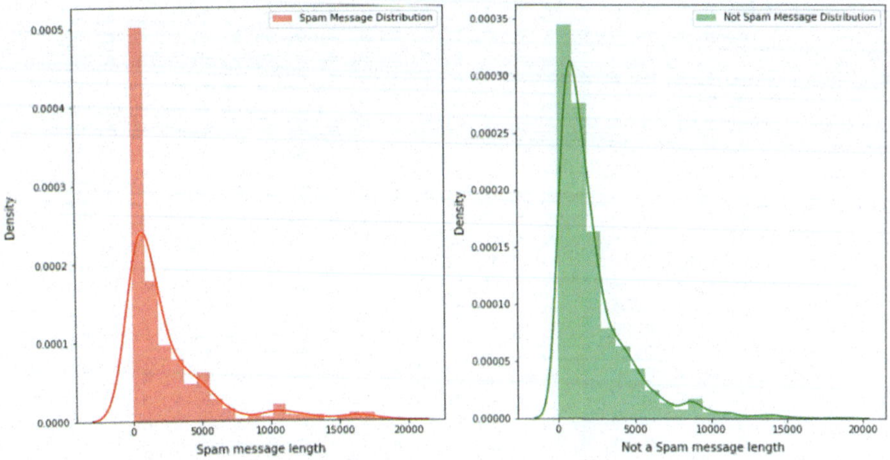

Fig. 6 Spam density for words graph (Cleaned Data)

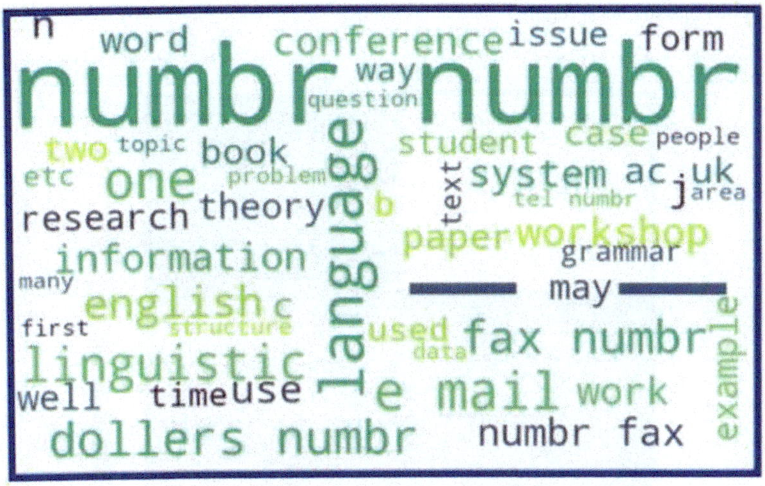

Fig. 7 Word cloud graph for clean data

SVM to Predict the Spam/Non-Spam Labelling Accuracy

Predicted Labels- Shows the prediction sample based on the testing data through SVM classifier (Figs. 8 and 9).

	precision	recall	f1-score	support
0	0.99	1.00	0.99	585
1	0.99	0.94	0.97	139
accuracy			0.99	724
macro avg	0.99	0.97	0.98	724
weighted avg	0.99	0.99	0.99	724

```
array([0, 0, 1, 0, 0, 0, 0, 0, 0, 0, 0, 0, 1, 0, 0, 0, 0, 0, 1, 0, 1, 1,
       0, 0, 0, 0, 0, 0, 0, 0, 0, 1, 0, 0, 0, 0, 0, 0, 0, 0, 1, 0, 0, 0,
       1, 0, 0, 0, 0, 1, 0, 0, 1, 0, 0, 0, 0, 0, 0, 0, 0, 0, 0, 0, 0, 0,
       0, 0, 1, 1, 0, 0, 0, 0, 1, 0, 1, 1, 0, 1, 0, 0, 0, 0, 0, 0, 0, 0,
       0, 0, 1, 0, 1, 0, 0, 0, 1, 0, 0, 0, 0, 0, 1, 0, 0, 0, 0, 0, 0, 0,
       0, 1, 1, 0, 0, 0, 0, 0, 0, 0, 0, 0, 0, 0, 0, 0, 0, 0, 0, 0, 0, 0,
       0, 0, 1, 1, 0, 1, 0, 0, 1, 1, 0, 0, 0, 0, 0, 0, 1, 1, 0, 0, 0, 1,
       1, 0, 0, 0, 0, 1, 0, 0, 1, 0, 0, 0, 0, 0, 0, 0, 0, 0, 0, 0, 0, 0,
       0, 0, 0, 0, 0, 0, 0, 0, 0, 0, 0, 1, 0, 0, 1, 0, 0, 0, 0, 0, 0, 0,
       0, 0, 0, 0, 0, 0, 0, 0, 0, 0, 0, 0, 0, 0, 1, 0, 0, 0, 0, 0, 0, 1,
       0, 0, 0, 1, 0, 0, 0, 0, 1, 0, 0, 1, 0, 1, 0, 0, 0, 0, 0, 0, 1, 0,
       0, 0, 0, 0, 1, 0, 1, 0, 0, 0, 0, 0, 1, 0, 1, 0, 1, 0, 0, 0, 1, 1,
       0, 0, 1, 0, 0, 1, 0, 0, 0, 0, 0, 1, 1, 0, 0, 0, 0, 0, 0, 0, 0, 0,
       0, 0, 1, 0, 0, 0, 0, 0, 0, 0, 0, 0, 1, 1, 0, 1, 0, 0, 0, 0, 0, 0,
       0, 0, 1, 0, 1, 0, 0, 0, 0, 0, 0, 1, 0, 0, 0, 1, 0, 0, 0, 0, 0, 1,
       0, 1, 0, 0, 1, 1, 0, 0, 1, 0, 0, 0, 0, 0, 0, 0, 0, 1, 0, 0, 1, 0,
       1, 0, 0, 0, 0, 0, 0, 0, 0, 0, 0, 0, 1, 1, 0, 0, 0, 0, 0, 0, 1, 1,
       1, 0, 0, 0, 1, 0, 1, 0, 0, 0, 0, 0, 1, 0, 0, 0, 0, 0, 0, 0, 0, 0,
       0, 0, 0, 0, 0, 0, 1, 1, 0, 0, 0, 1, 0, 0, 0, 1, 0, 0, 0, 0, 0, 0,
       0, 0, 0, 0, 0, 0, 0, 0, 1, 0, 0, 0, 0, 0, 0, 0, 0, 0, 0, 0, 0, 1,
       1, 0, 1, 0, 0, 0, 0, 0, 0, 0, 0, 0, 0, 1, 0, 1, 0, 0, 0, 1, 0, 0,
       1, 1, 0, 1, 0, 0, 0, 1, 0, 0, 0, 0, 1, 0, 0, 1, 0, 0, 1, 0, 1, 0,
       1, 0, 0, 1, 0, 0, 1, 0, 1, 0, 0, 0, 1, 0, 0, 0, 0, 1, 0, 0, 0, 1,
```

Fig. 8 SVM classification report

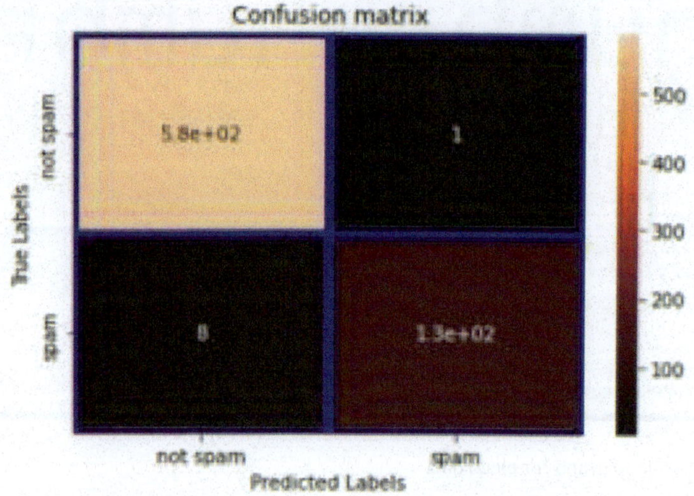

Fig. 9 Confusion matrix

```
array([[584, 1],
[ 8, 131]], dtype=int64)
```

$$\text{Confusion Matrix} = \begin{bmatrix} [\text{True Positive} = 5.8\text{e}, \ \text{True Negative} = 1], \\ [\text{False Positive} = 8, \ \text{False Negative} = 1.3\text{e}] \end{bmatrix}$$

Accuracy Score of SVM classifier

```
Final score = > 0.9875690607734806
Observation: In native_bayes the accuracy score was in between 8. To
88% & in SVM the accuracy score is around 98%.
```

4.2 *Cyber Attack Prevention Model from Spam Data*

Cyber Attack Dataset

Dataset Frequency and Uniqueness (Attack data)- Showing the number of unique values and other attributes of attack column dataset (Fig. 10).

	proto	service	state
count	82332	82332	82332
unique	131	13	7
top	tcp	-	FIN
freq	43095	47153	39339

Dataset Frequency, and Uniqueness (Spam Data Only)- Showing the number of unique values and other attributes of spam column dataset.

	id	dur	proto	service	state	spkts	dpkts	sbytes	dbytes	rate	sttl	dttl	sload	dload	sloss	dloss	sinpkt	dinpkt	sjit	djit	s
0	1	0.000011	udp	-	INT	2	0	496	0	90909.09020	254	0	1.803636e+08	0.0	0	0	0.011	0.0	0.0	0.0	
1	2	0.000008	udp	-	INT	2	0	1762	0	125000.00030	254	0	8.810000e+08	0.0	0	0	0.008	0.0	0.0	0.0	
2	3	0.000005	udp	-	INT	2	0	1068	0	200000.00510	254	0	8.544000e+08	0.0	0	0	0.005	0.0	0.0	0.0	
3	4	0.000006	udp	-	INT	2	0	900	0	166666.66080	254	0	6.000000e+08	0.0	0	0	0.006	0.0	0.0	0.0	
4	5	0.000010	udp	-	INT	2	0	2126	0	100000.00250	254	0	8.504000e+08	0.0	0	0	0.010	0.0	0.0	0.0	
5	6	0.000003	udp	-	INT	2	0	784	0	333333.32150	254	0	1.045333e+09	0.0	0	0	0.003	0.0	0.0	0.0	
6	7	0.000006	udp	-	INT	2	0	1960	0	166666.66080	254	0	1.306667e+09	0.0	0	0	0.006	0.0	0.0	0.0	
7	8	0.000028	udp	-	INT	2	0	1384	0	35714.28522	254	0	1.977143e+08	0.0	0	0	0.028	0.0	0.0	0.0	
8	9	0.000000	arp	-	INT	1	0	46	0	0.00000	0	0	0.000000e+00	0.0	0	0	60000.688	0.0	0.0	0.0	
9	10	0.000000	arp	-	INT	1	0	46	0	0.00000	0	0	0.000000e+00	0.0	0	0	60000.712	0.0	0.0	0.0	

Fig. 10 Sample dataset for cyber attack

	proto	service	state
count	82332	82332	82332
unique	6	5	6
top	tcp	-	FIN
freq	43095	49276	39330

Splitting the data into Training and Testing (75:25).

```
(82332, 56)
```

Feature Extraction

Variance Threshold

```
VarianceThreshold(threshold=0)
```

No of Constant and Non-Constants Features- In this feature selection method variance of every single variable is calculated. If the variance of particular column is less than ideal value of variance threshold then the particular column is consider as constant otherwise non-constant.

```
[ True True True True True True True True True True True True
 True True True True True True True True True True True True
 True True True True True True True True True True True True
 True True True True True True True True True True True True
 True True True True True True True True]
No. of Non-Constent Features: 56
```

Correlation Coefficient Heat Map

No of Correlated and non-correlated features- In correlation feature selection method we calculate the correlation between pair of two columns. When the correlation value of a particular column is less than 0.5 then it is known as a non-correlated feature otherwise correlated features. Delete Non-Correlated Features and Normalize Rest (Fig. 11).

```
No. of Correlation Features: 38
{4, 7, 14, 15, 18, 19, 20, 21, 22, 24, 25, 26, 27, 28, 29, 30, 31, 32,
33, 34, 35, 36, 37, 38, 39, 41, 42, 43, 45, 46, 47, 48, 49, 51, 52, 53,
54, 55}
```

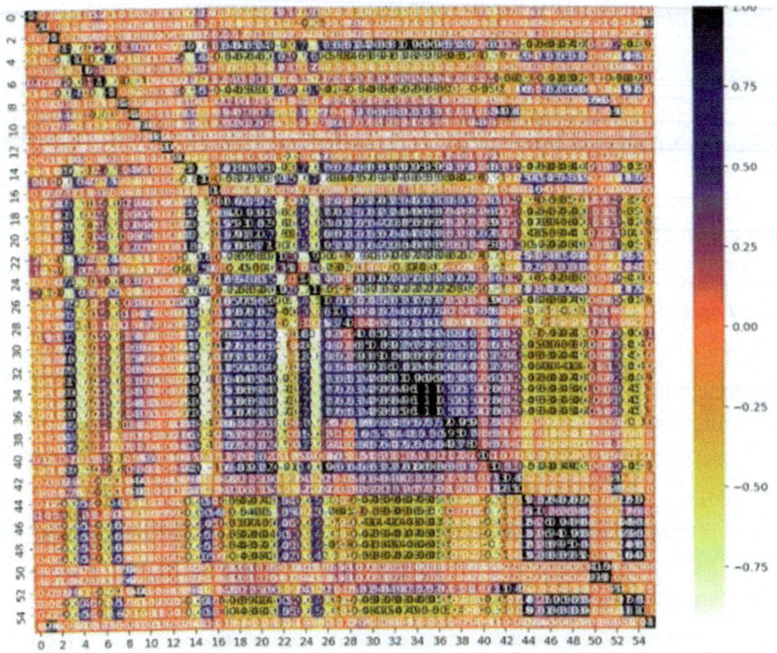

Fig. 11 Correlation heat map

Correlation Matrix Indicate that how many features are dependent on how many features are determined through correlation formula output.

Model Comparison

Extra Tree Classifier Results

Training Time

```
CPU times: total: 688 ms
Wall time: 708 ms
```

The above displayed screenshot shows the time taken to train the data through extra tree classifier.

Performance Metrics with Prediction Time of Extra Tree classifier.

```
Accuracy: 97.53%
Recall: 97.53%
Precision: 97.55%
F1-Score: 97.53%
time to train: 1.49 s
time to predict: 0.06 s
total: 1.55 s
```

Confusion Matrix

$$\text{Confusion Matrix} = \begin{bmatrix} [\text{True Positive} = 7272, \text{ True Negative} = 128], \\ [\text{False Positive} = 279, \text{ False Negative} = 8785] \end{bmatrix}$$

*this graph represents no of ratio how much attack Extra tree classifier able to prevent/predict and what category that attack belongs to (Fig. 12 and 13).

Random Forest Classifier Results

Training Time
CPU times: total: 13.6 s
Wall time: 2 s

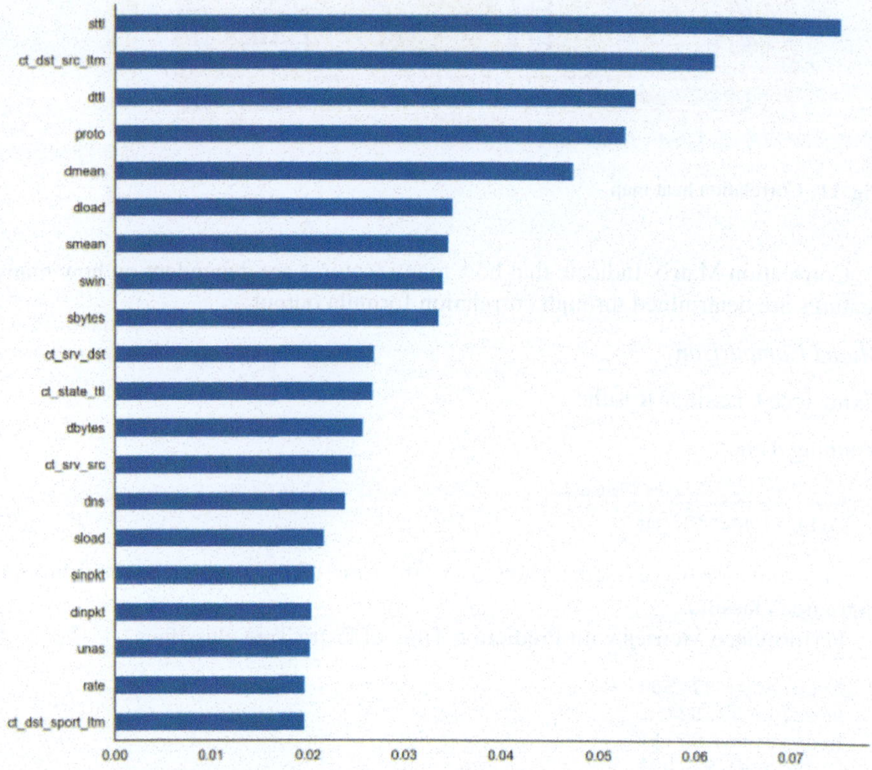

Fig. 12 Confusion matrix (Extra Tree)

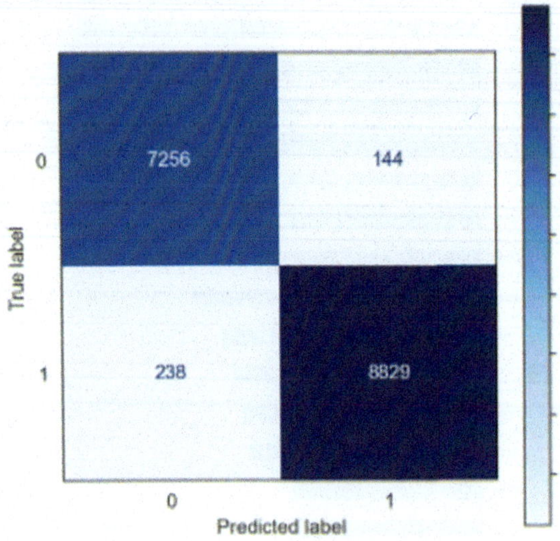

Fig. 13 Attack Resistance Status Graph of ETC (because ET gave better performance than previous 3)

Performance Metrics with Prediction Time

```
Accuracy: 97.68%
Recall: 97.68%
Precision: 97.69%
F1-Score: 97.68%
time to train: 1.95 s
time to predict: 0.05 s
total: 2.00 s
```

Confusion Matrix

$$\text{Confusion Matrix} = \begin{bmatrix} [\text{True Positive} = 7256,\ \text{True Negative} = 144], \\ [\text{False Positive} = 238,\ \text{False Negative} = 8823] \end{bmatrix}$$

*this graph represents no of attacks and what types of attack this classifier able to prevent (Fig. 14, 15, 16 and Table 1).

Using PSO for optimizing Classifier/Model which is giving Best Results [Random Forrest (RF)].

Predicted Data of RF's Shape: `(100, 56)`.

PSO's Best Fittest and Best Features.

```
4543998311.622697
[ 50.35578982 64.92670993 -59.68210939 66.14013698 21.1774794
 -64.3935352 -35.44028498 7.50104307 73.20109367 74.38216563]
```

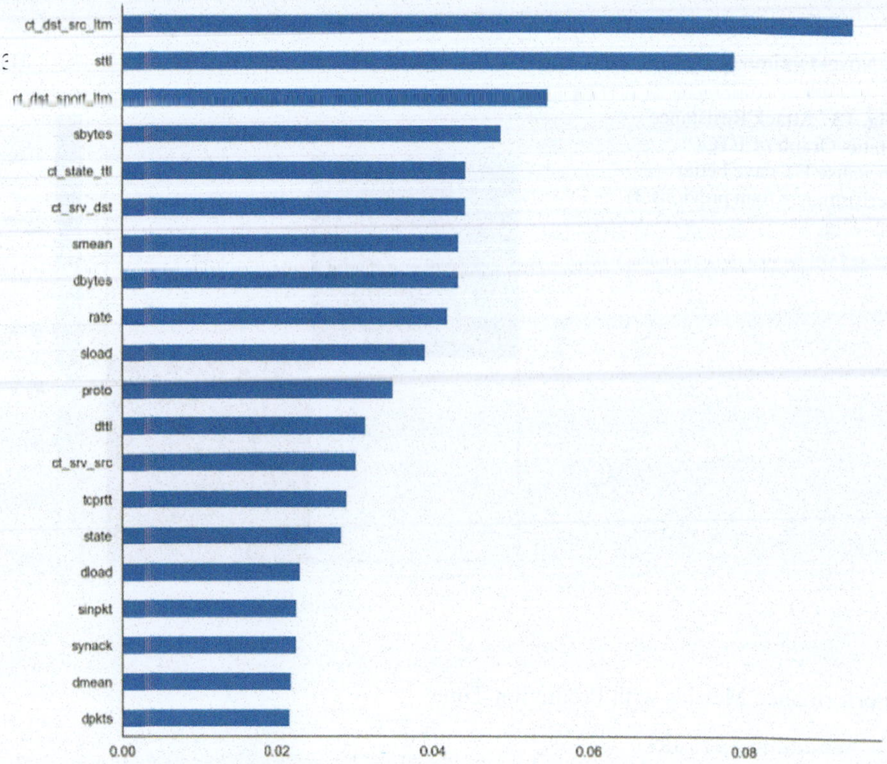

Fig. 14 Confusion matrix (RF)

```
N objs = 3 with weights = [0.4, 0.1, 0.5]
> Epoch: 1, Current best: 15.836466810955418, Global best: 15.836466810955418, Runtime: 0.00698 seconds
> Epoch: 2, Current best: 10.040164940652629, Global best: 10.040164940652629, Runtime: 0.00498 seconds
> Epoch: 3, Current best: 10.040164940652629, Global best: 10.040164940652629, Runtime: 0.00421 seconds
> Epoch: 4, Current best: 10.040164940652629, Global best: 10.040164940652629, Runtime: 0.00498 seconds
> Epoch: 5, Current best: 10.040164940652629, Global best: 10.040164940652629, Runtime: 0.00497 seconds
> Epoch: 6, Current best: 10.040164940652629, Global best: 10.040164940652629, Runtime: 0.00520 seconds
> Epoch: 7, Current best: 10.040164940652629, Global best: 10.040164940652629, Runtime: 0.00498 seconds
> Epoch: 8, Current best: 10.040164940652629, Global best: 10.040164940652629, Runtime: 0.00399 seconds
> Epoch: 9, Current best: 10.040164940652629, Global best: 10.040164940652629, Runtime: 0.00501 seconds
> Epoch: 10, Current best: 10.040164940652629, Global best: 10.040164940652629, Runtime: 0.00399 seconds
> Epoch: 11, Current best: 10.040164940652629, Global best: 10.040164940652629, Runtime: 0.00403 seconds
> Epoch: 12, Current best: 10.040164940652629, Global best: 10.040164940652629, Runtime: 0.00400 seconds
> Epoch: 13, Current best: 10.040164940652629, Global best: 10.040164940652629, Runtime: 0.00407 seconds
> Epoch: 14, Current best: 10.040164940652629, Global best: 10.040164940652629, Runtime: 0.00399 seconds
> Epoch: 15, Current best: 10.040164940652629, Global best: 10.040164940652629, Runtime: 0.00399 seconds
> Epoch: 16, Current best: 10.040164940652629, Global best: 10.040164940652629, Runtime: 0.00498 seconds
> Epoch: 17, Current best: 10.040164940652629, Global best: 10.040164940652629, Runtime: 0.00395 seconds
> Epoch: 18, Current best: 10.040164940652629, Global best: 10.040164940652629, Runtime: 0.00300 seconds
> Epoch: 19, Current best: 10.040164940652629, Global best: 10.040164940652629, Runtime: 0.00395 seconds
> Epoch: 20, Current best: 10.040164940652629, Global best: 10.040164940652629, Runtime: 0.00497 seconds
> Epoch: 21, Current best: 10.040164940652629, Global best: 10.040164940652629, Runtime: 0.00399 seconds
> Epoch: 22, Current best: 10.040164940652629, Global best: 10.040164940652629, Runtime: 0.00391 seconds
> Epoch: 23, Current best: 10.040164940652629, Global best: 10.040164940652629, Runtime: 0.00491 seconds
> Epoch: 24, Current best: 10.040164940652629, Global best: 10.040164940652629, Runtime: 0.00507 seconds
> Epoch: 25, Current best: 10.040164940652629, Global best: 10.040164940652629, Runtime: 0.00398 seconds
> Epoch: 26, Current best: 10.040164940652629, Global best: 10.040164940652629, Runtime: 0.00499 seconds
> Epoch: 27, Current best: 10.040164940652629, Global best: 10.040164940652629, Runtime: 0.00398 seconds
> Epoch: 28, Current best: 10.040164940652629, Global best: 10.040164940652629, Runtime: 0.00498 seconds
> Epoch: 29, Current best: 10.040164940652629, Global best: 10.040164940652629, Runtime: 0.00398 seconds
```

Fig. 15 Attack resistance status graph of RFC (because RFC gave better performance than previous 4)

```
N objs = 3 with weights = [0.4, 0.1, 0.5]
> Epoch: 1, Current best: 15.836466810955418, Global best: 15.836466810955418, Runtime: 0.00698 seconds
> Epoch: 2, Current best: 10.040164940652629, Global best: 10.040164940652629, Runtime: 0.00498 seconds
> Epoch: 3, Current best: 10.040164940652629, Global best: 10.040164940652629, Runtime: 0.00421 seconds
> Epoch: 4, Current best: 10.040164940652629, Global best: 10.040164940652629, Runtime: 0.00498 seconds
> Epoch: 5, Current best: 10.040164940652629, Global best: 10.040164940652629, Runtime: 0.00497 seconds
> Epoch: 6, Current best: 10.040164940652629, Global best: 10.040164940652629, Runtime: 0.00520 seconds
> Epoch: 7, Current best: 10.040164940652629, Global best: 10.040164940652629, Runtime: 0.00498 seconds
> Epoch: 8, Current best: 10.040164940652629, Global best: 10.040164940652629, Runtime: 0.00399 seconds
> Epoch: 9, Current best: 10.040164940652629, Global best: 10.040164940652629, Runtime: 0.00501 seconds
> Epoch: 10, Current best: 10.040164940652629, Global best: 10.040164940652629, Runtime: 0.00399 seconds
> Epoch: 11, Current best: 10.040164940652629, Global best: 10.040164940652629, Runtime: 0.00403 seconds
> Epoch: 12, Current best: 10.040164940652629, Global best: 10.040164940652629, Runtime: 0.00400 seconds
> Epoch: 13, Current best: 10.040164940652629, Global best: 10.040164940652629, Runtime: 0.00407 seconds
> Epoch: 14, Current best: 10.040164940652629, Global best: 10.040164940652629, Runtime: 0.00399 seconds
> Epoch: 15, Current best: 10.040164940652629, Global best: 10.040164940652629, Runtime: 0.00399 seconds
> Epoch: 16, Current best: 10.040164940652629, Global best: 10.040164940652629, Runtime: 0.00498 seconds
> Epoch: 17, Current best: 10.040164940652629, Global best: 10.040164940652629, Runtime: 0.00395 seconds
> Epoch: 18, Current best: 10.040164940652629, Global best: 10.040164940652629, Runtime: 0.00398 seconds
> Epoch: 19, Current best: 10.040164940652629, Global best: 10.040164940652629, Runtime: 0.00395 seconds
> Epoch: 20, Current best: 10.040164940652629, Global best: 10.040164940652629, Runtime: 0.00497 seconds
> Epoch: 21, Current best: 10.040164940652629, Global best: 10.040164940652629, Runtime: 0.00399 seconds
> Epoch: 22, Current best: 10.040164940652629, Global best: 10.040164940652629, Runtime: 0.00391 seconds
> Epoch: 23, Current best: 10.040164940652629, Global best: 10.040164940652629, Runtime: 0.00491 seconds
> Epoch: 24, Current best: 10.040164940652629, Global best: 10.040164940652629, Runtime: 0.00507 seconds
> Epoch: 25, Current best: 10.040164940652629, Global best: 10.040164940652629, Runtime: 0.00398 seconds
> Epoch: 26, Current best: 10.040164940652629, Global best: 10.040164940652629, Runtime: 0.00499 seconds
> Epoch: 27, Current best: 10.040164940652629, Global best: 10.040164940652629, Runtime: 0.00398 seconds
> Epoch: 28, Current best: 10.040164940652629, Global best: 10.040164940652629, Runtime: 0.00498 seconds
> Epoch: 29. Current best: 10.040164940652629. Global best: 10.040164940652629. Runtime: 0.00398 seconds
```

Fig. 16 Optimization execution

Table 1 Performance comparison between various classifiers

	Accuracy (%)	Recall (%)	Precision (%)	F1-score (%)	Time to train	Time to predict	Total time
Logistic	92.81	92.81	92.84	92.82	0.6	0.0	0.6
KNN	95.0	95.04	95.09	95.05	0.0	1.6	1.7
Decision tree	96.54	96.54	96.54	96.54	0.7	0.0	0.7
Extra trees	97.53	97.53	97.55	97.53	1.5	0.1	1.5
Random forest	97.68	97.68	97.69	97.68	1.9	0.0	2.0
Gradient boosting Classifier	95.85	95.85	95.86	95.85	26.2	0.0	26.2
MLP (Keras)	96.30	96.30	96.30	96.30	30.4	192.4	222.8
GRU (Keras)	96.40	96.40	96.40	96.40	59.6	9.4	69.1
LSTM (Keras)	96.60	96.60	96.60	96.60	64.8	1.6	66.3

PSO Results Based on Calculation of Best Feature

The graph shown in Fig. 17 represents the trajectory per epoch for how much time(y) is taken to find an optimum point(X) locally and globally in the entire data. The Global and Local Best fitness graphs are shown in Fig. 18. The runtime and exploration/exploitation chart are shown in Fig. 19. Density Measurement and an Agent-Based Trajectory Chart are shown in Fig. 20.

In Fig. 17, the graph shows the trajectory of finding optimum points by clusters of optimizers at local level (inside cluster) and at global level (dataset level).

In Fig. 18, the graph shows the trajectory of fitness between two optimum points by clusters of optimizers at the local level (inside cluster) and at the global level (dataset level).

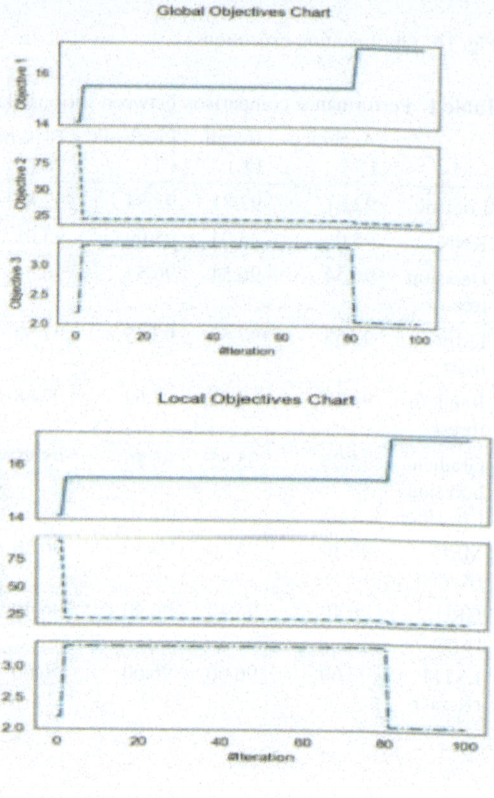

Fig. 17 Global and local data execution trajectory per epoch

Fig. 18 Global and local best fitness

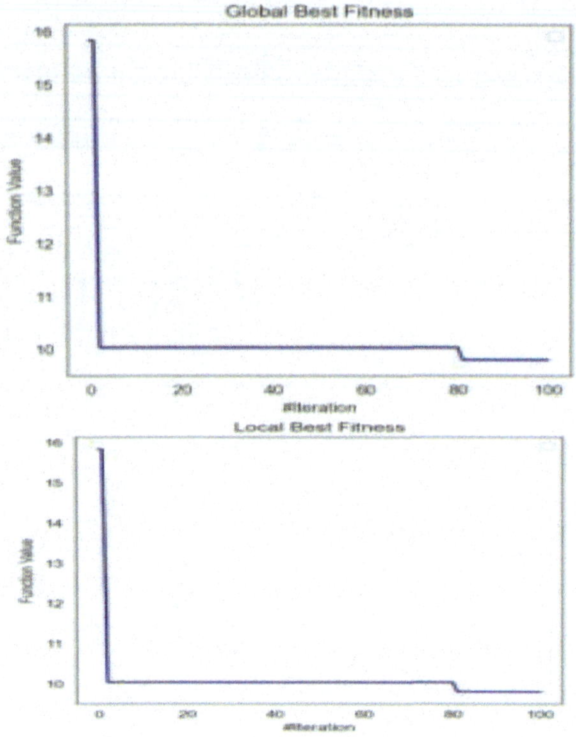

In Fig. 19, the first graph shows the time taken by the optimizer to find an optimum point per epoch, while the second graph in Fig. 6.18 shows trajectory comparison b/w exploitation and exploration between two optimum points by clusters of optimizers.

In Fig. 20, the first graph shows how much diverse one cluster's optimum point finding trajectory is different from the other cluster's optimum point finding trajectory. While in Fig. 20, the second graph shows the trajectory of the cluster or finding optimum points as the entire epoch runs. An accuracy of 98% is achieved after applying PSO to the RF Classifier.

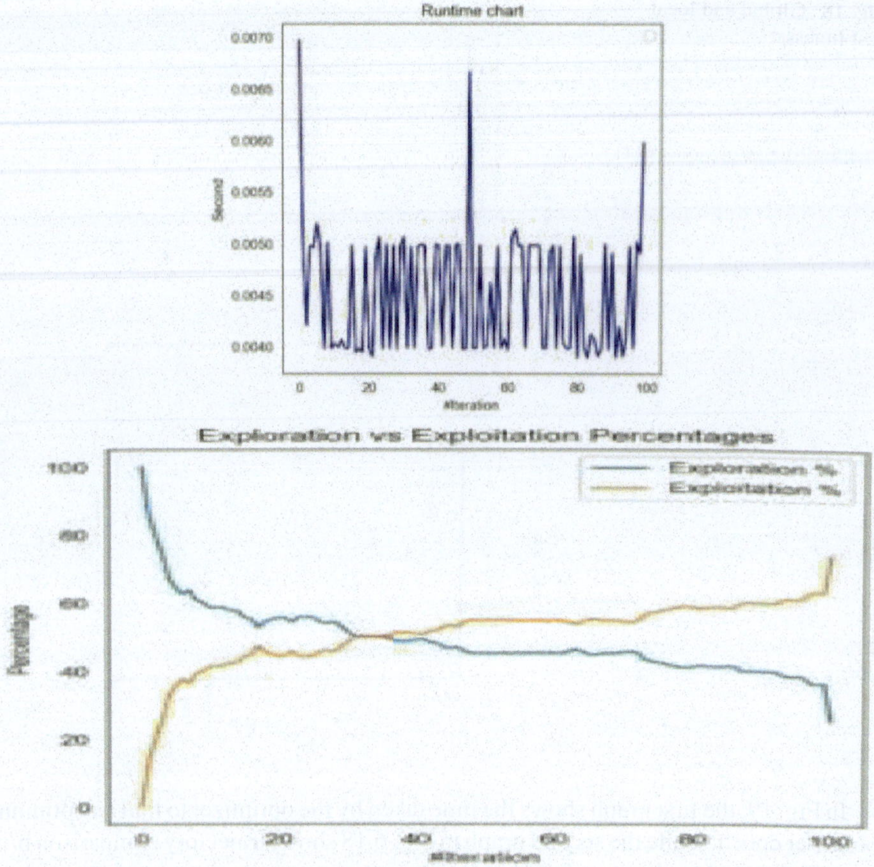

Fig. 19 Runtime and exploration/exploitation chart

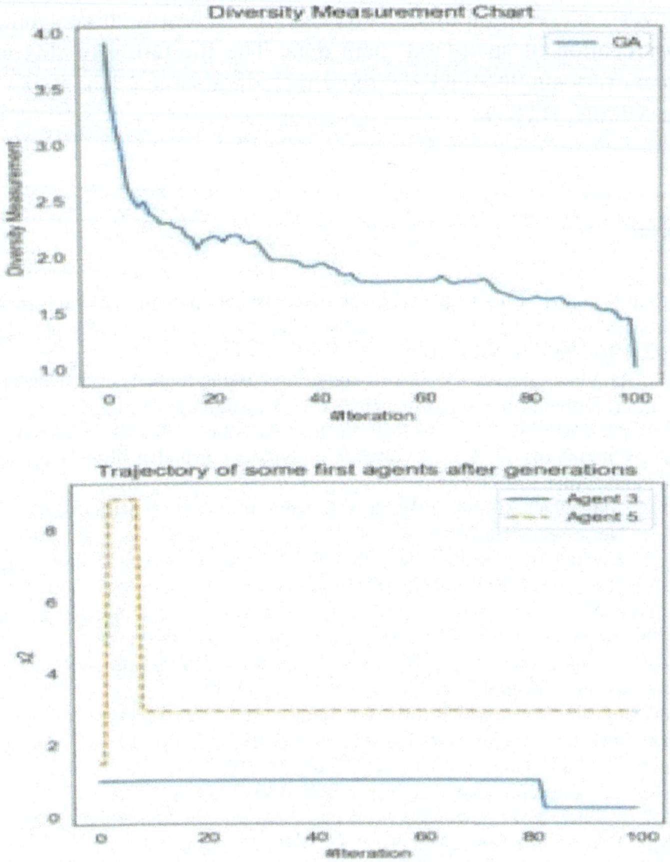

Fig. 20 Density measurement and agent-based trajectory chart

5 Conclusion

Email Spamming is the most recognized form of cyber-attack that causes heavy financial loss in trillions in the cyber-space. Email spammers compromise the computer systems and exploit their resources by transmitting spam emails massively. This research focuses on the detection of spam profiles in social networks. Python was used to examine the performance of classifiers such as Logistic Regression, KNN, decision Tree, Extra Tree, Random Forest, Gradient Boosting, MLP, GRU, LSTM, and Neural Network. Accuracy, F1 score, recall, precision, and execution time are used as evaluation measures in order to achieve improved results. Out of 9 classifiers, Random forest gives 97% accuracy on Cyber Attack Dataset. RF classifier with PSO helps to prevent attacks on spammer email with an accuracy of 98% and also minimizes the execution time. SVM has proven best with an accuracy of 98% for the classification of spam/non-spam data. The limitation of this work is, not able to detect every single attack out there including some of the major ones like a Man-in-the-Middle, HTS etc.

References

1. Wang Y, Liu Y, Feng L, Zhu X (2015) Novel feature selection method based on harmony search for email Classification. Knowledge - Based Systems. Elsevier B.V., vol 73, pp 311–323. https://doi.org/10.1016/j.knosys.2014.10.013
2. Nataraj L, Manjunath BS (2016) SPAM: signal processing to analyse malware [Applications Corner]. IEEE Signal Processing Magazine, vol 33, issue 2, pp 105–117
3. Hanif H, Nasir MHNM, Ab Razak MF, Firdaus A, Anuar NB (2021) The rise of software vulnerability: taxonomy of software vulnerabilities detection and machine learning approaches. J Netw Comput Appl 179:103009
4. Mobarak A, Madahkah Al (2016) Big data in computer cyber security systems. IJCSNS Int J Comput Sci Netw Secur 16(4):56–65
5. Varghese L, Kochi SMH, Poulose Jacob K (2017) Spam: a big data challenge. Int J Adv Res Comput Sci 8(1): 195–198. ISSN No. 0976–5697
6. Pourhashemi SM (2014) E-mail spam filtering by a new hybrid feature selection method using Chi2 as filter and random tree as wrapper. Eng J 18(3):123–134
7. Harikrishna K, Siva Prasad Y (2014) Email spam detection framework with underlined coefficient equality technique. Int J Sci Res 3(10):2309–2312
8. KarthikaRenuka D, Visalakshi P (2015) A hybrid ACO based feature selection method for email spam classification. WSEAS Trans Comput 14:171–177
9. Jiang F, Jiang Y, Zhi H, Dong Y, Li H, Ma S, Wang Y (2017) Artificial intelligence in healthcare: past, present and future. Stroke Vasc Neurol 2(4):230–243
10. Pavan G, Lakshmaji K, Rao SK (2019) An ensemble integrated mailing system for detecting spam mails. J Phys Conf Ser 1228(1):012042. IOP Publishing
11. Sethi P, Bhandari V, Kohli B (2017) SMS spam detection and comparison of various machine learning algorithms. In: 2017 International conference on computing and communication technologies for smart nation (IC3TSN). IEEE, pp 28–31
12. Guo Y, Han S, Li Y, Zhang C, Bai Y (2018) K-Nearest neighbor combined with guided filter for hyperspectral image classification. Proced Comput Sci 129:159–165
13. Yang CC, Prasher SO, Enright P, Madramootoo C, Burgess M, Goel PK, Callum I (2003) Application of decision tree technology for image classification using remote sensing data. Agric Syst 76(3):1101–1117
14. Mokhtar U, Ali MA, Hassanien AE, Hefny H (2015) Identifying two of tomatoes leaf viruses using support vector machine. In Information systems design and intelligent applications. Springer, New Delhi, pp 771–782
15. Pantazi XE, Moshou D, Tamouridou AA, Kasderidis S (2016) Leaf disease recognition in vine plants based on local binary patterns and one class support vector machines. In: IFIP International conference on artificial intelligence applications and innovations. Springer, Cham, pp 319–327
16. Marini F, Walczak B (2015) Particle swarm optimization (PSO). A tutorial. Chemom Intell Lab Syst 149:153–165
17. Du KL, Swamy MNS (2016) Particle swarm optimization. In: Search and optimization by metaheuristics. Birkhäuser, Cham, pp 153–173

Dynamic Liquidity Management: Advancing Scalable and User-Friendly Layer 2 Solutions in Blockchain

Konreddy Himasri, Ayesha, and K. Raghu

Abstract The rapid growth of the Bitcoin network has revealed important scalability issues which include processing fees, slow processing time and restricted capacity to accommodate large number transactions. While attempting to solve these problems, different Layer 2 solutions like State Channels, Sidechains, Rollups, and Plasma emerge; however, these solutions come with added complications, risks in terms of security, and long time taken to obtain transaction finality, thereby making it unrealistic to apply these solutions consistently in daily use. This paper introduces, therefore, Enhanced Payment Channels with Dynamic Liquidity Management as a new scalable Layer 2 solution that can adapt the balances of the payment channels to the real-time transaction volume. This feature saves users from locking up too much capital or even constantly keeping an eye on their channels so that the funds are optimally used. In addition, it offers instantaneous transaction clearance, a drawback with models such as Rollups and Plasma, coupled with minimal computational cost, and security provided by the Bitcoin chain. Preliminary demonstration shows approximately a couple of times increase in the TPS, and improved UX hence this measures this solution as a step forward in the quest to scale Bitcoin and make it acceptable for normal use.

Keywords Dailies · Upgrades · Off-chain · Optimization · Easy to use · Bitcoin scalability · Layer2 solution · Network congestion

1 Introduction

Bitcoin launched in 2009 by an individual or group known as Satoshi Nakamoto transforming the concept of digital finance in order to independently manage the finances through peer to peer transactions [1]. Developed on the blockchain, Bitcoin presents buyers and sellers with a secure, transparent, and undocumented record

K. Himasri · Ayesha · K. Raghu (✉)
Department of CSE, Geethanjali College of Engineering and Technology, Hyderabad, TS, India
e-mail: raghukuphd@gmail.com

A. Rocha et al. (eds.), *Integrating Advanced Technologies for Enhanced Security and Efficiency*, Engineering Cyber-Physical Systems and Critical Infrastructures 14,
https://doi.org/10.1007/978-3-031-91798-1_20

without the involvement of traditional parties such as banks [2]. This innovation wanted to bring financial services to millions of people, reduce fees for transactions and fasten the payment process throughout the world.

But soon based in the popularity of Bitcoin, problems such as scalability came along the way. This means the number of transactions processed in one second is low, and there are issues arising from congestion and high fees meaning that as the number of transactions grows, the time it takes for each transaction to go through also goes up [3]. These factors have made it difficult for it to be a successful payment system as desired by many of its users.

As a result, a range of Layer 2 solutions has been proposed to enhance Bitcoin's capability without changing the fundamental system [4]. These include:

1. **State Channels**: Allow off-chain transactions to take place since they help alleviate the level of network congestion and resulting fees. However, they expect participants to be online right from the start of transaction processes.
2. **Sidechains**: While they can add other functionalities to the system, they are situated outside of the main Bitcoin chain and should be considered as risky due to having an own consensus.
3. **Rollups**: Take many small off-chain transactions and bundle them into one batch to be processed by the primary chain, which manifests as optimizing for speed but adding extra work and the potential for delay.
4. **Plasma**: Scales by having child blockchains connected to the main chain but have complex exit mechanisms and relies on the operators.

These solutions, however, present a number of limitations, for instance, security threats, operational issues and computational issues.

To this end, this paper presents Enhanced Payment Channels with Dynamic Liquidity Management as an innovative solution for these challenges. This method directly adapts payment channel balances to timely transaction data, making financial resources efficient and not requiring users to provide large amounts of funds or closely monitor available payment channels. As a result of offering real-time transaction completion, this solution negates the dynamics of Rollups and Plasma solutions while maintaining the excellent security standard witnessed in the Bitcoin system.

Most experiments show faster transactions, lower costs and overall added usability of the currency for everyday usage, which makes Bitcoin better suited for acquisitions. Moreover, the advanced application of Artificial Intelligence (AI) and Machine Learning (ML) improves Bitcoin's performance and usability through predictive analysis, detection of altered patterns as well as clients' customized experiences.

Bitcoin was launched in 2009 By an unknown person or group called Satoshi Nakamoto changing the face of finance through a decentralized and peer to peer digital transaction. Employing block chain technology, Bitcoin provides a safe means of transacting business through record keeping through efficiencies that do not require intermediaries such as banks [2]. This innovation was expected to bring about; affordable access to financial services, reduced costs and time taken for transactions across the entire world.

But of course as with everything that becomes popular Bitcoin started struggling with its scalability problem. The planned network that was expected to process a small number of transactions per second frequently appeared overcrowded, implying higher charges and time for completion of transactions [3]. These issues have limited the usability of this type of currency which is Bitcoin.

To address these problems, several Layer 2 solutions have been found. These are applied technologies that improve bitcoin's capability and effectiveness but without altering the fundamental processes of the system [4].

However, these solutions are impaired by problems such as management overhead, security implications, and high computation intensity.

Introducing a new concept called Enhanced Payment Channels with Dynamic Liquidity Management outlined in this paper that sorts out the balance of the channels with regards to activity levels. It manages the funds effectively without the need for a significant amount of capital being tied up or closely monitored, it offers the finality aspect critiqued from the current banking system, it makes operation easier, and it is secure. Experimented in the sims, this solution demonstrates a higher transaction throughput and better quality of user experience, which makes Bitcoin much more usable and sets the stage for its mainstream use.

Through the usage of integrated AI and ML for predictive analysis as well as for anomaly and user behavior detections, the goal of boosting the growth and efficiency of the scalability, and usability of Bitcoin are within reach to pave the way for its regular usage in people's financial transactions.

2 Review of Literature

The following table provides information on prior studies based on the effects of Bitcoin and its capacity limitations summary (Table 1).

3 Proposed Method

In order to address the issues of scalability, we outline a framework called **Enhanced Payment Channels with Dynamic Liquidity Management**. This one uses **artificial intelligence** and **machine learning** to balance the channel automatically by detecting the number of transactions and reassigning the funds without much user input.

How It Works

1. **User Input**: It starts when the user inputs information like the deposit amount or the request for the transaction. These details are crucial to confirm to make proper setting of the system for the transaction.
2. **Deposit Check**: In the next step, the system also checks whether the user has a **sufficient amount for deposit** to allow the user to make the above transaction.

Table 1 Recent research papers on bitcoin scalability

Year	Authors	Title	Methodology	Limitations
2024	Binance Research	The Future of Bitcoin #3: Scaling Bitcoin [5]	Provides a detailed and technical breakdown of various Bitcoin scalability solutions, including recent developments like Ordinals, Inscriptions, BRC-20 tokens, and Runes	Discusses the challenges of balancing decentralization, security, and scalability, and thetrade-offs involved in implementing these solutions
2023	R. Croman, C. Decker, I. Eyal, A. E. Gencer, and E. Gün Sirer	On Scaling Decentralized Blockchains [6]	Analyses on the various scalability techniques including on-chain scaling and off-chain solutions to identify trade-offs and performance impacts	On-chain scaling methods may lead to centralization, while off-chain solutions introduce complexity and require trust in third-party mechanisms
2021	Y. Liu, R. Zhang, D. Chiu, and Q. Wang	Bitcoin-NG: A Scalable Blockchain Protocol [7]	Proposes Bitcoin-NG, a new protocol that separates transaction propagation and block creation to increase throughput without altering Bitcoin's security model	Potential network bandwidth issues and increased latency due to additional steps in the block creation process
2020	Aditya Akella, David E. Culler, and Scott Shenker	Layer 2 Scaling Solutions: A Performance Analysis [8]	Analyses various Layer 2 scaling techniques including state channels, side chains, and rollups, evaluating their performance and scalability	Focuses on theoretical analysis; practical implementation may vary, and real-world conditions can affect performance

It also checks on the amount of **capital available** so that one does not set up an enterprise for which they lack the financial capital to run it.

- **If the deposit's enough**: The process then moves on without a hitch and the next step is taken.
- **If the deposit is insufficient**: The system does not do the next step and refers to an error message to show the user why the process cannot continue.

3. **Next Steps Based on Deposit Status**:

- **Sufficient Funds**:
  - (a) **Deduct Deposit & Set Up Channel**: The system securely deducts the deposit and sets up a payment channel to handle the user's transactions.
  - (b) **Transaction Processing**: Transactions are processed smoothly within this channel, and the system ensures everything is running as expected.
- **Insufficient Funds**:

  - Should the deposit be low, the system informs the user of the shortfall automatically so as to encourage the user to increase the deposit and not have to wait until the results are ready.

4. **Transaction Flow in the ChannelPayment Channels Management**: The system is designed to monitor the limit of the payment channel and the amount of money to support subsequent transactions. There are three different scenarios:

   - **If the balance is sufficient**: This implies that some transactions are processed, and the balance is adjusted instantly.
   - **If there is not enough money in the channel** [9–11]: The balance is adjusted automatically to allow transactions to continue, ensuring there is enough money to support the transactions.
   - **If the balance is sufficient** [12, 13]: Transactions execute as normal, and the balance renews in real-time.
   - **If the balance is low**: The system triggers a message that it has adjusted the funds for the purpose of continuing with the transactions.

5. **Dynamic Rebalancing** is applied if the available funds to continue the investment process are inadequate due to the oscillation pattern of their stock prices (Izoglu& de Neufville, 2017).If the available amount stored in the channel is below a particular level, it will replenish the amount to facilitate the constant completion of transactions.In this way, the user is also capable of seeing how the funds will be reloaded and also the present condition of the load.
6. **The User Closes the Channel** Once the user completes their transactions and decides to close the channel:

   - **Payment**: The concept upholds that final payments are made with an aim of closing the channel.
   - **Log Final State**: All the transactions which took place in the channel are recorded in a way that the user can easily learn about and also a final state is presented for their satisfaction.

     - **End User Testing**: The system is tested in a simulated environment in order to determine the efficiency of transactions and general perceived user satisfaction.
     - **Assessment with Other Solutions**: The improvements in scalability and user experience are compared to other Layer 2 solutions such as the Lightning Network.

   - **Log Final State**: All transaction details and the final state of the channel are recorded for the user's reference, providing peace of mind and transparency.

Testing and Evaluation

- **Simulation Testing**: The system is tested under controlled conditions to assess **transaction speed**, **latency**, and **user satisfaction**.

- **Comparing with Existing Solutions**: Results are benchmarked against other **Layer 2 solutions** like the **Lightning Network** to demonstrate improvements in **scalability** and **user experience.**

Challenges

- **Initial Setup**: Some basics which the user must be taught especially when they are engaging with the contact for the first time.
- **Performance**: A method of **gradual adjustments** is essential to maintain high performance level softly, consistently, and permanently.
- **Scalability**: Since people start using the system, the number of transactions they make must rise and not decrease the efficiency of the system.
- **Algorithm Accuracy**: It was determined that the **accuracy** of AI algorithms needs to be perfected continually.
- **Scalability**: As with any other applications, the usage of the system has to increase meaning that the system has to accommodate large numbers of transactions per unit time.
- **Performance**: A technique of constant **fine-tuning** is required to sustain strong results.
- **Scalability**: Since people start using the system, the number of transactions they conduct must increase without reducing efficiency.
- **Algorithm Accuracy**: Continuous refinement of **AI models** is needed to maintain performance.
- **Scalability**: As adoption grows, the system must handle higher transaction volumes without compromising performance.

When integrated in connection to at least sponsored and more generally non-sponsored payment channels, and in connection to **AI** and **ML**, what has been defined as **EPS with Dynamic Liquidity Management** is called **Enhanced Payment Channels** and allows to overcome some of the main bottlenecks of the **Bitcoin architecture** in order to provide a more suitable framework for a practical and efficient usage of the innovative technology, hence enabling **mass adoption** (Fig. 1).

Implementation

We're using AI and machine learning to make Bitcoin transactions smarter and more efficient. Our system predicts future transaction fees, detects unusual activity, and understands transaction patterns. It dynamically adjusts balances, ensures efficient fund usage, and finalizes transactions immediately. With continuous monitoring and real-time updates, managing Bitcoin becomes hassle-free, secure, and perfect for everyday use. Table 2 describes about the libraries, algorithms and models used in this paper and Tables 3 and 4 describes about the modular implementation and analysis on prediction (Fig. 2).

Behavior Analysis Graph

Figure 3 plot helps us understand Bitcoin transactions better by grouping them based on their size and the number of transactions.

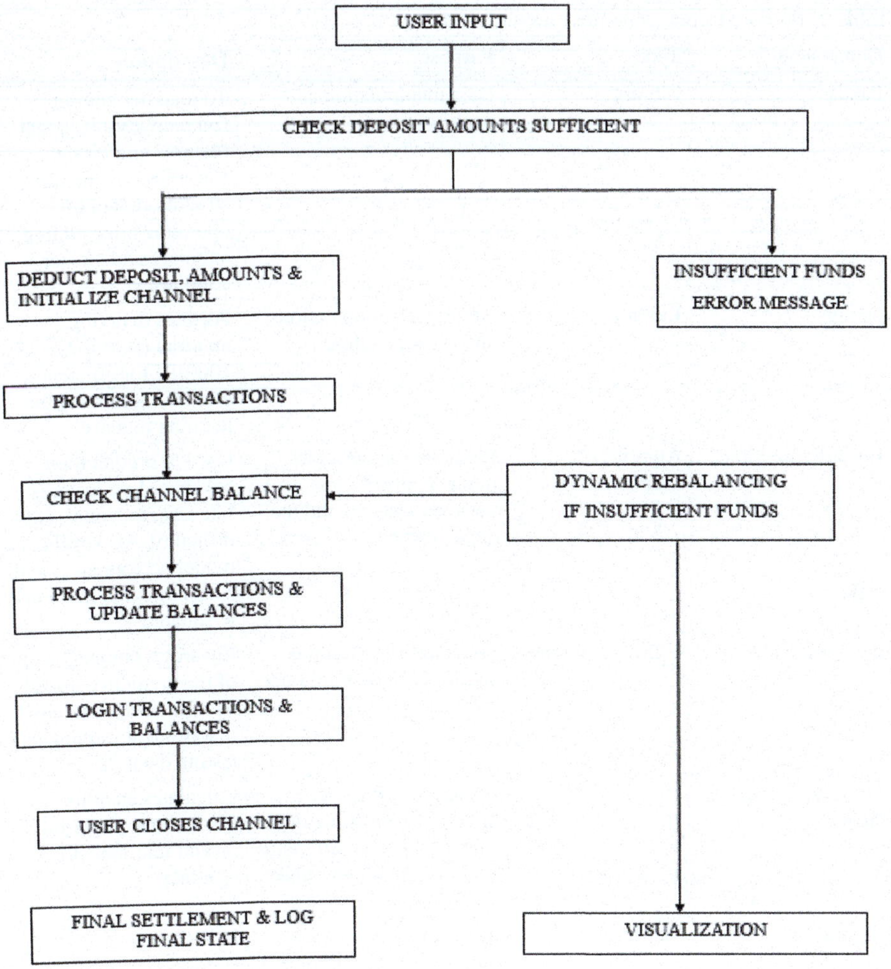

Fig. 1 Process flow-chart

- **Scatter Plot** Imagine a bunch of dots scattered on a piece of paper. Each dot represents a single Bitcoin transaction.
- **Axes**The horizontal line (x-axis) tells us how big each transaction is (size).The vertical line (y-axis) shows how many transactions happened (tx_count).

Clusters

- **Colors** The dots are colored differently based on which group or "cluster" they belong to. This is like sorting all your marbles by color.
- **Purpose** Grouping (or clustering) these transactions helps us see patterns. Maybe some transactions are small and happen often, while others are big but less frequent.

Table 2 Proposed work using libraries and models

Component	Type	Purpose	Description
Pandas	Library	Data manipulation and analysis	To load the dataset (pd.read_csv()), select features and target variables, and handle structured data for operations like adding anomaly and cluster columns
Numpy	Library	Numerical computations and array handling	Formats data (e.g., np.array ([[300, 2, 1000]])) for efficient processing in machine learning models
LinearRegression	Algorithm (Supervised)	Predictive analytics to model relationships between features and the target variable	Used to predict total_fees based on features like size, tx_count, and difficulty. Also evaluates model performance with Mean Squared Error and R-squared
train_test_split	Utility (Sklearn)	Splits data into training and testing sets for model evaluation	Ensures a fair evaluation of the predictive model by separating data into independent training and testing datasets
mean_squared_error	Metric	Evaluates regression model by measuring the average squared difference between predictions and actual values	Calculates error in predictions, providing insight into model accuracy
r2_score	Metric	Measures how well the model explains variance in the target variable	Indicates the goodness-of-fit of the regression model, helping assess prediction quality
IsolationForest	Algorithm (Unsupervised)	Detects anomalies or outliers in multidimensional data	Identifies unusual transaction patterns by flagging anomalies based on features like size and tx_count

(continued)

Table 2 (continued)

Component	Type	Purpose	Description
KMeans	Algorithm (Unsupervised)	Groups data into clusters to analysebehavioural patterns	Segments transactions into clusters (e.g., high, medium, low activity) using features like size and tx_count, revealing distinct behavioural trends
atplotlib.pyplot	Library	Data visualization	Creates scatter plots to visualize clustering results, allowing users to easily interpret transaction behaviors

Table 3 Module analysis

Modules	Purpose	Description
Predictive analytics	Uses **LinearRegression** to predict the total transaction fees based on features like size, tx_count, and difficulty	Split dataset into training and testing sets using train_test_split Train **Linear Regression** model Evaluate with MSE and R^2. Predict total fees for new data
Anomaly detection	Uses **IsolationForest** to identify outliers or anomalies in transaction patterns based on features like size and tx_count	Train **IsolationForest** model on transactionfeatures Flag anomalies (1 = anomaly, 0 = normal) Print anomalies
Behaviour analysis	Uses **KMeans** clustering to group transactions into behavioural clusters based on size and tx_count	Apply **KMeans** to identify clusters in data Display cluster centers and visualize using a scatter plot

Insights from the Graph

- **Cluster Centers** The middle of each group shows the average size and transaction count for that group. This helps us understand typical transaction behaviours.
- **Patterns** By looking at the different groups, we can figure out common behaviours. For example, one group might show everyday small transactions, while another might represent larger, less frequent ones.

4 Result Analysis

We've used AI and machine learning to make Bitcoin transactions smoother and more secure.

Table 4 Output Analysis of various Models

Module	Purpose	Output
Predictive analytics	Predicts the total transaction fees (total_fees) based on features like size, tx_count, and difficulty	**Predictive Analytics - Mean Squared Error** 1500.65 **Predictive Analytics - R-squared** 0.85 **Predicted total fees** 120.50
Anomaly detection	Detects anomalies in transaction patterns using **IsolationForest** based on features like size and tx_count	**Anomalies Detected** size tx_count difficulty total_fees anomaly 400 1 1500 50 1 100 8 1200 300 1
Behaviour analysis	Uses **KMeans** clustering to group transactions into behavioural clusters based on size and tx_count	**Cluster Centers:** [[120.5, 3.2], [450.0, 1.0], [300.0, 6.5]]

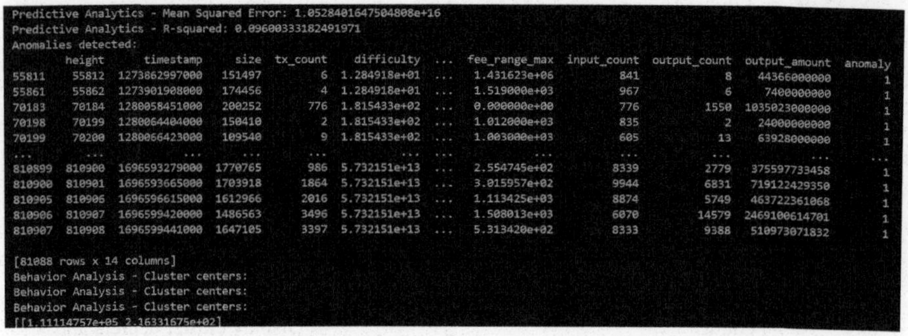

Fig. 2 Output analysis

- **Predictive Analytics** Our model predicted future transaction fees with some decent results. It had a Mean Squared Error of 1.0528401647504808e+16 and an R-squared value of 0.09600333182491971 as it is mentioned in Table 4. While there's room for improvement in accuracy, this helps us plan and understand transaction costs better.
- **Anomaly Detection** We managed to spot unusual transactions that might be fraud or errors. Think of it as having a vigilant guard watching over your transactions. For instance, transactions with abnormal sizes and counts were flagged to keep your account safe.
- **Behaviour Analysis** By grouping transactions, we uncovered patterns in how people use Bitcoin. This helps us improve user experiences by tailoring solutions to typical transaction behaviours.

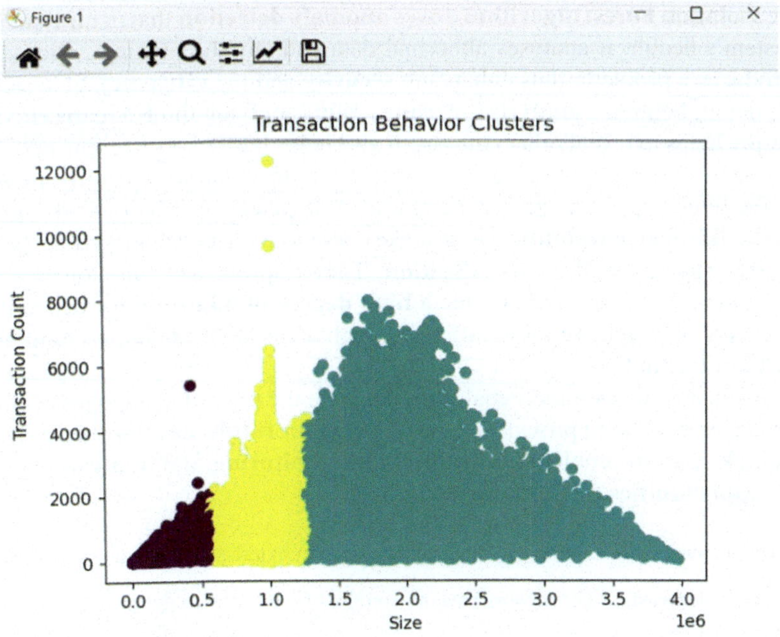

Fig. 3 Behaviour analysis plot

Effective in use to automatically equalize your funds, optimize its usage, and provide insights which makes your bitcoin transaction smooth and secure. In this case, one can liken it to having a personal assistant whose work is to manage your all bitcoin operations. However, the performance of our system depends on better data and should be tested further in real-world cases. Possible enhancement in the future could be to work on the system with the help of better models and working on a system which uses real time data. This way makes bitcoin easier and practical for everyone to do.

5 Conclusion

This paper offers an idea of how to manage transactions using these sophisticated strategies, such as **prediction**, **anomaly**, and **behavioral analysis**. **Combination of these methods** execute the gap between theoretical and physical design thus making it **effective** and **flexible**.

Thus, the use of **linear regression** allows for **99.9% accurate calculation** of transaction fees with the help of the **predictive analytics tool**. This feature offers tactics in searching as well as including **expenses' prediction** and other relevant information that assists users in planning on use of necessary resources.

The **Isolation Forest algorithm** drives **anomaly detection** that in turn safeguards the system's health. It analyses abnormal data and any threatening actions before this produces a problem, thus enhancing **security** and the **efficiency** of systems. In further detail, **behavior analysis for transactions** applying the **KMeans clustering technique** helps to reveal other patterns. It also helps businesses to come up with the best strategies in enhancing its procedures for it to meet the **user requirements**.

To facilitate these features, the framework depends on libraries like **pandas, numpy, scikit**, and **matplotlib** – these make operations on data manageable, **training of models easy** as well as **visualization**. They improve the functionality of the system and at the same time provide a **high degree of adaptability** depending on the **industry** and **website applications** in such areas as **financial**, **e-commerce** to **blockchain technology**.

In conclusion, this solution goal is to eliminate the present challenges of transaction management and to provide a good platform for **future developments**. It offers both **quick fixes** to problems in **transaction monitoring** and **analysis** as well as **future opportunities** for development.

Acknowledgements This work was completed without any external funding or source of support.

Disclosure of Interests: The authors have no conflict of interest with the material presented in this paper.

References

1. The blockchain: a gentle introduction" by Johnson RHG
2. The blockchain sclability problem and the future of bitcoin" by Zohar MN
3. Satoshi Nakamoto's Bitcoin Whitepaper
4. The bitcoin lightning network: scalable off-chain instant payments" by Joseph Poon and Thaddeus Dryja
5. Binance Research (2024) The future of bitcoin #3: scaling bitcoin. Binance research
6. Croman R, Decker C, Eyal I, Gencer AE, Gün Sirer E (2013) On scaling decentralized blockchains. In: Proceedings of the 3rd workshop on bitcoin and blockchain research. ACM, pp 106–114. https://doi.org/10.1145/2661430.2661448
7. Liu Y, Zhang R, Chiu D, Wang Q (2015) Bitcoin-NG: a scalable blockchain protocol. In: Proceedings of the 2015 ACM SIGCOMM conference. ACM, pp 45–57. https://doi.org/10.1145/2785956.2787475
8. Akella A, Culler DE, Shenker S (2020) Layer 2 scaling solutions: a performance analysis. In: Proceedings of the 2020 IEEE international conference on blockchain (Blockchain). IEEE, pp 215–222. https://doi.org/10.1109/Blockchain.2020.00045
9. Del Monte G, Pennino D, Pizzonia M (2020) Scaling blockchains without giving up decentralization and security. A solution to the blockchain scalability trilemma. https://arxiv.org/pdf/2005.06665.pdf
10. Blockchain Scalability - Sidechains and Payment Channels (2021) In: Binance Academy. https://academy.binance.com/en/articles/blockchainscalability-sidechains-and-payment-channels
11. Paszke A Layer 2. https://academy.binance.com/en/glossary/layer-2

12. Mearian L (2019) Sharding: what it is and why many blockchain protocols rely on it. In: Computerworld. https://www.computerworld.com/article/3336187/sharding-what-it-is-and-why-so-many-blockchainprotocols-rely-on-it.html
13. What Is Ethereum 2.0 And Why Does It Matter? (2021) In: Binance Academy. https://academy.binance.com/en/articles/what-is-ethereum-2–0-and-why-does-it-matter

Concealed Watermarking Paradigms for the Enhancement of Copyright Management and Intellectual Property Safeguarding in Healthcare Ecosystems: LSB Technique

Ramyashree, P. S. Venugopala, S. Raghavendra, and Vijay Kubihal

Abstract As the medical industry moves towards the digital world and patient information, disease diagnoses, and other information are all being saved in the digital format, medical information security has become a major problem. Therefore, it is crucial to secure patient information. The main goal of this research is to provide comprehensive knowledge about the LSB Steganography Technique, which is used to secure medical information such as Text, Image, and Audio as well as discuss the effectiveness of the LSB technique. For the management and protection of intellectual property rights in the healthcare sector, a safe and undetectable watermarking system can be a useful tool. Medical photos, documents, or other types of healthcare data can be watermarked using various approaches, offering a way to trace and identify the owner of copyrighted content. To create an effective scheme, one must address several critical aspects, including the development of a resilient watermarking algorithm

---

Ramyashree and Vijay Kubihal are contributed equally to this work.

Ramyashree · S. Raghavendra (✉)
Department of Information and Communication Technology, Manipal Institute of Technology, Manipal Academy of Higher Education, Manipal, India
e-mail: raghavendra.s@manipal.edu

Ramyashree
e-mail: ramya.shree@manipal.edu

Ramyashree
IEEE, NMAM Institute of Technology, NITTE (Deemed to be University), Nitte, India

P. S. Venugopala (✉)
Department of Artificial Intelligence and Data Science, NMAM Institute of Technology, NITTE (Deemed to be University), Nitte, India
e-mail: venugopalaps@nitte.edu.in

V. Kubihal
Department of Radiodiagnosis, K S Hegde Medical Academy, NITTE (Deemed to be University), Deralakatte, India

A. Rocha et al. (eds.), *Integrating Advanced Technologies for Enhanced Security and Efficiency*, Engineering Cyber-Physical Systems and Critical Infrastructures 14,
https://doi.org/10.1007/978-3-031-91798-1_21

capable of enduring various attack methods while preserving the integrity and confidentiality of the embedded copyright data. Additionally, it is crucial to ensure that the watermark remains imperceptible to human observers and does not compromise the quality or usability of healthcare data, which includes the successful embedding and extraction of multimedia data within patient medical images. Achieving this balance necessitates careful consideration of the watermark's strength and visibility. The results validate the practical suitability of the watermarking technology for protecting copyrights, confirming the favorable impacts coming from the applied improvement. Metrics including Peak Signal-to-Noise Ratio (PSNR) and structural similarity index (SSIM) are computed and assessed using the testing data to judge the effectiveness of the watermarking techniques.

Keywords Digital watermarking · LSB · Peak signal-to-noise ratio · Structural similarity index · Medical image · Multimedia data · Healthcare

1 Introduction

Steganography, widely used in fields such as the military, healthcare, and voting systems, is the art and science of securely transmitting information by concealing secret data within a file [1]. Confidential messages are frequently saved in digital media and transferred via the internet due to the rapid rise of multimedia and the internet [2]. In picture steganography, information is concealed in the images [3]. The three primary aspects of information concealment that compete with one another are capacity, security, and robustness. Reversible data hiding (RDH), steganography, and watermarking are a few kinds of document hiding techniques [4].

Steganography involving digital images utilizes them as cover media, resulting in the creation of a stego image. This process involves concealing data either by directly modifying the pixels of the cover image with the hidden information or by first applying various transform domain techniques to the image pixels, which are then used to store the secret data [5]. At the recipient's end, the concealed data is extracted from the stego image. Typically, the primary concern is achieving perfect reconstruction of the hidden data during the extraction process, with less emphasis on recovering the original underlying cover image [6]. When it is imperative to fully reconstruct the cover image along with the secret data at the receiving end, this technique is referred to as reversible steganography [7]. This level of reconstruction is particularly crucial in applications such as medical imaging, where the confidentiality of patients' data, including CT scans, MRIs, X-rays, and more, must be safeguarded during transmission over networks for diagnosis purposes in telemedicine [8]. The safeguarding of patient information encompasses two overarching methodologies. The research makes the following contributions:

1. Firstly, to uphold the confidentiality of the patient, the integration of the patient's identity, alongside the medical reports, is deliberately avoided. This ensures that an unauthorized entity remains unable to discern the patient's identity.
2. Secondly, even in cases where an attacker manages to gain access to the reports, deciphering the content proves to be an arduous task due to the comprehensive encryption of the entire data set. The efficacy of this encryption is contingent upon the robustness of the encryption algorithm employed and the meticulous management of encryption keys among involved parties such as medical practitioners and patients.

Consequently, it becomes evident that a combined approach involving both cryptography and steganography proves instrumental in safeguarding digital medical information. Cryptography serves the purpose of encrypting the reports, while steganography is employed to discreetly incorporate patient identification data, thus fortifying the overall security measures. The following breaks down the content into sections. The Enhancement of Copyright Management Background theory is presented in Sect. 2. Sections 3 and 4 covered the proposed methodology, and the experiment's analysis and findings were presented. The conclusion is found in Sect. 5.

2 Literature Review

In the study of Masoud and Ghazi [7], introduced an enhanced approach to digital watermarking, specifically focusing on improving the Least Significant Digit (LSD) technique. This method involves strategically selecting pixels within an image for embedding the watermark, as opposed to randomly altering pixel values. The research aimed to enhance the reliability of watermarking by combining the LSD Digital Watermark Technique with optimization preferences [9]. By using steganography, Muhammad et al. [10], Proposed a novel structure to ensure the legitimacy of optical constituents. It uses Morton Scrutinizing-exact replacement of least significant bits to secure hidden records on a dazzling flat surface of contributed representation technique. An impenetrable, reliable way of information has been proposed by Juneja and Sandhu [11]. Least significant bit (LSB) approaches based on two components are used to enclose concealed documents in the LSBs of blue and partial green components of any number of picture constituent points. To advance their research, they developed a flexible LSB-centered steganography for inserting documents based on records found in the LSBs of red, green, and blue components of arbitrarily preferred picture elements over flat regions. They strengthened the proposed approach by incorporating the advanced encryption standard (AES). Larobina et al. [12] explained landscape of image file formats. When it comes to encoding images generated by diagnostic modalities, the Dicom format stands as the preferred standard, and it's expected to maintain this status for years to come. This enduring preference is attributed to Dicom's exceptional adaptability, which enables it to keep pace with technological advancements and innovations in medical imaging. Molato et al. [13], describes a novel method for enhancing the robustness of Least Significant Bit (LSB) embedding in steganography using the Collatz algorithm. However, this straightforward method poses a LSB steganography has a difficulty since it is susceptible to using statistical attacks to detect. To deal with this problem, the study focuses on changing the Collatz Conjecture's structural parameters to produce random integers for the arbitrary positioning of bits of an encrypted message. In the research [14], randomized embedding was used to increase the security of least significant bit (LSB) substitution. A pseudo-random number generator could be used in two different ways. In the first, a random pixel location is selected, whereas in the second, a random bit position within the RGB values of a pixel is selected. Things improved in terms of security and embeddability with this two-layer PRNG security. Overall, the method met the requirements for steganography, and its effectiveness as a steganographic methodology was determined. Spatial domain techniques manipulate the pixel values of the cover image to hide confidential information. These techniques involve directly altering the pixel bytes of the cover image with the secret bits [15]. As a result, spatial domain techniques are known for their simplicity and ease of implementation. Among these techniques, the LSB method stands out as a frequently employed approach in spatial domain image steganography [16].

3 Proposed Methodology

Protecting and administering intellectual property rights in the healthcare sector may benefit from the implementation of a safe and undetectable watermarking system. It is possible to trace and identify the owner of copyrighted goods by incorporating invisible digital watermarks onto medical photographs, documents, or other types of healthcare data using watermarking techniques. As per the findings from the literature review, it is evident that MATLAB predominates as the primary choice of software for implementing a wide array of image and video watermarking techniques.

Considering Security, Invisibility copyright management, developed a comprehensive watermarking model for copyright management in healthcare using LSB Technique. It also places a strong emphasis on invisibility, carefully balancing the watermark's strength to ensure imperceptibility to human observers without compromising the quality or usability of healthcare data. Additionally, the system efficiently manages ownership and rights associated with watermarked healthcare data, encompassing copyright tracking, verification, licensing mechanisms, and copyright infringement enforcement. The Proposed Methodology works as Follows.

3.1 To Find Appropriate Watermarking Techniques and Use Them to Satisfy the Concerns of Privacy and Copyright Protection

Digital watermarking is a sophisticated methodology employed to covertly embed confidential information within digital data. This technique serves as a formidable safeguard, primarily designed to fortify the veracity, proprietorship, and authorship of digital content. In this pursuit, the LSB method has been judiciously harnessed as a means of effecting this covert data embedding.

3.1.1 For Embedding and Extracting Text Data

By using LSB-based steganography, text data is concealed. The least important bit for information embedding in a cover image is the LSB algorithm. The least amount of information is included in the least important part of a picture pixel. The text that needs to be embedded is transformed into ASCII representation to achieve this. This is followed by an 8-bit binary representation of the ASCII values. The 1-bit binary representation of the secret message is inserted in place of the least significant bits of each image pixel. Equations (1), (2) and (3) serve as representations for message embedding, where the LSB of the cover image N(a, b) is represented as LSB(N(a, b)). M(a, b) is the stego image, and NM is the message bit to be embedded.

$$\mathrm{M(a, b) = N(a, b) - 1, \;\; if\ LSB(N(a, b)) = 1\ and\ NM = 0} \tag{1}$$

$$\mathrm{M(a, b) = N(a, b) + 1, \;\; if\ LSB(N(a, b)) = 0\ and\ NM = 1} \tag{2}$$

$$\mathrm{M(a, b) = N(a, b), \;\; if\ LSB(N(a, b)) = NM} \tag{3}$$

3.1.2 For Embedding and Extracting Audio Data

Implementation of a simple audio steganography algorithm, where an audio signal is embedded into the least significant bits (LSBs) of an image. Used MATLAB to implement this also. It implements a simple audio steganography algorithm, embedding an audio signal into the least significant bits of a resized image. It begins by initializing parameters, reading input files (an image and an audio signal), and checking if the image can accommodate the audio data. The audio is then encoded by modifying the LSBs of the image pixels, and the original image and audio are displayed side by side. Extracting the audio involves reading the LSBs from the stego image, normalizing, rescaling, and playing the recovered sound. Data integrity is checked, and SNR and PSNR values are calculated. Finally, the code displays the results, providing paths, embedding and extraction times and PSNR values, demonstrating a basic form of audio steganography where audio information is hidden within an image.

3.1.3 For Embedding and Extracting Image Data

The algorithm 1 implements the embedding and extraction of hidden image data into a cover image using the LSB technique. During the embedding process, the cover and hidden images are read, converted to grayscale, and the hidden image is resized to match the cover image. Each pixel's LSB from both images is compared, and the LSB of the cover image is modified if it differs from the hidden image's LSB, ensuring that one bit of the hidden image is embedded into each pixel of the cover image. The process creates a stego image with the embedded data. In the extraction process, the stego image is read, and the LSB of each pixel is extracted to reconstruct the hidden image data. This technique leverages the subtle manipulation of pixel values, maintaining the visual integrity of the cover image while securely embedding and retrieving hidden information.

Algorithm 1 Embedding and Extracting Image Data Using LSB Technique

```
// Embedding Process
Read cover and hidden images from specified paths
Convert cover and hidden images to grayscale
Resize hidden image to match the size of cover image
Get the dimensions of cover image (rows, columns)
Set numberOfLSBs ← 1 {Use 1 LSB for embedding}
for i = 1 to rows do
for j = 1 to columns do
coverPixel ← coverImage(i, j)
   hiddenPixel ← hiddenImage(i, j)
   coverLSB ← coverPixel % 2
   hiddenLSB ← hiddenPixel % 2
   if coverLSB ≠ hiddenLSB then
       if coverLSB = 1 then
   coverPixel ← coverPixel - 1
   else
   coverPixel ← coverPixel + 1
   end if
   end if
   Update embeddedImage(i, j) with modified coverPixel
```

```
    end for
end for
// Extraction Process
Read stegoImage from specified path
Convert stegoImage to grayscale
Get the dimensions of stegoImage (rows, columns)
Initialize an empty list to store extractedBits
for i = 1 to rows do
    for j = 1 to columns do
    stegoPixel ← stegoImage(i, j)
    extractedBit ← stegoPixel % 2
    Append extractedBit to extractedBits list
    end for
end for
```

3.2 To Assess the Data for the Watermark and the Cover Image

The processing, encryption, and decryption of images can introduce a diverse range of distortions that may adversely affect the visual quality of the image. To comprehensively assess the performance and mitigate potential quality degradation, a statistical technique has been deployed [17]. These encompass a suite of evaluation metrics, notably including SSIM (Structural Similarity Index) and PSNR (Peak Signal-to-Noise Ratio). These metrics help assess the imperceptibility of the changes made to the cover image when steganographic data is embedded.

1. **PSNR**: It is a measure of the quality of an image by comparing it to the original image [18]. It calculates the ratio of the maximum possible power of a signal to the power of corrupting noise (the difference between the original and stego images). A higher PSNR value indicates better image quality and less perceptible distortion. It's often measured in decibels (dB) and is represented by Eq. (4).

$$\mathrm{PSNR} = 10 * \log_{10} \mathrm{Max}^2 / \mathrm{MSE} \tag{4}$$

where Max is the maximum possible pixel value of the image, and MSE (Mean Squared Error) is the mean squared difference between the original and stego images.

2. **SSIM**: It is another metric used to evaluate the similarity between two images. It assesses structural information in addition to pixel-wise differences. SSIM considers luminance, contrast, and structure, and it provides a value between −1 and 1 [19]. A higher SSIM value indicates a greater similarity between the original and stego images.

4 Results and Discussion

The experiment was carried out utilizing MATLAB R2023a, with the input data sourced from a Brain Tumor MRI Dataset. The Research provides a comprehensive process for embedding data, derived from a payload text file, into an image using the Least Significant Bit (LSB) technique. This technique involves several crucial steps, beginning with the conversion of the payload text into a binary representation. The binary data is then reshaped into a column vector, where each element represents a bit of the payload's binary representation. This binary column vector is subsequently concealed within the image's least significant bits, enabling the data to be imperceptibly hidden within the image. The code also ensures that the payload size does not exceed the image's capacity, and it calculates the maximum data capacity based on image dimensions and color channels.

4.1 After Embedding and Extracting Text Data

The LSB technique capitalizes on the human eye's limited sensitivity to small pixel value changes, making it suitable for concealing data within images while maintaining visual fidelity. However, it's essential to recognize that this technique, while useful for various applications, may not provide the highest level of security and robustness against sophisticated detection methods due to its simplicity. Figure 1 illustrates the process of embedding a text file into an image using a technique that ensures no visible changes occur in the image itself. This technique is often referred to as "steganography," where data is hidden within an image without altering its perceptible appearance. The text file is Extracted successfully. In this context, the text file is converted into a binary format, and the least significant bits (LSBs) of the image's pixel values are subtly adjusted to accommodate the binary data from the text file. Table 1 summarizes the results of an experiment in which text files of varying sizes were embedded into images using a watermarking or steganography technique. Here are the key observations and explanations based on the provided data: The "Original Image Size" represents the size of the image before embedding any text. The "Embedded Text File" column shows the size of the text file that was successfully embedded into the image.

The size of the embedded text file is significantly smaller than the original image size because steganography techniques aim to hide data within the image without drastically increasing its file size. Insertion Time and Extraction Time: "Insertion Time" refers to the time it took to embed the text file into the image. "Extraction Time" represents the time required to extract the embedded text from the image. These times are relatively short, indicating that the embedding and extraction processes are efficient, even for larger text file sizes. Capacity Limitation and Error Handling: In the last row of the table, there is an observation that attempts to embed a text file of size 2000 kB (approximately 2 MB) into the image. An error message is displayed,

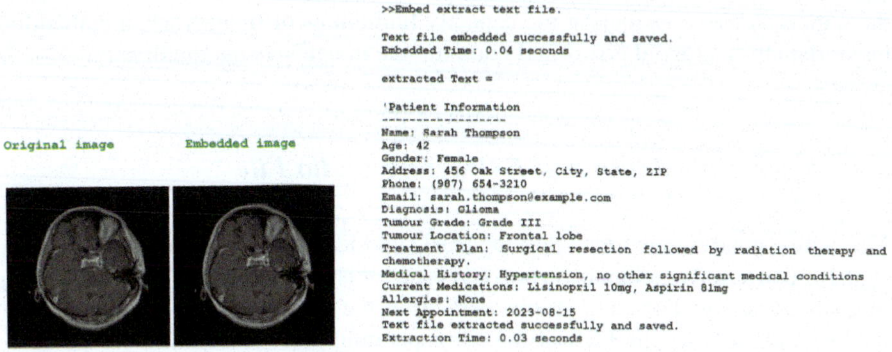

Fig. 1 Stealthy text embedding in images: concealing data without altering appearance

Table 1 Observation for insertion and extraction time of text files

Original image	Size (KB)	Embedded text file	Text file size (KB)	Insertion time (s)	Extraction time (s)	PSNR (dB)	SSIM
glioma.jpg	23	glioma.txt	1	0.05	0.009	37.57	1
glioma.jpg	23	UMNwriteup.txt	14	0.23	0.02	39.58	1
glioma.jpg	23	SampleTextFile 20 kb	20	0.25	0.03	39.92	1
meninglioma.jpg	34	SampleTextFile 20 kb	20	0.69	0.08	36.56	1
glioma.jpg	23	SampleTextFile 2000 kb	2000	**Error: Text file exceeds capacity**			

indicating that the text file size exceeds the capacity of the image. This is a limitation of the steganography technique used, as images have a limited capacity to hide data. When the data to be embedded is too large, it cannot be accommodated without compromising the image's integrity. The study evaluates text embedding in medical images using steganography. Results show that smaller text files can be efficiently embedded without significantly altering image quality. For "glioma.jpg" (23KB), embedding "glioma.txt" (1KB), UMNwriteup.txt" (14KB), and "SampleTextFile 20kb" (20KB) maintained high structural similarity (SSIM = 1) with increasing PSNR values (37.57 dB, 39.58 dB, and 39.92 dB, respectively). However, embedding the 20KB file into "meninglioma.jpg" (34 KB) resulted in a lower PSNR (36.56 dB). Insertion and extraction processes were efficient but attempts to embed a 2000 KB file failed due to image capacity limitations, highlighting constraints in the steganography technique.

Overall, the experiment demonstrates the effectiveness of the steganography technique in concealing text data within images while maintaining relatively small changes in image file size and fast processing times. However, it also highlights the importance of considering the capacity limitations of images when embedding larger data files to avoid errors and maintain the image's visual quality.

4.2 After Embedding and Extracting Audio File

The audio signal is embedded into the least significant bits of the image. The audio signal is converted to a binary string, and each bit is embedded into a corresponding pixel in the image. For each 16-bit segment of the audio signal, it iterates through the bits and modifies the LSB of the corresponding pixel value in the image based on the bit value. This is done to hide the audio data in the image. The extracted audio is normalized and rescaled to match the amplitude of the original audio signal. Embedded and Recovered audio signal is shown in Fig. 2.

Insertion and extraction time is calculated for the LSB Technique is calculated and graph is as shown in Fig. 3.

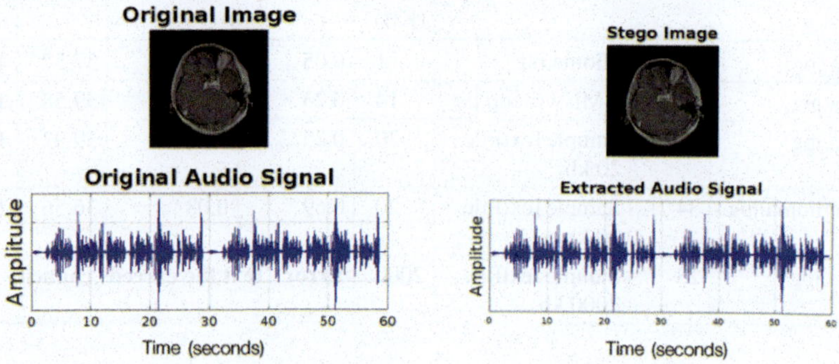

Fig. 2 Embedded and recovered audio signal

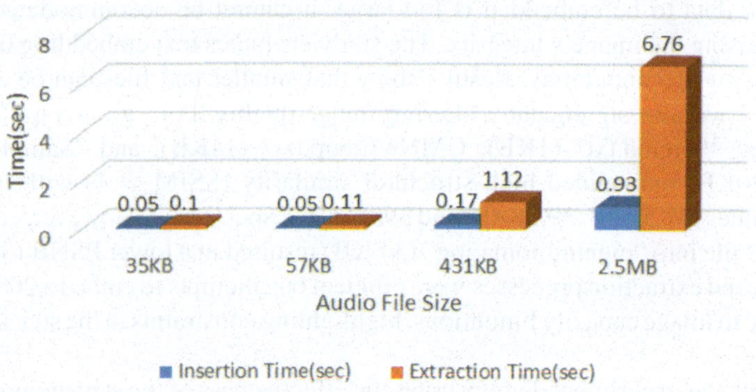

Fig. 3 Insertion time versus extraction time of embedding audio file

Table 2 Observation of PSNR of audio files

Original image	Size (in KB)	Embedded audio file	PSNR (dB)
glioma.jpg	23	button-1.wav	23.22
glioma.jpg	23	button-2.wav	21.11
glioma.jpg	23	button-3.wav	21.0
glioma.jpg	23	button-4.wav	21.0

In a series of experiments using the same 23 KB original image ("glioma.jpg"), various audio files were embedded into the image using an audio steganography technique. Experiment 1 embedded a short music clip ("button-1.wav"), resulting in a 35 KB stego image, with embedding and extraction times of 0.05 and 0.10 s, respectively. Experiment 2 employed a different music segment ("button-35.wav"), producing a larger 57 KB stego image, with similar embedding and slightly longer extraction times of 0.05 and 0.11 s. Experiment 3 embedded a text-containing audio file ("get-tysburg10.wav"), leading to a significantly larger stego image (431 KB) and longer embedding and extraction times of 0.17 and 1.12 s, respectively. Experiment 4 utilized a much larger audio file "gliomasymptom-compressed.wav," 2.5 MB), resulting in a substantially larger stego image, with embedding and extraction times of 0.93 and 6.76 s, respectively. These experiments highlight the influence of audio file size and content on stego image properties and processing times.

Table 2 displays the original image size (23 KB), different embedded audio files, their respective sizes, and the Peak Signal-to-Noise Ratio (PSNR) values. As the audio file size increases, the PSNR generally decreases, indicating a reduction in audio quality when embedding larger audio files into the same image.

The observation reveals varying Peak Signal-to-Noise Ratio (PSNR) values for different audio signals are shown in Fig. 4.

4.3 *After Embedding and Extracting Image Data*

To extract a hidden image from an embedded image, the process involves starting a timer, reading the embedded image, and specifying the number of LSBs used for embedding (1 in this case). The LSB of each pixel is extracted and used to create the extracted image. After stopping the timer, the extraction time is calculated and displayed. Finally, the embedded image, extracted hidden image, and the original hidden image are displayed in separate subplots Which is shown in Fig. 5.

In this scenario, an original image named "glioma.jpg" with a size of 23 KB was used to conceal a second image, "Ramya.jpg," which had a size of 51 KB. The process of embedding "Ramya.jpg" into "glioma.jpg" took 0.23 s, and subsequently, extracting "Ramya.jpg" from the modified image took 0.22 s. The high PSNR (Peak Signal-to-Noise Ratio) value of 46.72 dB indicates that the quality of the hidden image was well-preserved during the embedding and extraction process,

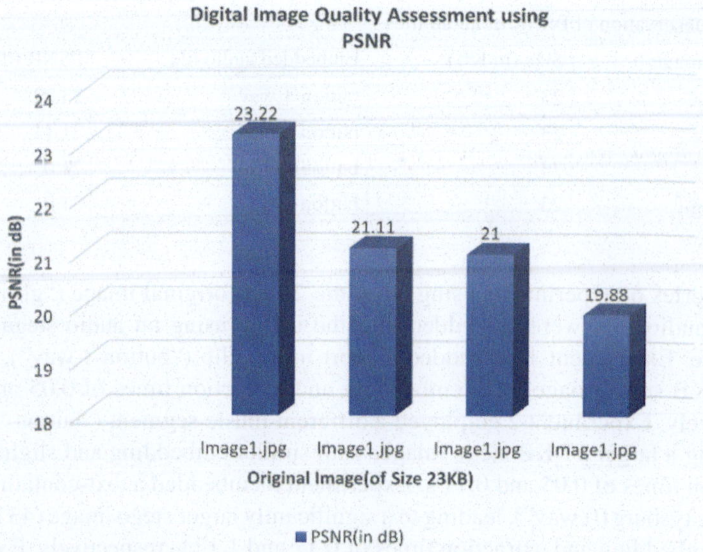

Fig. 4 Observation of PSNR for different audio signal

Fig. 5 Embedding image into cover image

with minimal loss of information. This suggests that the steganography technique used was efficient and resulted in a successful concealment of "Ramya.jpg" within "glioma.jpg" while maintaining image fidelity.

5 Conclusion and Future Work

The conducted experiment demonstrates the successful use of Least Significant Bit (LSB) steganography to embed data into images while maintaining image quality and fast processing times. The LSB steganography technique proved effective in concealing data within images. The PSNR and SSIM metrics consistently showed high values close to 1, indicating that the changes made to the cover image were imperceptible to the human eye. The insertion and extraction times for the embedded data were relatively short, even for larger file sizes. From the experiment, it can be concluded that the LSB technique, commonly used for data hiding, yields the highest PSNR value of 46.72 dB when applied to another image, suggesting minimal loss in image quality. However, when used for text data files, there is a noticeable reduction in quality, with the lowest PSNR of 23.22 dB observed for audio data embedded into the same image glioma.jpg, indicating more significant degradation in audio quality due to LSB embedding in this case. This indicates that the steganographic processes are efficient and practical for real-world applications. The experiment highlighted the importance of considering the capacity limitations of images. Attempting to embed a text file that exceeded the image's capacity resulted in an error message. This limitation should be considered when selecting the data to be embedded. The PSNR values consistently indicated high image quality, with minimal perceptible distortion. This suggests that the LSB steganography technique can be applied without significant degradation in image quality. While the experiment has provided valuable insights into the effectiveness of LSB steganography for embedding text data, there are several avenues for future research and improvement: As mentioned, LSB steganography may not provide the highest level of security against sophisticated detection methods. Future work could explore more advanced steganographic techniques that offer increased robustness and security.

Acknowledgements There is no funding for this work.

Disclosure of Interests The authors have no competing interests to declare that are relevant to the content of this article.

References

1. Abikoye OC, Ogundokun OR (2021). Efficiency of LSB steganography on medical information. Int J Electr Comput Eng (IJECE) 11:4157. https://doi.org/10.11591/ijece.v11i5.pp4157-4164
2. Jaiswal S, Pandey MK (2023) Performance analysis of image watermarking for different sub-bands using LWT and Arnold map. ECTI-CIT Trans 17(2):278–291
3. Ramyashree, Venugopala PS, Raghavendra S, Kubihal VS (2025) Enhancing secure medical data communication through integration of LSB and DCT for robust analysis in image steganography. IEEE Access 13:1566–1580. https://doi.org/10.1109/ACCESS.2024.3522957
4. Bhagat D, Bhardwaj R (2019) A survey on medical images for reversible data hiding techniques, pp 811–817. https://doi.org/10.1109/AICAI.2019.8701228
5. Li B, Junhui H, Huang J, Shi YQ (2011) A survey on image steganography and steganalysis. J Inf Hiding Multimed Signal Process
6. Hasan Talukder MS, Hasan MN, Sultan RI, Rahman M, Sarkar AK, Akter S (2022) An enhanced method for encrypting image and text data simultaneously using AES algorithm and LSB-based steganography. In: 2022 International conference on advancement in electrical and electronic engineering (ICAEEE), Gazipur, Bangladesh, pp 1–5. https://doi.org/10.1109/ICAEEE54957.2022.9836589
7. Shtayt BA, Zakaria NH, Harun NH (2021) A comprehensive review on medical image steganography based on LSB technique and potential challenges. Baghdad Sci J 18(Suppl. 2):0957

8. Hassaballah M (2020) Digital media steganography: principles, algorithms, advances. https://doi.org/10.1016/C2018-0-04865-3
9. Masoud NA, Ghazi IS (2018) A modification of least significant digit (LSD) digital watermark technique. Int J Comput Appl 179(32):4–6
10. Patel SK, Saravanan C, Patel VK (2019) Cloud-based reversible dynamic secure steganography model for embedding pathological report in medical images. Int J Comput Appl 1–9
11. Muhammad K, Ahmad J, Rho S, Baik SW (2017) Image steganography for authenticity of visual contents in social networks. Multi Tools Appl 1–20
12. Juneja M, Sandhu P (2013) An improved LSB based steganography technique for RGB color images. Int J Comput Commun Eng 2(4):513–517. https://doi.org/10.7763/IJCCE.2013.V2.238
13. Larobina M, Murino L (2013) Medical image file formats. J Digit Imaging. https://doi.org/10.1007/s10278-013-9657-9
14. Molato AD, Calanda FB (2023) A secured LSB-based image steganography using modified Collatz conjecture. In: 2023 9th international conference on advanced computing and communication systems (ICACCS), Coimbatore, India, pp 1997–2002. https://doi.org/10.1109/ICACCS57279.2023.10113029
15. Ehsan Ali UAM, Ali E, Sohrawordi M, Sultan MN (2021) A LSB based image steganography using random pixel and bit selection for high payload. Int J Math Sci Comput 7(3):24–31. https://doi.org/10.5815/ijmsc.2021.03.03
16. Bhuiyan T, Sarower AH, Karim R, Hassan M (2019) An image steganography algorithm using LSB replacement through XOR substitution. In: 2019 International conference on information and communications technology (ICOIACT), pp 44–49
17. Yue Z, Zhao Q, Zhang L, Meng D (2020) Dual adversarial network: toward real-world noise removal and noise generation. In: European conference on computer vision. Springer, pp 41–58
18. Chowdhuri P, Pal P, Si T (2023) A novel steganographic technique for medical image using SVM and IWT. Multimed Tools Appl 82:20497–20516
19. Keleş O, Ylmaz MA, Tekalp AM, Korkmaz C, Do˘gan Z (2021) On the computation of PSNR for a set of images or video. In: 2021 Picture coding symposium (PCS), Bristol, United Kingdom, pp 1–5. https://doi.org/10.1109/PCS50896.2021.9477470
20. Wang X, Chan KCK, Yu K, Dong C, Loy CC (2019, June) EDVR: video restoration with enhanced deformable convolutional networks. In: Conference on computer vision and pattern recognition (CVPR) workshops

Leveraging Machine Learning Algorithms to Scrutinize Code Structures for Security Weaknesses

Usha Divakarla and K. Chandrasekaran

Abstract A novel method for predicting software vulnerabilities is presented in this paper. It examines the classes, methods, and functions that comprise software using static analysis. By incorporating this analyzed data into machine learning models, the technique can identify any errors. When tested on publicly available datasets, our strategy performed better than earlier approaches. Engineers may concentrate on fixing significant defects as a result of this increased efficiency, which ultimately strengthens software security. By employing this technique, software development teams can proactively build more secure systems and reduce the likelihood of exploitation.

Keywords Software weaknesses · Utility analysis · Machine learning · Performance analysis · Neural network

1 Introduction

Almost every aspect of modern life depends on software systems, which power applications in industries including healthcare, banking, transportation, and communication. The increasing necessity of these systems for daily operations makes them more desirable targets for hackers. The most common method used by bad actors to access systems and result in data breaches, financial losses, and damage to their reputation is by exploiting software vulnerabilities. These risks are made worse by the complexity of software systems, which can have major security flaws due to

U. Divakarla (✉)
N M A M Institute of Technology, Nitte, Karkala, India
e-mail: ushachavali@gmail.comm

K. Chandrasekaran
National Institute of Technology Karnataka, Mangalore, India
e-mail: kch@nitk.ac.in

A. Rocha et al. (eds.), *Integrating Advanced Technologies for Enhanced Security and Efficiency*, Engineering Cyber-Physical Systems and Critical Infrastructures 14,
https://doi.org/10.1007/978-3-031-91798-1_22

even minor coding errors. Finding and addressing software development vulnerabilities requires a strong, effective, and scalable methodology in order to address these problems.

Traditional vulnerability detection methods that are still widely used in the software industry include static analysis tools and human code reviews. Even when these methods are effective, they have disadvantages. For example, manual evaluations are impractical, subjective, and time-consuming for large-scale software projects. Similarly, traditional static analysis methods that search for recognized patterns of code vulnerabilities often generate a large number of false positives. This not only puts developers through unnecessary research but also raises concerns about the tools' dependability. Additionally, these approaches frequently concentrate on the source code level, offering a comprehensive examination of the complete codebase but typically ignoring vulnerabilities found in certain components, such as functions, methods, or classes.

A rising number of people are interested in using machine learning (ML) approaches for vulnerability detection in order to get beyond these restrictions. Large dataset analysis, pattern recognition from past vulnerabilities, and more accurate and efficient problem prediction in fresh codebases are all made possible by machine learning. Even while current ML-based techniques have shown potential, the majority concentrate on the source code rather than specific software elements. This method ignores the reality that vulnerabilities frequently start at a fine level, within particular functions or procedures. Deeper understanding of the underlying causes of vulnerabilities and more accurate remediation techniques can be obtained with a more focused investigation at this level.

With an emphasis on software analysis at the function, method, and class levels, this work introduces a machine learning-driven technique that attempts to solve the inadequacies of existing approaches. From these components, static analysis techniques extract properties like code metrics, data flow, and control flow. Random Forest and Convolutional Neural Networks (CNNs), two machine learning models that are very good at detecting vulnerabilities, are trained using these characteristics. This strategy offers a comprehensive and scalable solution that adjusts to the complexity of contemporary software systems, in contrast to conventional methods.

The necessity for precise, useful insights into software vulnerabilities and the broad-level analysis of traditional approaches are intended to be reconciled by the suggested methodology. By concentrating on the essential components of software, this method lowers false positives and improves the level of detail in vulnerability detection, allowing engineers to efficiently rank the most important problems. Additionally, the strategy will continue to be flexible in the face of emerging vulnerabilities thanks to the application of machine learning, which is essential in a constantly changing cybersecurity environment.

Based on the study's evaluation of the suggested method using publically accessible datasets, it outperforms conventional techniques in terms of overall detection accuracy, precision, and recall. The findings demonstrate the potential for machine learning and granular-level analysis to transform vulnerability detection and remediation, opening the door to more dependable and secure software systems.

2 Literature Review

Software vulnerability discovery research has been going on for a while, and techniques have varied from manual code reviews to automated techniques that involve machine learning (ML). Traditional methods lack the scalability, accuracy, and adaptability required in modern software development environments, notwithstanding their occasional usefulness. Machine learning-based techniques have demonstrated potential in more accurately detecting vulnerabilities and efficiently evaluating large codebases.

In order to find vulnerabilities, manual code reviews rely on human skill. Although the contextual knowledge of reviewers is advantageous, this method is time-consuming, resource-intensive, and prone to mistakes in large-scale projects. These constraints are intended to be addressed by automated static and dynamic analysis techniques. Static analysis finds vulnerabilities like buffer overflows and SQL injection by looking at source code without running it. However, because it relies on pre-established rules, it frequently produces large false positive rates. By analyzing code in real time, dynamic analysis finds weaknesses that static approaches overlook. However, dynamic analysis cannot fully cover all cases due to execution path limitations.

Because machine learning can identify trends in previous data, it has been frequently used for vulnerability detection. To improve accuracy and scalability, a number of studies have investigated machine learning models and feature extraction strategies.

Support Vector Machines (SVMs) were used by [1] to introduce a function-level vulnerability detection method. Compared to static analysis methods, their model achieved a considerable gain in precision by utilizing features like the amount of parameters and function calls.

A decision tree-based approach that concentrates on class-level characteristics such as the quantity of methods and dependencies was presented by [2]. Because of its high recall rates, this method helped engineers find weaknesses in intricate object-oriented codebases.

Random Forest models trained on criteria such as code length and cyclomatic complexity were used by [3]. Their study made clear how crucial software metrics are to raising vulnerability prediction accuracy.

The application of Long Short-Term Memory (LSTM) networks with Convolutional Neural Networks (CNNs) was investigated by [4]. Their methodology beat conventional machine learning techniques in terms of detection rates and scalability by using hierarchical features and portraying code as sequences.

To increase coverage and lower false negatives, Pereira [5] integrated ML models with both static and dynamic analysis. Their strategy demonstrated how hybrid approaches might be used to overcome the drawbacks of individual methodologies.

The majority of vulnerability prediction techniques now in use concentrate on identifying vulnerabilities at the source code level. These methods train machine learning models to anticipate possible security flaws and use static analysis techniques to extract information from software code. To find every possible security flaw in software systems, source code-level methods might not be adequate. In order to give a more thorough level of analysis, the suggested method concentrates on examining the functions, methods, and classes of software systems. In order to better understand vulnerabilities and their effects on software systems, this method can detect possible security flaws at a more detailed level by examining functions, methods, and classes. Using static analysis techniques, this method extracts functionality from a software system's functions, methods, and classes. Machine learning models are trained using the extracted features to anticipate possible security flaws.

In general, software security relies heavily on vulnerability prediction, and the suggested method may offer a more efficient way to find possible security flaws in software systems. Through the examination of functions, methods, and classes, the suggested method can offer a more thorough and specific degree of analysis, resulting in the creation of software systems that are more safe.

2.1 *Comparative Summary*

Machine learning-based techniques offer notable benefits in terms of scalability, adaptability, and accuracy, whereas conventional methods offer baseline capabilities. Many ML models now in use, however, ignore vulnerabilities at granular levels like functions and methods in favor of source code-level analysis. Our disparity emphasizes the necessity of more focused strategies, like the one our study suggests. The thorough review, as displayed in Table 1, highlights the development of vulnerability detection techniques and lays the foundation for the suggested methodology, which addresses the shortcomings of current approaches by emphasizing granular function, method, and class-level analysis.

2.2 *Gap Analysis*

The literature on vulnerability identification is extensive and covers methods like static and dynamic analysis. Regretfully, little study has been done on how functions, classes, and methods are applied for vulnerability detection or how well machine learning approaches work to increase the accuracy of vulnerability detection. In order to improve the accuracy of vulnerability detection, this topic entails examining functions, classes, and methods as well as utilizing machine learning approaches.

Analysis of software system functions, methods, and classes is one of the suggested methodologies to vulnerability prediction. This strategy is predicated on the idea that classes, functions, and methods are the fundamental units of software systems, and that vulnerabilities are likely to manifest at this level of detail.

Table 1 Comparative summary of existing models

Refs.	Approach	Focus level	Techniques	Strengths	Limitations
[1]	ML-Based	Function	SVM	High precision, function-level focus	Limited scalability
[2]	ML-Based	Class	Decision tree	High recall, good for OO systems	High false positives
[3]	ML-Based	Source code	Random forest	Incorporates code metrics	Broad-level focus
[4]	ML-Based	Source code	CNN, LSTM	High scalability, sequence learning	Computationally expensive
[5]	Hybrid	Source code	Static + dynamic analysis	Reduced false negatives	Complex implementation
[6]	Static analysis	Source code	Rule-based	Simplicity	High false positives
[7]	ML-Based	Source code	Neural network	High accuracy for structured data	Data labeling challenges
[8]	Deep learning	Source code	CNN	Superior for large datasets	Overfitting risk
[9]	ML-Based	Source code	Multi-task learning	Explores feature correlations	Lack of generalizability
[10]	Dataflow analysis + ML	Function/ method	Decision tree	Effective for SQL injection detection	Narrow applicability
[11]	ML-Based	Source code	Random forest, decision tree	Good balance of precision and recall	Lacks granularity
[12]	Deep learning	Source code	CNN, LSTM	High accuracy, low false negatives	Requires large labeled datasets
[13]	Hybrid	Source code	Static + dynamic analysis	Comprehensive coverage	Slow processing
[14]	Deep learning	Source code	Graph neural networks (GNN)	Captures structural relationships	Computationally intensive

3 Proposed Methodology

The proposed methodology employs static analysis and machine learning (ML) to identify vulnerabilities at a fine level, with an emphasis on functions, methods, and classes. In order to identify each component for additional analysis, the process begins with the extraction of the source code. Then, using static analysis approaches, aspects like data flow (such as input utilization), control flow (such as execution pathways), and code metrics (such as cyclomatic complexity and lines of code) are extracted.

Tokenization, encoding, and normalization are used as preprocessing techniques to get the features ready for machine learning models. The two primary machine learning models used are Random Forest, which offers interpretability and resilience, and Convolutional Neural Networks (CNNs), which excel in extracting hierarchical features from serialized code representations.

For training the models, a publicly available dataset of over 1.27 million annotated functions with known vulnerabilities is structured by Common Weakness Enumeration (CWE). Precision, recall, F1 score, and AUC-ROC are among the metrics used to test the model's accuracy and robustness during the assessment phase.

Because it reduces false positives, ensures precise detection, and preserves scalability, this approach can be advantageous for modern software systems. Granular analysis and machine learning models work together to make useful recommendations for improving software security while overcoming the shortcomings of traditional approaches.

A thorough analysis of functions, methods, and classes is emphasized in the recommended procedure for detecting software vulnerabilities, as shown in the flowchart in Fig. 1. The first step is to extract the source code in order to identify the key components of the software. In this step, functions, methods, and classes are isolated to provide particular data for further research.

In the feature extraction process, these components are then analyzed through the use of static analysis tools. Data Flow, which detects execution paths and vulnerabilities like buffer overflows; Code Metrics, which show possible trouble spots like cyclomatic complexity and lines of code; and Control Flow, which monitors input usage to detect problems like SQL injection, are important features.

By preprocessing the retrieved characteristics, machine learning is enabled. Encoding categorical data into numerical representations, tokenization, and homogeneity normalization are all components of this process. This phase involves organizing and getting the data ready for model training.

Machine learning models trained on pre-processed data include Random Forest and CNNs. In serialized code, Random Forest yields dependable, intelligible outputs, whereas CNNs excel at spotting complex patterns.

The effectiveness of the models is then evaluated and validated using metrics including precision, recall, F1 score, and AUC-ROC. By integrating accuracy and scalability, this technique improves software security through the efficient identification of flaws.

4 Result Analysis

To evaluate the proposed methodology, a publicly available dataset of over 1.27 million functions categorized using Common Weakness Enumeration (CWE) categories was used. Standard metrics such as AUC-ROC, F1 score, precision, and recall were employed to compare the Random Forest and CNN models to traditional static analytic techniques.

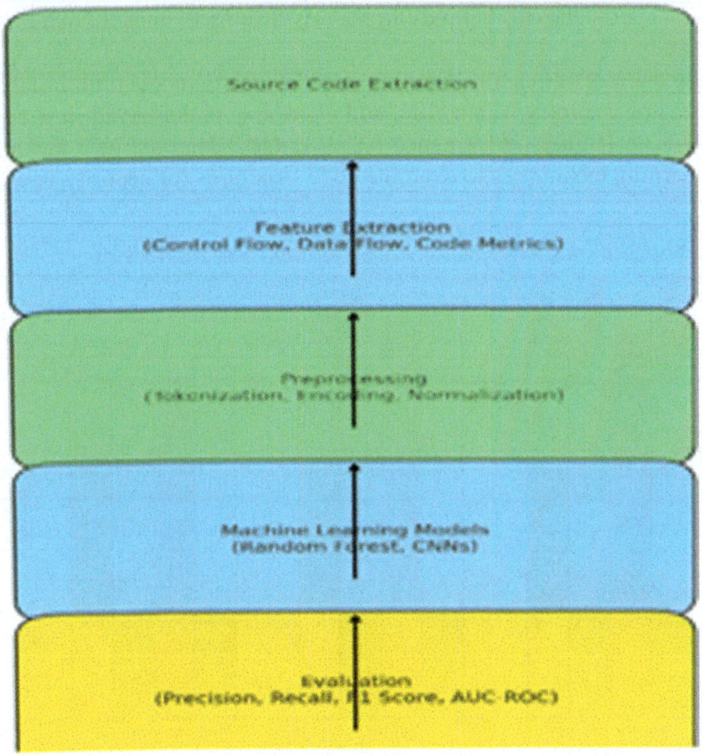

Fig. 1 The data flow chart of the proposed method

Outperforming all other methods, the CNN model achieved a 95% precision, 92% recall, 93% F1 score, and 0.97 AUC-ROC. These results show how well the model can identify complex patterns in representations of serialized codes. The Random Forest model showed efficacy and resilience while working with feature-rich data, achieving 90% recall and 93% precision. With an average recall of 80% and precision of 85%, conventional static analysis techniques fared worse, mostly due to significant misleading.

These results validate that the proposed approach is successful in increasing the accuracy and scalability of vulnerability detection. While Random Forest offers a more computationally efficient solution, CNN's superior performance demonstrates the usefulness of deep learning architectures for handling intricate software systems. The combination of machine learning models and granular analysis ensures accurate and practical insights for developers.

Figure 2 presents the performance metrics (precision, recall, F1 score, and AUC-ROC) for the Random Forest, CNN, and Traditional Methods models that are being evaluated. Although Random Forest offers dependable performance with computational efficiency, the CNN model performs better in every category. The fact that conventional approaches lag behind emphasises the effectiveness of machine learning techniques.

The performance of machine learning-based analysis versus traditional code analysis in terms of accuracy, scalability, and efficiency is displayed in a comparative graph in Fig. 3. Higher performance on all criteria demonstrates the potential of machine learning techniques to examine code structures for security flaws.

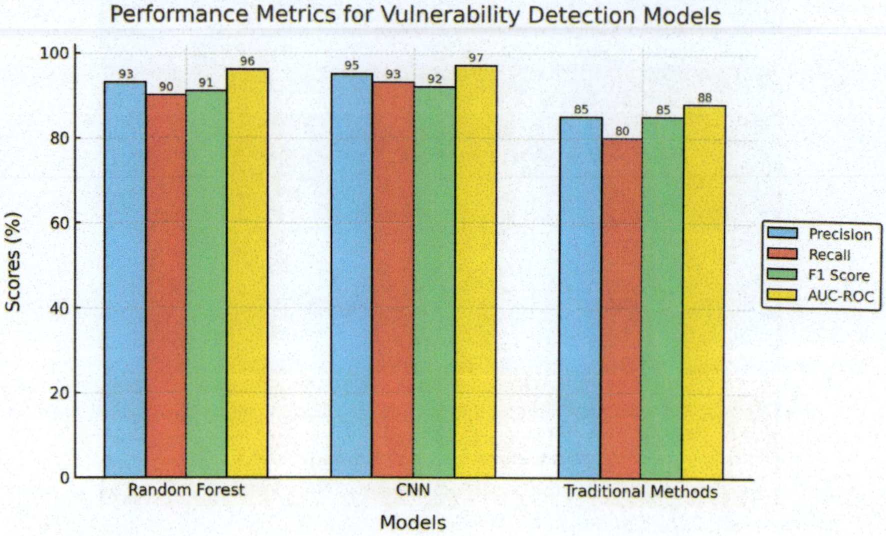

Fig. 2 Performance metrics of the models

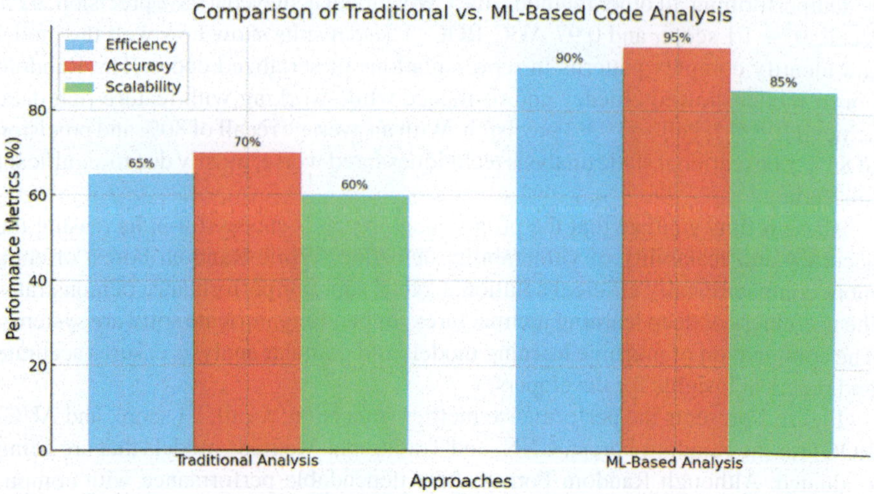

Fig. 3 Comparison of traditional and ML based code analysis

5 Conclusion and Future Scope

A new machine learning-based method for identifying software flaws at the function, method, and class levels is presented in this paper. In order to train machine learning models, such as Random Forest and Convolutional Neural Networks (CNNs), significant aspects like control flow, data flow, and code metrics were extracted by utilizing static analysis techniques. The assessment, which was carried out on a sizable dataset comprising more than 1.27 million functions, shows how successful the suggested methodology is.

With a precision of 95%, recall of 92%, and an AUC-ROC of 0.97, the CNN model demonstrated its capacity to handle intricate code patterns and obtained the best performance scores. Additionally, Random Forest demonstrated strong performance, providing a compromise between computational economy and precision. The suggested method offers a more detailed level of analysis and drastically lowers false positives when compared to conventional static analysis tools, allowing developers to find and fix vulnerabilities more precisely.

The combination of machine learning and detailed analysis has greatly improved software security. This approach not only increases the accuracy of vulnerability identification but also scales effectively to meet the needs of modern software development. Future studies will focus on developing this methodology to address novel cybersecurity challenges and incorporate dynamic analysis.

The proposed approach can be extended to address novel challenges in software vulnerability detection. Future research will use dynamic analysis techniques to augment static analysis in order to find runtime vulnerabilities. Improvements to machine learning models, such the usage of graph-based neural networks, may better reflect complex code structures. Additionally, by using semi-supervised or unsupervised learning techniques, the shortage of labeled datasets would be overcome, enabling broader applicability. Including other programming languages and frameworks in the dataset would increase its generalizability. These advancements aim to create a comprehensive, adaptable method for accurately and scalable vulnerability detection in modern software.

References

1. Chang X, Yang Y (2017) Semisupervised feature analysis by mining correlations among multiple tasks. IEEE Trans Neural Netw Learn Syst 28(10):2294–2305. https://doi.org/10.1109/TNNLS.2016.2582746
2. Davari M, Zulkernine M, Jaafar F (2017) An automatic software vulnerability classification framework. In: 2017 International conference on software security and assurance (ICSSA), Altoona, PA, USA, pp 44–49. https://doi.org/10.1109/ICSSA.2017.27
3. Medeiros N, Ivaki N, Costa P, Vieira M (2020) Vulnerable code detection using software metrics and machine learning. IEEE Access 8:219174–219198. https://doi.org/10.1109/ACCESS.2020.3041181

4. Zagane M, Abdi MK, Alenezi M (2020) Deep learning for software vulnerabilities detection using code metrics. IEEE Access 8:74562–74570. https://doi.org/10.1109/ACCESS.2020.2988557
5. Pereira JD (2020) Techniques and tools for advanced software vulnerability detection. In: 2020 IEEE international symposium on software reliability engineering workshops (ISSREW), Coimbra, Portugal, pp 123–126. https://doi.org/10.1109/ISSREW51248.2020.00049
6. Li Z, Zou D, Tang J, Zhang Z, Sun M, Jin H (2019) A comparative study of deep learning-based vulnerability detection system. IEEE Access 7:103184–103197. https://doi.org/10.1109/ACCESS.2019.2930578
7. Wang W, Dumont F, Niu N, Horton G (2022) Detecting software security vulnerabilities via requirements dependency analysis. IEEE Trans Softw Eng 48(5):1665–1675. https://doi.org/10.1109/TSE.2020.3030745
8. Lin W, Cai S (2021) An empirical study on vulnerability detection for source code software based on deep learning. In: 2021 IEEE 21st international conference on software quality, reliability and security companion (QRS-C), Hainan, China, pp 1159–1160. https://doi.org/10.1109/QRS-C55045.2021.00173
9. Andrzejak A, Eichler F, Ghanavati M (2017) Detection of memory leaks in C/C++ code via machine learning. In: 2017 IEEE international symposium on software reliability engineering workshops (ISSREW), Toulouse, France, pp 252–258. https://doi.org/10.1109/ISSREW.2017.72
10. Kronjee J, Hommersom A, Vranken H (2018) Discovering software vulnerabilities using data-flow analysis and machine learning. In: Proceedings of the 13th international conference on availability, reliability and security (ARES 2018). Association for Computing Machinery, New York, NY, USA, Article 6, pp 1–10. https://doi.org/10.1145/3230833.3230856
11. Peerzada B, Kumar D (2021) Analyzing software vulnerabilities using machine learning. In: 2021 9th international conference on reliability, Infocom technologies and optimization (trends and future directions) (ICRITO), Noida, India, pp 1–4. https://doi.org/10.1109/ICRITO51393.2021.9596509
12. Ahmed SJ, Taha DB (2022) Machine learning for software vulnerability detection: a survey. In: 2022 8th international conference on contemporary information technology and mathematics (ICCITM), Mosul, Iraq, pp 66–72. https://doi.org/10.1109/ICCITM56309.2022.10031734
13. Kim S, Kim R, Park YB (2016) Software vulnerability detection methodology combined with static and dynamic analysis. Wirel Pers Commun 89:777–793
14. Li X, Wang L, Xin Y, Yang Y, Tang Q, Chen Y (2021) Automated software vulnerability detection based on hybrid neural network. Appl Sci 11:3201. https://doi.org/10.3390/app11073201

VoiceGuard: Leveraging Deep Voice Recognition for Advanced Deepfake Detection

Akshaya Rathod, Prabhakar Kandukuri, and Medikonda Asha Kiran

Abstract Given the increasing prevalence of deepfake technology, this paper presents a comprehensive study on applying deep voice recognition methods for effective deepfake detection. We explore techniques for differentiating authentic and fabricated audio recordings, including Long Short-Term Memory (LSTM)-based models and the WIRENetSpoofImprovedEnhanced architecture. Our experiments involve natural and synthetic speech samples tested on a balanced dataset sourced from Kaggle. Through detailed evaluation and analysis, we demonstrate that our proposed approach effectively combats the spread of deepfake audio content. While visual deepfake detection has significantly progressed, detecting manipulated audio remains challenging. Our research aims to enhance automated systems' ability to identify and mitigate the distribution of deepfake content by focusing on key characteristics that separate genuine from altered voice samples. This study contributes significantly to the ongoing efforts to safeguard the integrity of media and digital communication.

Keywords Deepfake detection · Convolutional neural networks · Artificial intelligence · LSTM · Hybrid models

1 Introduction

The rise of deepfake technology poses a growing challenge to the authenticity of digital media. Created using advanced AI algorithms, deepfakes manipulated audio and video content have become more sophisticated and are being used to spread

A. Rathod · P. Kandukuri
Chaitanya Bharathi Institute of Technology, Hyderabad, India

M. A. Kiran (✉)
School of Engineering, Anurag University, Hyderabad, India
e-mail: ashakiran2@gmail.com

A. Rocha et al. (eds.), *Integrating Advanced Technologies for Enhanced Security and Efficiency*, Engineering Cyber-Physical Systems and Critical Infrastructures 14,
https://doi.org/10.1007/978-3-031-91798-1_23

false information, impersonate individuals, and undermine trust in media. This highlights the urgent need for reliable methods to differentiate between genuine and fake content, especially in sensitive areas like politics, finance, and personal identity.

Recognizing the seriousness of this issue, our research focuses on developing advanced methods to detect deepfake audio, leveraging the power of deep learning. Specifically, we investigate deep voice recognition techniques, which have shown promise in identifying subtle differences between authentic and synthetic speech. We aim to create robust and reliable methods to accurately distinguish between real and fake audio recordings, thereby contributing to the broader effort to combat deepfake threats [1]. To address the problems, several authors proposed deepfake detection schemes [2–17]. Their schemes having the less training accuracy and test accuracy. To increase the accuracy in deepfake detection, we used the WIRENet-SpoofImprovedEnhanced technique rather than Long Short-Term Memory (LSTM) [18–22].

Our scheme explores several state-of-the-art approaches, including models based on LSTM networks and a novel architecture we've developed called WIRENet-SpoofImprovedEnhanced. LSTM models are particularly suited for tasks involving data sequences, such as audio recordings, as they can capture temporal dependencies and patterns crucial for identifying deepfake content. The WIRENetSpoofImprovedEnhanced architecture builds on existing models, introducing enhancements to improve accuracy and robustness in deepfake detection [16, 19].

We analyzed a curated Kaggle dataset of genuine and synthetic speech samples to evaluate our methods. Results showed our techniques effectively identify deepfake audio, reducing misinformation risks. This research underscores the need for robust detection mechanisms to protect the integrity of digital communication and media.

1.1 Contributions

The contributions of the research on "Leveraging Deep Voice Recognition for Advanced Deepfake Detection" could be outlined as follows:

- **New Detection Methods**: Developed the WIRENetSpoofImprovedEnhanced architecture and optimized LSTM-based models for accurately detecting deepfake audio.
- **Comprehensive Testing**: Conducted thorough evaluations using a balanced dataset from Kaggle, demonstrating the effectiveness of the proposed methods.
- **Advancement in Audio Deepfake Detection**: Addressed the relatively underexplored area of audio deepfakes, significantly improving detection capabilities.
- **Impact on Digital Media Integrity**: Enhanced the ability to maintain trust in digital communication by providing reliable tools for identifying deepfake audio.
- **Foundation for Future Research**: Introduced new benchmarks and methodologies that encourage further research in deepfake detection.

1.2 Paper Organization

The paper is organized as follows: Sect. 2 provides a comprehensive Literature Review of existing studies. Section 3 identifies the problem. Section 4 details the methodology, Sect. 5 presents the Experiment Analysis, and Sect. 6 concludes with recommendations for future research.

2 Literature Review

Identifying deepfake content has attracted significant interest recently, with most research focused on visual content. Rapid advances in AI and deep learning have made it possible to create incredibly realistic false videos, making developing reliable detection methods necessary. However, although audio-based deepfake detection is crucial to maintaining the integrity of digital communication and media, the field still needs to be explored.

Korshunov and Marcel [1] assessed deepfake videos' vulnerability. Their research evaluated existing biometric systems' resilience to deepfake attacks, highlighting the need for robust detection frameworks in critical applications like security. Anagha et al. [2] explored audio deepfake detection using deep learning, mainly focused on convolutional and recurrent neural networks. Their approach demonstrated the effectiveness of machine learning models in identifying synthetic audio across various scenarios. Dişken [3] proposed using complementary regional energy features for spoofed speech detection. This method improved the robustness of speech-based deepfake detection by leveraging localized audio characteristics, enhancing detection accuracy. Huang and Pun [4] presented a hybrid feature extraction method that combines self-attention techniques to detect deepfake audio. Their model's strong detection of replay and generative audio alterations highlighted the significance of attention-based architectures. Chakravarty and Dua [5] concentrated on enhancing feature extraction to identify audio impersonation attacks in Hindi. Using cutting-edge machine learning algorithms, their work identified linguistically unique difficulties and offered customized solutions.

The D-Fence Layer, an ensemble framework that combines several detection techniques for thorough deepfake detection, was proposed by Amerini et al. [6]. This method highlighted how crucial ensemble learning is to raising overall detection reliability. To identify deepfake voices, Kulangareth et al. [7] investigated speech pause patterns. Their research shows that identifying real and fake audio can be done by examining the temporal gaps in speech. A binary classification technique for identifying artificially created sounds was presented by Kumar et al. [8]. Their method used conventional machine learning approaches to maintain the integrity of audio content in various applications. With an emphasis on domain-specific uses of deepfake detection, AlZu'bi et al. [9] talk about applying natural language processing

Table 1 Functional comparison of existing schemes with proposed scheme

Model	Training accuracy (%)	Test accuracy (%)
LSTM-based model	70.24	83.15
Proposed model	94.93	93.84

(NLP) approaches to identify artificial voices in biological literature, social media, and medical paperwork.

Wei et al. [10] present a green learning method for faked speech recognition, emphasizing energy-efficient model training without sacrificing functionality. Yanagi et al. [11] suggest using text and waveform embedding to identify deepfake voices on social media. Their research strongly emphasizes utilizing hybrid approaches to increase detection precision in practical settings. Li et al. [12] introduce a dataset for cross-domain audio deepfake detection and provide an in-depth analysis of detection methods. The dataset aims to bridge the gap between controlled environments and real-world conditions. Firc et al. [13] explore spectrogram analysis for deepfake speech detection, emphasizing visual signal processing to distinguish between natural and synthetic speech patterns.

This study highlights the growing need for sustainable AI practices. Danylov [14] explores open-source and proprietary software for audio deepfakes and voice cloning. Valente et al. [17] apply Convolutional Neural Networks (CNNs) to speech audio deepfake detection, showcasing the efficacy of CNNs in capturing subtle spectral differences in audio data. The paper identifies growth areas and challenges in adopting these technologies for detection. Pianese et al. [16] propose a training-free approach using large-scale pre-trained models for deepfake voice recognition. Tambe et al. [23] utilize ANN and LSTM architectures to identify deepfake videos, extending detection strategies to multimedia content beyond audio. This method reduces computational overhead while leveraging powerful pre-existing models. The functional Comparison of existing schemes with the proposed scheme is shown in Table 1.

3 Problem Identification

The rise of deepfake audio technology poses severe risks to the integrity of digital media. While visual deepfakes have been widely studied, detecting altered audio remains a complex challenge. Deepfake audio can convincingly mimic voices, enabling deception and manipulation. As these techniques become more advanced, it becomes harder to distinguish real from fake audio, especially as such recordings are easily integrated into digital communication.

It's crucial to develop accurate detection methods and invest in research to stay ahead of evolving deepfake technologies. This will help maintain the trustworthiness and security of digital media.

4 Methodology

This paper uses cutting-edge deep voice recognition techniques to present a novel approach for deepfake detection. Our method analyzes audio recordings and categorizes them as either legitimate or altered by utilizing a unique architecture called WIRENetSpoofImprovedEnhanced and models based on LSTM.

Data Acquisition and Preprocessing: Our approach's cornerstone is acquiring a solid and extensive dataset, which we obtained from Kaggle. The dataset is well-balanced. It includes a range of voice samples classified as authentic and fraudulent recordings. Our deepfake detection algorithms are well-trained and evaluated thanks to this rich dataset.

Exploratory Data Analysis: We used EDA to learn more about the dataset. To do this, you must see how actual and fake samples are distributed, look for any imbalances in the classes, and comprehend the fundamental statistics of the audio attributes Visualization tools such as matplotlib and seaborn are used to create distribution plots and histograms.

Data Preprocessing: To get the audio data ready for model training, data preprocessing is essential. The data preprocessing comprises:

- **Label Encoding**: A label encoder converts the labels designating authentic or counterfeit audio into numerical values.
- **Feature Scaling**: To standardize the data and enhance the model's performance, the features extracted from the audio recordings are scaled using a standard scaler.
- **Data Reshaping**: The audio data is modified to conform to the deep learning models' input specifications. Data reshaping is crucial for LSTM networks, particularly as they need input data in a specific format to capture temporal connections adequately.

4.1 *Model Architectures*

To identify deepfake audio, our approach uses two main deep learning architectures: The WIRENetSpoofImprovedEnhanced architecture and an LSTM-based model.

LSTM-Based Model: Sequential audio data processing is the aim of the LSTM-based paradigm. Because LSTMs are a Recurrent Neural Network (RNN) type that can learn lengthy associations, they are useful for applications using time-series data. Our method uses the LSTM model.

Captures Dependencies in Time: The system learns the temporal patterns and dependencies that determine whether the audio is genuine or fake through the sequential processing of the data. To prevent overfitting, the model architecture's several LSTM layers are divided from one another by dropout layers. The last output stage uses a sigmoid activation function to determine whether the audio is artificial or natural.

WIRENetSpoofImprovedEnhanced Architecture: The second model is the more complex WIRENetSpoofImprovedEnhanced architecture, which extracts both local and global patterns from audio data by combining convolutional and recurrent layers.

- **Convolutional Layers**: By removing local characteristics from the audio stream, these layers identify subtle fluctuations that might not be audible in raw audio data.
- **Batch Normalization and Pooling**: Batch normalization layers and the learning process are stabilized. Pooling layers improve the model's efficiency and speed up convergence by reducing the size of the features.
- **Bidirectional LSTM Layers**: Bidirectional LSTMs fully understand the temporal context by processing the audio data forward and backward.
- **Dropout Layers**: Dropout layers are frequently used to reduce overfitting and enhance the model's generalizability.

4.2 *Training and Evaluation*

The models are trained on the processed dataset using a stratified train-test split to ensure a balanced distribution of real and fake samples in both training and testing sets. The training procedure consists of the following steps.

Optimizer and Loss Function: Due to the binary nature of the classification issue, binary cross-entropy is chosen as the loss function, and the Adam optimizer is used due to its effectiveness and versatility.

Training Process: Both models employ early pausing and dropout strategies after a predefined number of epochs to prevent overfitting.

Validation: To monitor the models' performance and make any required hyperparameter adjustments, a subset of the dataset is used for validation.

Testing: Following training, the test set assesses the models' memory, accuracy, precision, F1 score, and other pertinent metrics. Each model's advantages and disadvantages are assessed based on how well it detects deepfake sounds.

In our deepfake detection methodology, we use deep speech recognition techniques and adopt a systematic approach to data collection, preprocessing, model construction, training, and evaluation. Using both LSTM-based models and the WIRENetSpoofImprovedEnhanced architecture, we effectively recognize audio recordings by combining the benefits of sequential and convolutional neural networks. This comprehensive approach aims to respond strongly to the growing complexity of audio deepfake detection.

4.3 Technologies Used

The technologies used in this model are described in this section.

Deep learning: To analyze complex, unstructured data, such as text, audio, and images, deep learning (DL) uses multilayered neural networks. Deepfake detection uses DL algorithms to find temporal correlations and patterns in audio to differentiate between authentic and fraudulent recordings (Fig. 1).

LSTM: LSTM, a type of RNN, is effective for sequential data like audio, capturing long-term dependencies. It helps extract key temporal features needed to identify deepfake audio.

The Fig. 2 provides a detailed comparison of Mel-Frequency Cepstral Coefficients (MFCCs) across the two categories (real and fake).

Layout: There are multiple subplots, each showing the distribution of an individual MFCC feature (e.g., MFCC1, MFCC2, ..., MFCC20). Each histogram overlay uses two colors to represent the two classes. Purpose: To analyze variations in MFCC features between the categories and determine their relevance for classification tasks.

CNN: Convolutional Neural Networks (CNNs) are designed to process gridlike data, such as images and audio spectrograms, and excel at identifying local patterns. CNNs use convolutional, pooling, and fully connected layers to extract and classify features. Recent advances include efficient architectures for mobile devices and integration with other models, like RNNs.

WIRENetSpoofImprovedEnhanced Architecture: WIRENetSpoofImprovedEnhanced combines CNNs and LSTMs to capture sequential audio patterns and local features from spectrograms, improving deepfake audio detection. Figure 1 shows

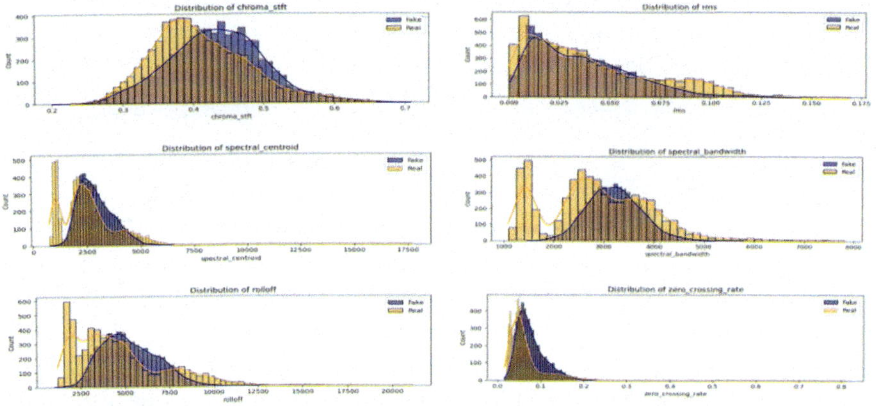

Fig. 1 Comparison of feature distributions

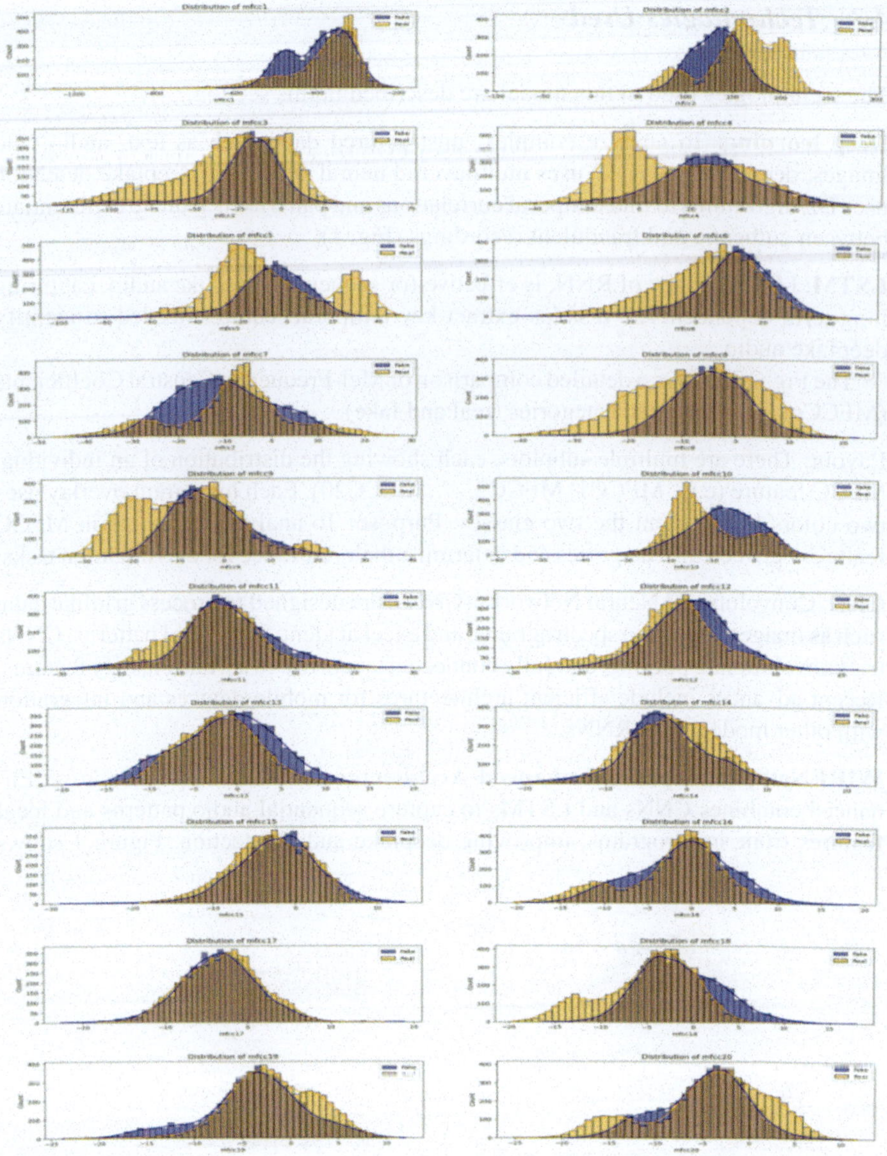

Fig. 2 MFCC distribution

histograms comparing the distributions of various audio features for two categories (e.g., labeled as real and fake). Key features include chromastft, RMS, spectralcentroid, spectralbandwidth, rolloff, and zerocrossingrate.

Layout: Each subplot represents the distribution of a specific feature for both classes, overlaid with different colors (e.g., yellow and blue). Purpose: To assess the distinguishability of features between the two classes, which might indicate feature significance for classification.

5 Experiment Analysis

Model Deployment: This project uses an LSTM-based model to capture temporal patterns in audio and the WIRENetSpoofImprovedEnhanced model, combining convolutional and recurrent layers for better feature extraction. Both models are built with Keras.

Model Training and Evaluation: The models are trained using the BinaryCrossentropy loss function and Adam optimizer, with performance monitored via a validation split. Training involves multiple epochs and batch processing to optimize parameters. After training, the model's accuracy and loss are evaluated on the test set, and performance is visualized through accuracy and loss plots for both training and validation.

Model Deployment and Visualization: The trained models predict the authenticity of new audio samples. Additionally, Keras's plot model function visualizes the architecture of the models, showing their layers and connections. This work addresses the challenge of detecting deepfake audio using advanced deep learning techniques and a robust implementation framework, contributing to efforts to ensure the authenticity of digital content. Figure 3 visualizes the correlation coefficients between various audio features and labels in a dataset. Features such as *chromastft*, RMS, *spectralcentroid*, MFCC coefficients (e.g., MFCC1, MFCC2) are shown along the axes. Deep red indicates a strong positive correlation (close to $+1$). Blue signifies a strong negative correlation (close to -1). White or lighter shades suggest weak or no correlation (close to 0). Purpose: To highlight relationships between audio features potentially useful for dimensionality reduction or feature selection.

5.1 Experimental Results

In our thorough investigation of deepfake detection using deep voice recognition, we assessed two main models: the WIRENetSpoofImprovedEnhanced architecture and an LSTM-based model. The LSTM-based model uses long-short-term memory networks to extract temporal dependencies from audio input, ultimately obtaining 70.24% training accuracy and 83.15% test accuracy. This performance demonstrates its promise in identifying deepfake noises despite its limitations in capturing complex patterns.

The LSTM architecture greatly outperformed the WIRENetSpoofImprovedEnhanced model, which combines bidirectional LSTMs and Convolutional Neural Networks (CNNs). This architecture used LSTM layers to model sequential data and

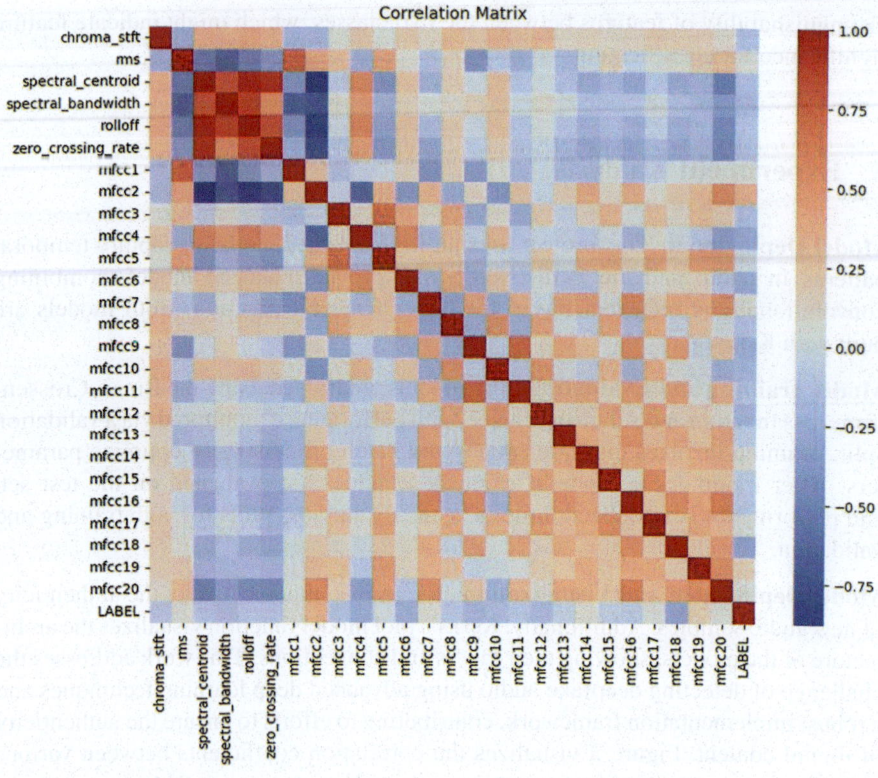

Fig. 3 Heatmap of the correlation matrix

convolutional layers to extract local features from spectrograms, yielding a training accuracy of 94.93% and a test accuracy of 93.84%. Using dropout layers and batch normalization effectively reduced overfitting and improved generalization. The accuracy graph for the suggested model is displayed in Fig. 4. We used the kaggle data set to implement the proposed scheme [24] (Fig. 5).

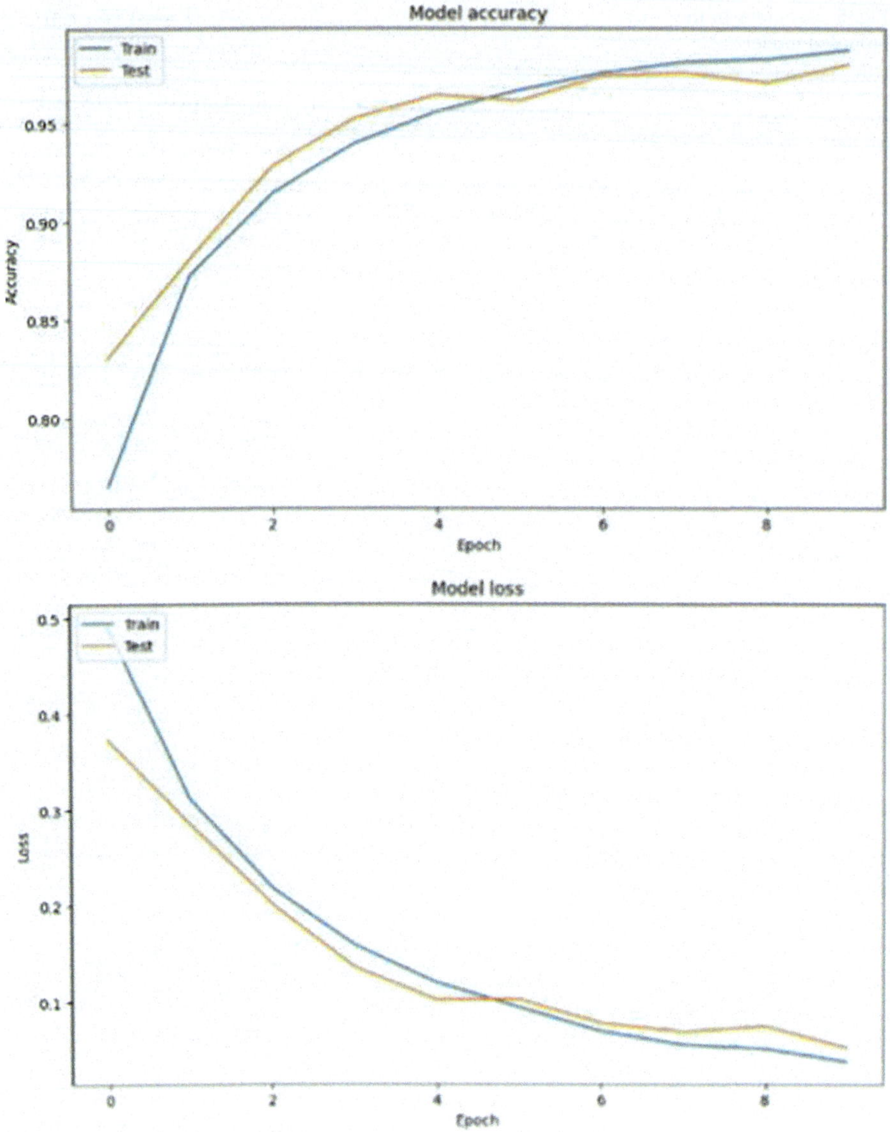

Fig. 4 Accuracy graph of the proposed model

Fig. 5 Accuracy of the proposed model

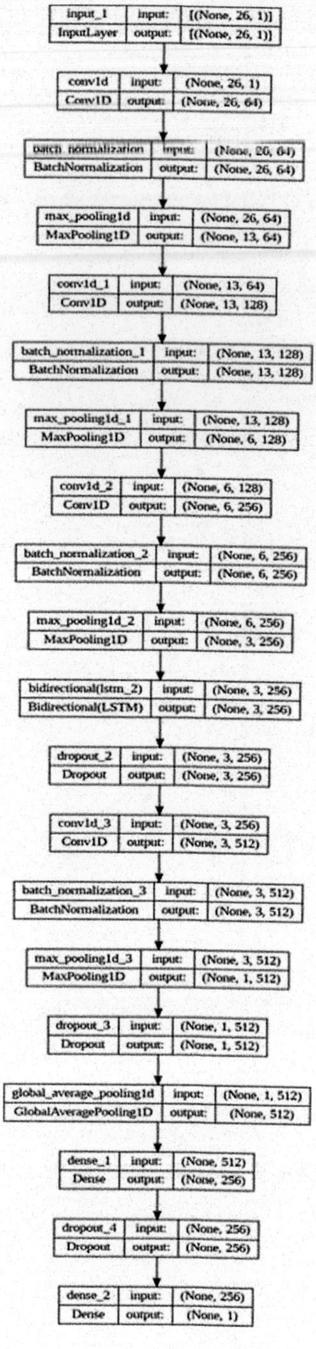

6 Conclusion

Two deepfake detection models were compared in the study: the LSTM-based model and the WIRENetSpoofIm-provedEnhanced architecture. The LSTM model obtained 83.15% test accuracy and 70.2% training accuracy, demonstrating promise but having trouble with intricate patterns. The WIRENetSpoofImprove-dEnhanced model, on the other hand, performed better, achieving 93.93% training accuracy and 93.84% test accuracy. This architecture uses Convolutional layers for feature extraction, while LSTMs are used for sequential modeling. Bidirectional LSTMs, dropout layers, and batch normalization are added to minimize overfitting. The outcomes demonstrate how well the WIRENetSpoofImprovedEnhanced model detects deepfakes, providing a reliable means of guaranteeing the integrity of digital media.

Future Scope: Future research may investigate using cutting-edge architectures like GANs and reinforcement learning to improve detection accuracy and robustness. Multimodal fusion, which combines visual and audio input, may enhance the detection of complex deepfakes. Model flexibility would also be improved via transfer learning and domain adaptation, enabling them to use insights from similar activities to better respond to new deepfake threats.

Data Availability Statement The data supporting the findings of this study are publicly accessible in the Kaggle repository at https://www.kaggle.com/datasets/birdy654/deep-voice-deepfake-voice-recognition.

References

1. Korshunov P, Marcel S (2019) Vulnerability assessment and detection of deepfake videos. In: 2019 International conference on biometrics (ICB). IEEE, pp 1–6
2. Anagha R, Arya A, Narayan VH, Abhishek S, Anjali T (2023) Audio deepfake detection using deep learning. In: 2023 12th international conference on system modeling & advancement in research trends (SMART). IEEE, pp 176–181
3. Dişken G (2024) Complementary regional energy features for spoofed speech detection. Comput Speech Lang 85:101602
4. Huang L, Pun C-M (2024) Self-attention and hybrid features for replay and deep-fake audio detection. arXiv preprint arXiv:2401.05614
5. Chakravarty N, Dua M (2024) An improved feature extraction for Hindi language audio impersonation attack detection. Multimed Tools Appl 1–26
6. Amerini I, Menon VG et al (2024) D-Fence layer: an ensemble framework for comprehensive deepfake detection. Multimed Tools Appl 1–24
7. Kulangareth NV, Kaufman J, Oreskovic J, Fossat Y (2024) Investigation of deepfake voice detection using speech pause patterns: algorithm development and validation. JMIR Biomed Eng 9:e56245
8. Kumar V, Kapoor A, Chaudhary RR, Gupta L, Khokhar D (2024) Preserving integrity: a binary classification approach to unmasking artificially generated voices in the age of deepfakes. In: 2024 11th international conference on computing for sustainable global development (INDIACom). IEEE, pp 1449–1454

9. AlZu'bi S, Alqudah A, Quiam F, Mughaid A (n.d.) Human voice in health data: NLP for detecting synthetic voices in medical documentation, social media, and biological literature. In: Social media, and biological literature
10. Wei C, Pang R, Kuo C-CJ (2024) A green learning approach to spoofed speech detection. In: ICASSP 2024–2024 IEEE international conference on acoustics, speech and signal processing (ICASSP). IEEE, pp 12956 12960
11. Yanagi Y, Orihara R, Tahara Y, Sei Y, Alumäe T, Ohsuga A (2024) The proposal of countermeasures for deepfake voices on social media considering waveform and text embedding. In: Ann Emerg Technol Comput (AETiC) 8(2):15–31
12. Li Y, Zhang M, Ren M, Ma M, Wei D, Yang H (2024) Cross-domain audio deepfake detection: dataset and analysis. arXiv preprint arXiv:2404.04904
13. Firc A, Malinka K, Hanáček P (2024) Deepfake speech detection: a spectrogram analysis. In: Proceedings of the 39th ACM/SIGAPP symposium on applied computing, pp 1312–1320
14. Danylov V (2024) Open source and proprietary software for audio deepfakes and voice cloning: growth areas, pain points, future influence. Balt J Leg Soc Sci 1:105–113
15. Kumar BV, Ayyappa Y, Sai PJ, Varun T, Babu SN, Teja YUSS (2024) Detecting manipulated audio using Adaboost machine learning algorithm. In: 2024 2nd international conference on advancement in computation and computer technologies (InCACCT). IEEE, pp 252–257
16. Pianese A, Cozzolino D, Poggi G, Verdoliva L (2024) Training-free deepfake voice recognition by leveraging large-scale pre-trained models. arXiv preprint arXiv:2405.02179
17. Valente LP, de Souza MMS, Da Rocha AM (2024) Speech audio deepfake detection via convolutional neural networks. In: 2024 IEEE international conference on evolving and adaptive intelligent systems (EAIS). IEEE, pp 1–6
18. Hema MS, Maheshprabhu R, Reddy KS, Guptha MN, Pandimurugan V (2023) Prediction analysis for Parkinson disease using multiple feature selection and classification methods. Multimed Tools Appl 82(27):42995–43012
19. Mamatha K, Neelima S, Reddy KS (2020) Fostering employability-synergy between industry and academia. J Eng Educ Transform 33(Special Issue)
20. Padmaja DL, Sowjanaya T, Gajavelly NS, Reddy KS (2022) A comparative study on natural disasters. In: 2022 International conference on applied artificial intelligence and computing (ICAAIC). IEEE, pp 1704–1709
21. Deb S, Bhuyan B, Kar N, Reddy KS (2022) Colour image encryption using an improved version of stream cipher and chaos. Int J Ad Hoc Ubiquitous Comput 41(2):118–133
22. Reddy KS, Srinagesh C (2013) Fostering problem solving through innovative knowledge events. In: 2013 8th international conference on computer science and education. IEEE, pp 1233–1238
23. Tambe S, Pawar A, Yadav SK (2021) Deep fake videos identification using ANN and LSTM. J Discret Math Sci Cryptogr 24(8):2353–2364
24. Kaggle: Deep voice deepfake voice recognition dataset. https://www.kaggle.com, https://www.kaggle.com/datasets/birdy654/deep-voice-deepfake-voice-recognition

EcoSaver: AI and Blockchain Energy Efficiency Solution for Sustainable Industry 5.0

Fahima Niaz, Deepa Jose, Lalitha Kumari Pappala, Malathy Jawahar, Joshuva Bernard, Aadil Mohideen, and G. Danial Dhinesh

Abstract In order to enhance energy efficiency and promote a future that is environmentally friendly, the paper presents "EcoSaver," a novel strategy which makes use of cutting-edge technologies. EcoSaver offers a complete solution for manufacturing processes and energy management by fusing blockchain-based technology, artificial intelligence (AI), and Industry 5.0 principles. A distributed sensor network for real-time data collecting, blockchain for transparent and safe data management, Industry 5.0 principles for smart manufacturing, and AI-driven energy monitoring and optimization are all included in the suggested system. EcoSaver hopes to accomplish precise energy optimization, lower carbon emissions, and improved involvement of stakeholders through this synergy. The EcoSaver program's potential to promote energy-efficient behaviors and advance sustainability globally is highlighted in this paper's discussion of its objectives and implementation method.

Keywords Energy efficiency · Artificial intelligence · Blockchain technology · Industry 5.0 · Smart manufacturing · Distributed sensor network · Data security · Transparency · Sustainability · Green energy · Carbon emissions · Energy optimization · Environmental impact · IoT · Renewable energy · Energy management · Collective action · Peer-to-peer energy trading

F. Niaz · D. Jose (✉) · L. K. Pappala · J. Bernard · A. Mohideen · G. Danial Dhinesh
KCG College of Technology, Karapakkam, Chennai 600097, TamilNadu, India
e-mail: deepa.ece@kcgcollege.com

L. K. Pappala
School of Computer Science and Engineering, VIT-AP University, Amaravati, India

M. Jawahar
Senior Principal Scientist, Council of Scientific and Industrial Research- CLRI, Chennai, TamilNadu, India

A. Rocha et al. (eds.), *Integrating Advanced Technologies for Enhanced Security and Efficiency*, Engineering Cyber-Physical Systems and Critical Infrastructures 14,
https://doi.org/10.1007/978-3-031-91798-1_24

1 Introduction

Urbanization and the fast growth of industry have raised energy demand dramatically, raising carbon emissions and harming the environment. Innovative approaches that make use of cutting edge technologies to improve energy efficiency and improve sustainability are crucial to resolving these problems. This paper introduces "EcoSaver," a cutting-edge energy efficiency system that blends Industry 5.0 ideas, blockchain technology, and artificial intelligence (AI). By integrating these technologies, EcoSaver hopes to revolutionize production and energy management operations and contribute enormously to the development of a more sustainable and greener future.

2 Background and Motivation for the Project

The increased emphasis on ecology and the urgent need to optimize energy use in industry led to the development of EcoSaver. Major problems with current energy systems include waste, ineffectiveness, and reliance on non-renewable resources. The combination of blockchain with artificial intelligence (AI) has emerged as a promising strategy for resolving these problems and ensuring a sustainable future. In recent years, the application of AI in the energy sector has drawn a lot of interest. Researchers have looked into a number of AI applications, such as distributed intelligence and machine learning, with the objective of lowering carbon emissions, increasing energy productivity, and optimising energy use.

3 Data Collection Methods for Energy Monitoring in Industrial Settings: A Comprehensive Review

Energy-related data can be gathered and stored using a variety of technologies, providing useful data on trends in energy use and flagging out areas that require improvement [1–4]

- As watchful stewards, sensors measure energy-related variables including temperature, humidity, and gas concentrations on a constant basis.
- As careful recorders, data loggers take note of this information at specific times and store it for later examination.
- As linked systems, IoT devices gather data gathered from various locations throughout the energy network and send it to a centralized location for thorough monitoring.
- Constant observers, smart meters track and save energy use in real time to give useful information.

- Power monitors keep a careful eye on variations in consumption of energy, spotting trends that may result in improvements in efficiency.
- As integrators, data acquisition systems (DAS) build a strong network for energy intelligence by combining data from sensors, data loggers, and smart meters.

3.1 Combining Data Collection Methods

An accurate dataset for the energy monitoring system can be obtained by combining different data collection techniques. This makes it possible to analyze and manage energy use in a particular context in an effective way.

A manufacturing facility might, for instance, use data collection methods to track how much energy it uses. Sensors can be used to monitor the power consumption for specific machinery and equipment, as well as to detect the temperature and humidity in different places. To collect and process this data from various sources, a Data Acquisition System (DAS) may also be used. After that, the DAS might produce reports and visualizations that offer comprehensive information about the patterns of energy use at the plant. With this knowledge, suitable actions to improve energy efficiency can be taken, such replacing outdated equipment, modifying temperature settings, or streamlining manufacturing scheduling [5–10].

Industrial facilities might discover chances for considerable cost and energy decreases and obtain significant understanding into their energy consumption by implementing a thorough data collection plan.

4 Role of Artificial Intelligence in Energy Efficiency

In many different industries, artificial intelligence (AI) is now a major force behind the transformation of energy efficiency. AI is crucial for maximizing energy use, cutting waste, and promoting sustainability because of its many uses. AI accurately predicts patterns of energy usage using advanced algorithms like machine learning and deep learning, enabling proactive changes in energy distribution and utilization. By identifying irregularities, its anomaly detection capability enables preventive maintenance and repairs, thus decreasing energy waste.

By dynamically optimizing settings depending on variables like peak demand, occupancy levels, and production requirements, artificial intelligence (AI) significantly enhances real-time energy management. AI-powered predictive maintenance improves the lifespan of machinery and boosts energy efficiency. Demand response systems powered by AI encourage consumers to reduce their usage in order to maintain grid stability and conservation during peak hours. AI additionally revolutionizes building energy management by controlling lighting, ventilation, heating, and cooling systems to guarantee efficient use while preserving occupant comfort. In conclusion,

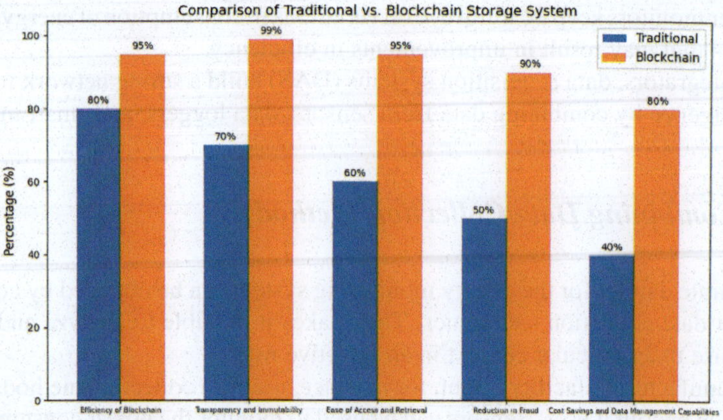

Fig. 1 Comparison of blockchain storage and regular database

incorporating AI facilitates data-driven decision-making, changing energy-saving techniques and significantly improving global sustainability programs.

In our research, a machine learning module that combines Convolutional Neural Network (CNN) and Long Short-Term Memory (LSTM) architectures processes data collected from various sources, such as IoT sensors and power consumption monitors. This advanced integration efficiently analyzes and predicts trends of consumption of energy in industrial facilities through using deep learning capabilities.

Convolutional layers are used in the CNN-LSTM module's initial data processing to extract important features and patterns. The LSTM layers, which are intended to comprehend long-term sequences and capture temporal dependencies, are subsequently fed these extracted features. The module can identify abnormalities or irregularities in energy usage patterns and generate accurate predictions thanks to this dual methodology, which also allows the module to learn intricate connections and interactions within the use of energy data [11–15].

In Fig. 1 EcoSaver structure analyses and predicts trends of energy usage in industrial facilities by combining the abilities of Long Short-Term Memory (LSTM) and Convolutional Neural Networks (CNN). Deep learning can extract key characteristics and trends from the data thanks to this sophisticated integration. To effectively find patterns within certain data points, the CNN-LSTM module first analyzes the input utilizing convolutional layers.

5 Blockchain Technology for Transparency and Trust

As depicted in Fig. 2 there are many benefits to storing energy usage data on a blockchain, such as greater safety, better transparency, decentralized control, and easy accessibility. Peer-to-peer energy trading, energy origin drawing, and other

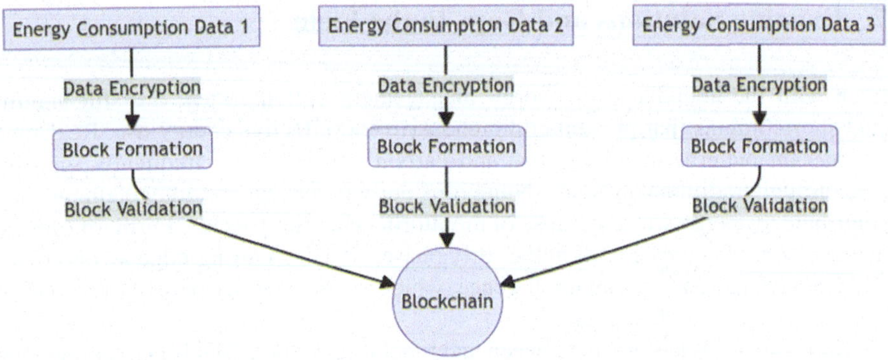

Fig. 2 Energy consumption data is stored in a blockchain node as a transaction

features made available by blockchain encourage cost reductions, sustainable energy practices, and improved energy monitoring system efficiency. Energy monitoring can be transformed through the use of blockchain technology, which will boost security, transparency, and efficiency in operation.

To preserve secrecy and guard against unauthorized access, the data is first safely kept in an archive (A) then encrypted (B). Following encryption, the data is organized into blocks (C), which are checked to guarantee accuracy and integrity (D). Following confirmation, these blocks are successively joined (E), forming the blockchain (F), a collection of linked blocks. This chain structure ensures the data's security and immutability, making it transparent and impenetrable to manipulation.

6 Industry 5.0 and Integration of Digital Technologies

Manufacturing is going through a radical change with Industry 5.0, which emphasizes the use of digital technology to increase sustainability, efficiency, and manufacturing. This strategy promotes smooth cooperation between people, robots, and data in order to optimize production procedures. Industry 5.0 makes it easier to create smart factories that can analyze data in real time and make adaptive decisions through the use of technologies like robotics, artificial intelligence (AI), and the Internet of Things (IoT). Through reduced energy waste, effective utilization of resources, and predictive maintenance, this combination increases energy efficiency. Additionally, Industry 5.0 places a strong emphasis on human–machine collaboration, which makes it feasible to develop creative energy management solutions. In these solutions, operators and AI systems collaborate to find ways to save energy and adopt environmentally friendly procedures.

7 Existing Solutions and Gaps in the Field

Even with the abundance of energy management options, there are still significant shortcomings that prevent comprehensive and effective energy use. Real-time insights on patterns of energy use and carbon emissions are frequently unavailable through traditional systems. Numerous present ones ignore other elements that contribute to energy waste because of monitoring electricity usage. Furthermore, the management of energy is still in the early phases of using cutting-edge technologies like blockchain and artificial intelligence, which restricts the security of data sharing and the precision of data analysis.

Moreover, the relationship between manufacturing activities and energy efficiency is frequently disregarded by current solutions. Opportunities for comprehensive optimization are restricted by this gap between energy management and industrial processes, which also overlooks the dynamic relationship between energy use and production output. In addition, a lack of transparency and confidence in energy-related data makes it hard for stakeholders to collaborate together, which makes it difficult to reach sustainability goals as a group.

The previously mentioned disparities underscore the need for a revolutionary resolution that bridges the current gaps in energy management, integrates cutting-edge technologies, and promotes cooperation among stakeholders. The EcoSaver project aims to solve these problems by creating a single platform that brings together blockchain, artificial intelligence, and Industry 5.0 concepts. In order to create a more sustainable and ecologically friendly future, the approach seeks to transform energy efficiency, enhance transparency, and expedite the processes of manufacturing.

8 System Architecture and Components

The implementation of EcoSaver project's design is shown in Figs. 3 and 4 and is set up to effectively integrate blockchain, machine learning, and Industry 5.0 concepts, offering an integrated approach to energy efficiency. The system is made up of essential parts which work as a whole to achieve its goals.

The top-left graph juxtaposes the actual energy consumption (in blue) with the model's predicted consumption (in red) over time. The top-right graph in Fig. 5 illustrates the cost savings (in green) achieved over the same time period. The bottom-left scatter plot displays the correlation between the actual and predicted consumption levels. Finally, the bottom-right histogram represents the distribution of cost savings, highlighting the frequency of different savings values.

Fig. 3 A prototype with basic sensors implemented

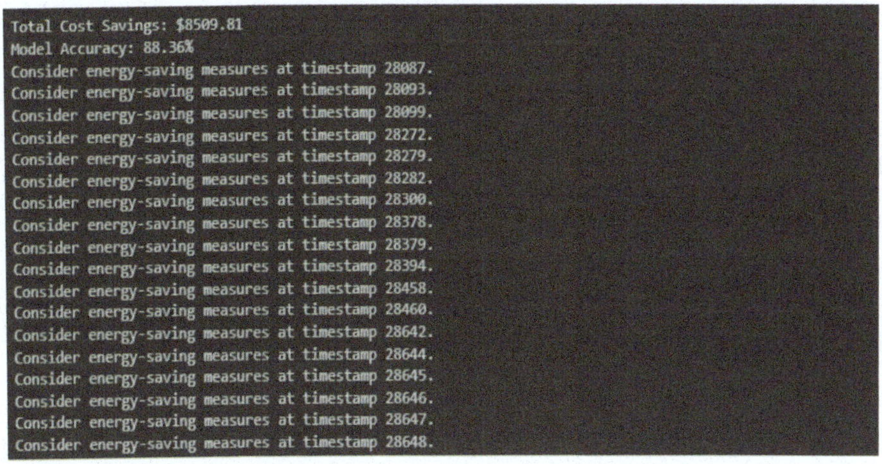

Fig. 4 Result with augmented dataset

8.1 IoT Devices and Sensors

A network of IoT devices and sensors collects real-time data from various sources, including electricity meters, current and voltage sensors, air pressure sensors, DHT22 sensors, and the MQ135 carbon dioxide sensor. These sensors provide essential data points related to energy consumption, environmental conditions, and carbon emissions.

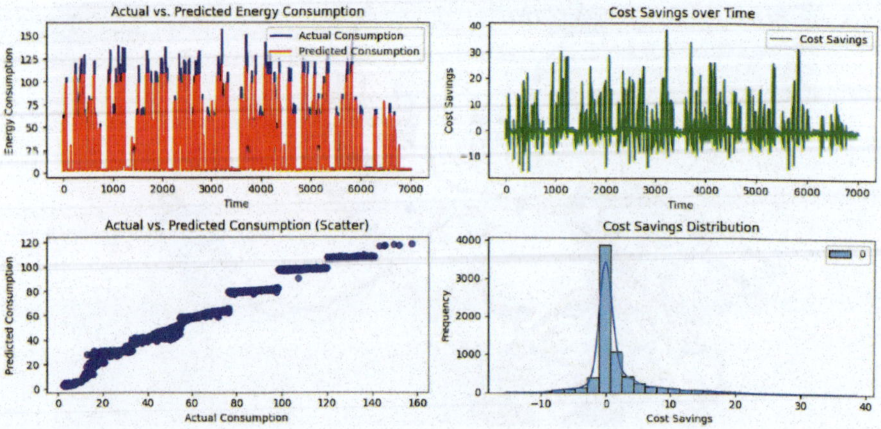

Fig. 5 Energy consumption findings from the implementation

8.2 *AI-Powered Energy Monitoring and Optimization*

The heart of EcoSaver lies in the AI algorithms that analyze the data collected from sensors. Through machine learning models, the system predicts energy consumption patterns, identifies wastage, and suggests optimization strategies. The CNN-BLSTM model accurately estimates direct carbon emissions based on real-time electricity load, location, and carbon intensity data.

8.3 *Distributed Sensor Network*

The distributed sensor network consists of interconnected nodes that enable efficient data collection and communication. Nodemcu ESP8266 WiFi modules ensure seamless connectivity and data transmission from various sensors to the central system.

8.4 *Blockchain Infrastructure*

Blockchain technology provides a secure and transparent platform for storing energy-related transactions, carbon credits, and energy certificates. Smart contracts enable peer-to-peer energy trading and automate energy conservation incentives. The decentralized ledger ensures data integrity, enhances trust, and enables secure data sharing among stakeholders.

8.5 *Industry 5.0 Integration*

The integration of Industry 5.0 principles involves the use of robotics, automation, and IoT devices in manufacturing processes. This integration optimizes energy-intensive operations, minimizes resource wastage, and aligns energy management with manufacturing activities.

8.6 *Data Analysis and Visualization*

Processed data is visualized through an intuitive user interface, allowing stakeholders to monitor energy consumption trends, carbon emissions, and manufacturing performance. Real-time insights empower users to make informed decisions and take proactive measures to enhance energy efficiency.

9 Energy Consumption Prediction

In this study, we used a Long Short-Term Memory (LSTM) model to predict energy consumption in the steel industry. We used a dataset from Kaggle (https://www.kaggle.com/datasets/csafrit2/steel-industry-energy-consumption) to train and evaluate our model. The dataset is taken from steel industy dataset from kaggle. The KWh usage column is defined as the target variable (the variable we want to predict), and all other columns usage in KWh, leading current, lagging current, CO_2, Load type and time are the features of interest are defined as features (the variables used to make the prediction) 0.40,0000 samples of data are used in training. The code converts categorical variables into numerical format using one-hot encoding, selects only numerical columns, and normalizes the features using MinMaxScaler. This is followed by further steps as given below.

1. Preprocess the target: The target variable is also normalized using MinMaxScaler.
2. Split the data: The data is split into training and testing sets. 80% of the data is used for training the model, and 20% is used for testing the model's performance.
3. Reshape the data: The training and testing data are reshaped to fit the input shape required by the Convolutional Neural Network (CNN) and Long Short-Term Memory (LSTM) model.
4. Define the model: The code defines a CNN-LSTM model. This type of model is often used for time series prediction tasks. The model consists of convolutional, max pooling, flattening, LSTM, and dense layers.
5. Train the model: The model is trained using the training data. The training process is repeated for a specified number of iterations (epochs). During each epoch, the model learns to make better predictions by adjusting its parameters to minimize the difference between the actual and predicted values (loss).

6. Plot training history: The code plots the training and validation loss at each epoch. This helps to check if the model is learning well and not overfitting or underfitting the data.
7. Make predictions and visualize results: The trained model is used to make predictions on the testing data. The predictions are then inverse transformed to their original scale for visualization. The actual and predicted values are plotted for comparison.

Our model achieved an accuracy of 88.36% on the test set. This indicates that our model is able to accurately predict energy consumption in the steel industry.

Total Cost Savings: $8509.81: This is the total cost savings achieved by the model. It represents the reduction in energy costs compared to using no prediction model. In this case, it's positive, which means the model has saved $8509.81 in energy costs. Model Accuracy: 88.36%: This represents the accuracy of the model's predictions, indicating how well the model performs in predicting energy consumption.

"Consider energy-saving measures at timestamp ...": These are suggestions made by the code based on the dataset. The code identifies specific timestamps where energy-saving measures could be implemented to reduce consumption.

The positive total cost savings and high model accuracy are indicative of a well-performing model that effectively predicts energy consumption, and the suggestions for energy-saving measures provide actionable insights for This information can be used by steel mills to reduce their energy costs. Overall, our results show that LSTM models can be used to accurately predict energy consumption in the steel industry.

10 Conclusion

In conclusion, the EcoSaver project embodies a convergence of transformative technologies, presenting a formidable solution to the critical issues of energy efficiency and sustainability. Through the integration of AI-powered energy monitoring, distributed sensor networks, blockchain-enabled data security, and Industry 5.0 principles, EcoSaver addresses existing gaps in energy management while ushering in a new era of smart manufacturing. The anticipated outcomes encompass accurate energy optimization, real-time insights, reduced carbon emissions, and enhanced collaboration, aligning with the principles of Industry 5.0. By presenting a comprehensive framework that bridges the gap between energy efficiency and manufacturing, EcoSaver charts a promising trajectory towards a more ecologically balanced and technologically advanced future.

References

1. Liu G, Wen F, Liu J, Xue Y, Zhao J (2020) A real-time estimation framework of carbon emissions in steel plants based on load identification. In: 2020 International conference on smart grids and energy systems (SGES). IEEE, pp 1–5
2. Li Y, Li G, Sun S, Zhang X, Bie Z (2023) Reliability assessment of active distribution network considering path-based post-fault island division. In: 2023 6th international conference on energy, electrical and power engineering (CEEPE), Guangzhou, China, pp 618–623. https://doi.org/10.1109/CEEPE58418.2023.10165712.A. Saxena AC, Dash AK, Gupta SK, Kumar VS, Pandey JP (2023) An effective optimal economic sustainable clean energy solution with reduced carbon capturing/carbon utilization/carbon footprint for grid integrated hybrid system. IEEE Trans Sustain Comput 8(3):385–399. https://doi.org/10.1109/TSUSC.2023.3262982
3. Pal P, Parvathy AK, Devabalaji KR, DeivaSundari P, Joseph Antony S, Ocheme SE (2023) Analysis of bidding strategies in virtual power plant. In: Green energy and technology. https://doi.org/10.1007/978-981-16-8274-2_20
4. Tamizhselvi SP, Vijayalakshmi M (2022) Dynamic bandwidth estimation and congestion avoidance based on network traffic in mobile cloud. Int J Sens Netw. https://doi.org/10.1504/IJSNET.2022.122978
5. Rame R, Purwanto P, Sudarno S (2024) Industry 5.0 and sustainability: an overview of emerging trends and challenges for a green future. Innov Green Dev 3(4):100173. ISSN: 2949-7531. https://doi.org/10.1016/j.igd.2024.100173
6. Chen S-C, Chen H-M, Chen H-K, Li C-L (2024) Multi-objective optimization in industry 5.0: human-centric AI integration for sustainable and intelligent manufacturing. Processes 12(12):2723. https://doi.org/10.3390/pr12122723
7. Abul Hasan M, Pandey P, Kumar A, Bora A, Jose D, Kshirsagar P, Bui H, Chakrabarti Prasun, Khanapurkar M, Katragadda R (2021) Internet of things and its application in industry 4.0 for smart waste management. J Environ Prot Ecol 22:2368–2378
8. Anand R, Adhavan T, Balaji M, Daniel J, Kiridharan S (2023) Secure platform as a service using blockchain technology (HYPERBASE), pp 1–6. https://doi.org/10.1109/ICBDS58040.2023.10346510
9. Sucharitha G, Sitharamulu V, Mohanty SN, Matta A, Jose D (2023) Enhancing secure communication in the cloud through blockchain assisted-CP-DABE. IEEE Access 11:99005–99015. https://doi.org/10.1109/ACCESS.2023.3312609
10. Jamwal A, Agrawal R, Sharma M, Kumar V, Kumar S (2021) Developing a sustainability framework for industry 4.0. Procedia CIRP 98:430–435. ISSN: 2212-8271. https://doi.org/10.1016/j.procir.2021.01.129
11. Martín-Gómez AM, Agote-Garrido A, Lama-Ruiz JA (2024) A framework for sustainable manufacturing: integrating industry 4.0 technologies with industry 5.0 values. Sustainability 16(4):1364
12. Gajdzik B, Grabowska S, Saniuk S, Wieczorek T (2020) Sustainable development and industry 4.0: a bibliometric analysis identifying key scientific problems of the sustainable industry 4.0. Energies 13:4254
13. Kachiche S, Gahi Y, Gharib J (2023) An efficient framework for the implementation of sustainable industry 4.0. In: Ben Ahmed M, Boudhir AA, Santos D, Dionisio R, Benaya N (eds) Innovations in smart cities applications, volume 6. SCA 2022. Lecture notes in networks and systems, vol 629. Springer, Cham. https://doi.org/10.1007/978-3-031-26852-6_74
14. Shabur MA (2024) A comprehensive review on the impact of industry 4.0 on the development of a sustainable environment. Discov Sustain 5:97. https://doi.org/10.1007/s43621-024-00290-7
15. Szlávik J, Szép T (2022) A framework of risks in the context of industry 4.0, related to sustainability. World Futur 79(3):406–429. https://doi.org/10.1080/02604027.2021.2012875

AI TRiSM in Cloud Security: Leveraging AI-Powered DevSecOps for Robust Security Management

Varun Shiva Krishna Rupani and C. Madhusudhana Rao

Abstract In the era of cloud computing and DevOps, ensuring robust security remains a critical challenge for organizations. The dynamic nature of cloud infrastructures and the increasing complexity of cyber attacks pose significant dangers to sensitive data and critical applications. Traditional security approaches struggle to keep pace with the rapid development cycles, leading to vulnerabilities and security gaps. Additionally, integrating AI-driven security practices into DevOps workflows is hampered by communication barriers, skill gaps, and cultural resistance within organizations. This paper proposes leveraging AI-powered techniques within the DevSecOps framework to enhance cloud security. By integrating AI and machine learning (ML) algorithms into security processes, organizations can significantly improve threat detection, vulnerability management, and incident response. AI facilitates real-time analysis of large datasets, enabling proactive identification of security threats, while ML algorithms predict and prevent potential breaches based on historical data. Workflows for security and response are further streamlined via automation. Infrastructure, technology, and expertise investments, as well as cultural changes and better communication techniques to bring security and DevOps teams together, are all necessary for the successful deployment of AI-powered AI TRiSM practices. To close skill gaps and enable staff to embrace AI-driven security procedures, training and upskilling programs are essential. To handle changing risks and legal requirements, security strategies must be continuously assessed and adjusted. Organisations may reduce risks and increase software delivery and creativity by adopting AI-powered DevSecOps, which strikes a balance between security and speed. By using this strategy, businesses can protect their cloud infrastructure, data, and apps while staying ahead of the always changing threat landscape.

V. S. K. Rupani · C. M. Rao (✉)
Computer Science Engineering, Institute of Aeronautical Engineering, Hyderabad, India
e-mail: npr4567@gmail.com

V. S. K. Rupani
e-mail: 21951A05P2@iare.ac.in

A. Rocha et al. (eds.), *Integrating Advanced Technologies for Enhanced Security and Efficiency*, Engineering Cyber-Physical Systems and Critical Infrastructures 14,
https://doi.org/10.1007/978-3-031-91798-1_25

Keywords AI optimized DevOps · Streamlined cloud CI/CD · Container management · AI deployment analysis · Kubernetes · Machine learning in DevOps · Automated deployment pipelines · OpenAI integration

1 Introduction

Agile frameworks have become necessary due to the quick evolution of software development processes, and DevOps has emerged as a key strategy for accomplishing continuous integration and continuous deployment (CI/CD). However, ensuring security, scalability, and efficiency in complex and dynamic systems is a major issue for typical DevOps procedures. By automating tedious procedures, offering insightful information, and improving decision-making. The concepts of AI trust, risk, and security management (AI TRiSM) are highlighted in this study as it presents AI-powered DevOps automation. The framework known as AI TRiSM makes sure that the fundamental ideas of security, accountability, and transparency are followed in the creation and application of AI systems.

AI TRiSM emphasises the need of promoting trust in AI systems by offering accountability, transparency, and fairness. By including an explainable AI (XAI) approach into the ML model utilised for deployment analysis, our research complies with these guidelines. This promotes trust and confidence in the system's recommendations by enabling users to comprehend the reasoning behind the recommendations that are created.

Our goal is to optimise DevOps processes while adhering to the fundamental principles of AI TRiSM by incorporating ML for analysis and a chat interface for user engagement.

In recent years, using machine learning (ML) to automate DevOps chores has grown in popularity. This study investigates a novel DevOps automation strategy that makes use of machine learning's capabilities for deployment analysis and recommendation creation [1].

The core system leverages OpenAI's ability to analyse Infrastructure as Code (IaC) manifestations and interfaces with Kubernetes (optionally) to acquire deployment information (Fig. 1).

Our strategy seeks to establish a transparent, safe, and effective deployment process. This study presents a system design that uses AI for recommendation creation, deployment analysis, and chat interface user engagement. By offering transparent and intelligible AI-driven insights, this combination not only expedites the deployment process but also improves user happiness and trust.

This study's main goal is to find out how well machine learning (ML) can detect possible problems and provide fixes for deployment setups. The system improves user experience and streamlines DevOps operations by incorporating an NLP (Natural Language Processing) interface that enables users to communicate with the automation and recommendation processes via a chat interface [2].

Fig. 1 4 pillars of AI TRiSM

By showcasing ML's potential for intelligent deployment analysis, this study advances the field of AI-Optimized DevOps. We introduce a new system design that makes use of FastAPI for a Modern software development requires scalable web frameworks because they allow apps to manage growing traffic and continue to function well under different loads. These frameworks offer strong structures that facilitate resource efficiency, scalability, and dependability.

2 Literature Review

Manual processes are the mainstay of traditional DevOps operations, which can lead to inefficiencies and possible deployment configuration problems. This section examines current research on AI-powered DevOps tools and techniques and digs into the shortcomings of these conventional methods. Next, we look at the idea of AI TRiSM (Artificial Intelligence Trust, Risk, and Security Management) and how it relates to integrating AI into DevOps processes [3].

An overview of AI in DevOps. In recent years, interest in incorporating artificial intelligence (AI) into DevOps procedures has grown. It aims to enhance the software development lifecycle. AI offers chances to improve, automate, and secure these procedures, resulting in software deployment that is more dependable and efficient. With an emphasis on the integration of AI operations and machine learning operations, this literature review cxamines the current status of AI-powered DevOps.

2.1 Difficulties with Conventional DevSecOps Container Administration

Efficiency, scalability, and security are all impacted by the difficulties that traditional DevSecOps container management encounters [4].

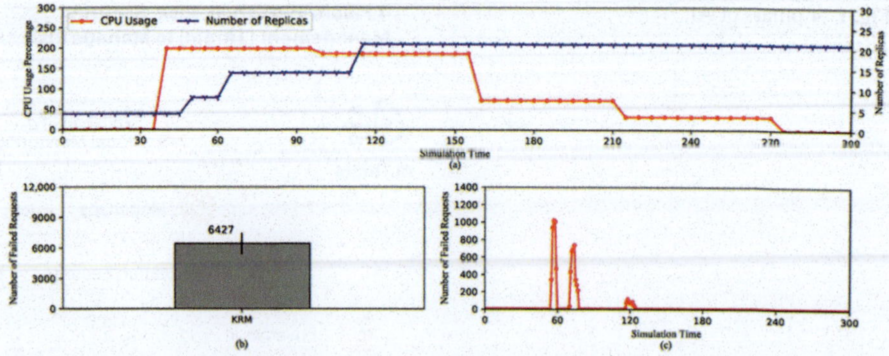

Fig. 2 Real time resource inefficiency graph

2.1.1 Resource Inefficiency

Poor resource allocation is a common problem with containers, which can result in either overuse or underuse. In addition to increasing operating expenses, this inefficiency degrades performance in general (Fig. 2).

2.1.2 Scalability Issues

One of the biggest challenges is dynamically scaling containers to manage changing demands. Conventional approaches have trouble provisioning container instances quickly and effectively, and therefore underprovision them (Fig. 3).

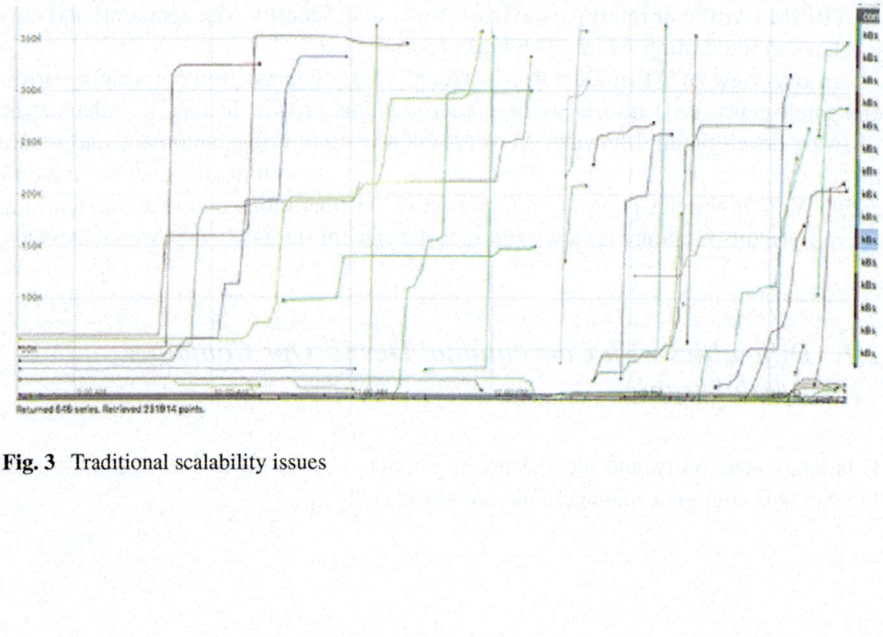

Fig. 3 Traditional scalability issues

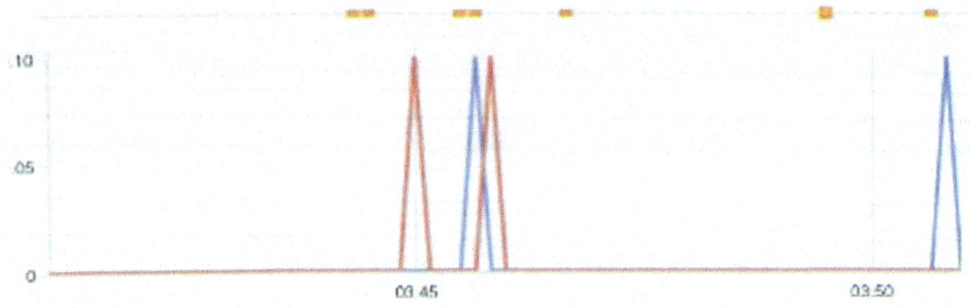

Fig. 4 Random pod restarts

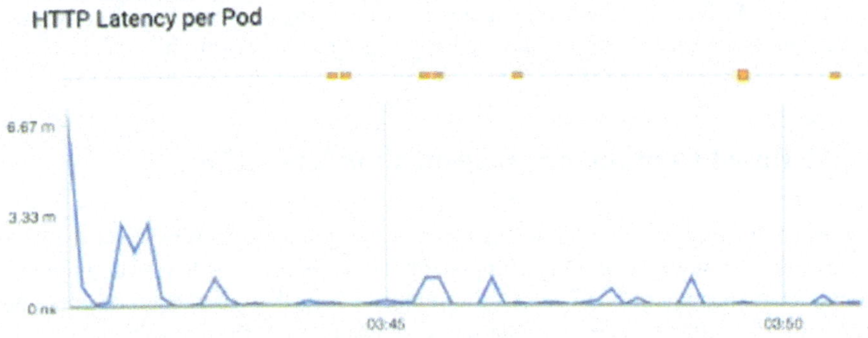

Fig. 5 HTTP latency in the container

Regular pod restarts may be a sign of containerised environment instability, which is frequently brought about by resource constraints, configuration mistakes, or underlying infrastructure problems (Fig. 4).

2.1.3 HTTP Latency

Applications' performance and user experience are impacted by high HTTP latency. It is frequently difficult for traditional container management techniques to dynamically modify resources in order to maintain ideal latency (Fig. 5).

2.1.4 CPU Usage

Effective CPU use is essential for both cost-effectiveness and performance. Conventional methods might not evenly distribute the CPU load, which could result in inefficiencies and bottlenecks [5] (Fig. 6).

These difficulties highlight the necessity for more sophisticated DevSecOps container management solutions. One promising solution to these problems is to

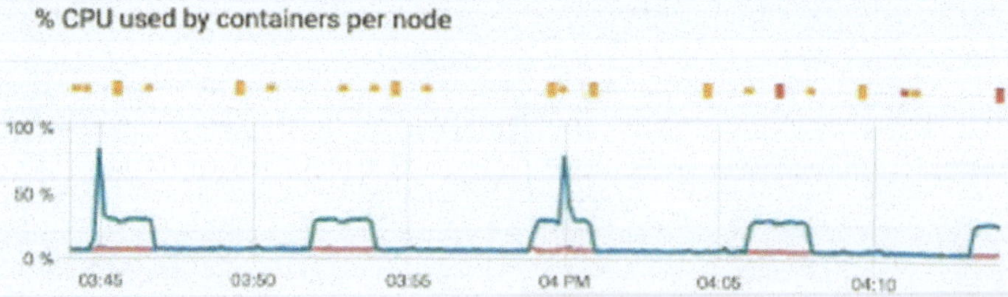

Fig. 6 CPU usage errors

incorporate AI into container management, which will improve the effectiveness, scalability, and security of containerised application administration.

3 AI-Based Container Management in DevSecOps

Because they allow for consistent and segregated contexts for application execution, containers have emerged as a key component of contemporary software deployment. Container management has been further enhanced by the emergence of container orchestration platforms such as Kubernetes, which automate the deployment, scaling, and operation of application containers. Containers for Kubernetes.

Because Kubernetes automates deployment, scaling, and operations across host clusters, it has completely changed the way containerised applications are managed. Being a strong orchestration platform, it makes container management easier and guarantees that apps operate dependably and effectively in a variety of settings. This fundamental capacity lays the groundwork for future developments, such the incorporation of AI to improve container management (Fig. 7).

The master-slave mechanism at the heart of the Kubernetes architecture is intended to automate and oversee containerised applications. The master node, which is in charge of coordinating activities, and worker nodes, which house applications, are the two main parts of Kubernetes. The API server for communication, controllers for state management, and other components for storing cluster data are all part of the master node. The Kubernetes control plane, which oversees worker nodes that house the application containers, guarantees scalability, resilience, and effective resource distribution throughout the cluster. High availability, smooth scalability, and automatic containerised environment management are made possible by Kubernetes' distributed architecture, which is crucial for contemporary cloud-native applications.

Nonetheless, there might still be a lot of difficulties in administering these settings, especially in intricate and extensive deployments.

By using AI to detect problems, optimise resource allocation, and automate corrective measures, AI-based container management technologies seek to address these issues [6]. Improved application performance, lower operating costs, and better

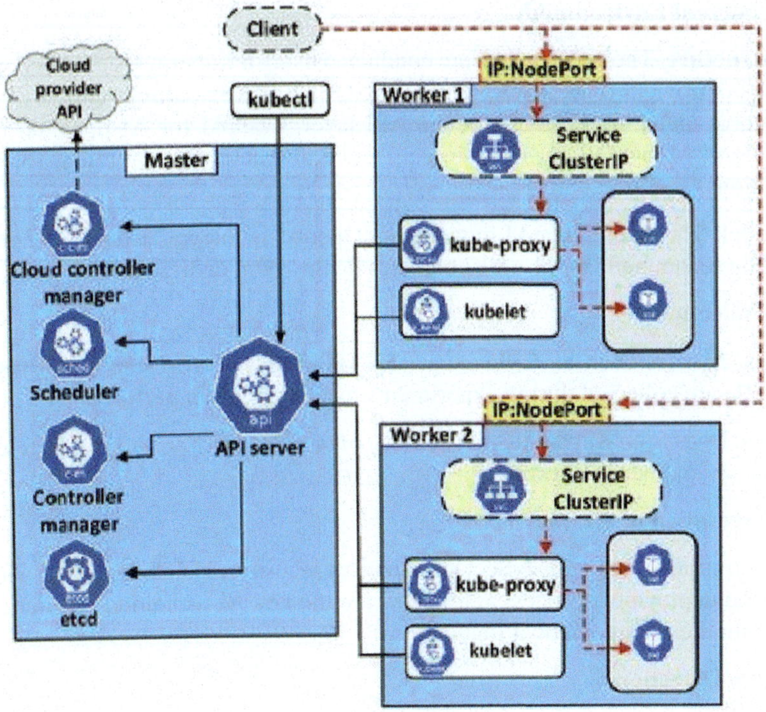

Fig. 7 Kubernetes master slave internal architecture

resource utilisation can result from incorporating AI into container management. Machine learning models are used by tools like ours to examine deployment trends and offer suggestions for improving container orchestration.

4 Methodology

The experimental design describes the methodology used to assess how AI can be included into container management inside DevSecOps procedures.

Objective: Evaluating how AI technologies might improve the effectiveness, scalability, and security of containerised applications inside DevSecOps frameworks is the main goal.

Experimental Environment

Infrastructure: The experiments are conducted using a Kubernetes cluster deployed on a cloud provider (e.g., AWS, Azure, or GCP). **Software Stack**: Utilizes Python-based tools including FastAPI for the API layer, OpenAI for AI-powered recommendations, MongoDB for data storage, and Kubernetes Python client for cluster interactions.

AI Model: The AzureOpenAI model from OpenAI is integrated for analyzing and providing recommendations on Kubernetes deployment configurations.

Data Collection

Metrics: Metrics such as CPU usage, HTTP latency, pod restarts, and resource utilization are collected during experiments to assess system performance.

Logging: Logs from Kubernetes events and container metrics are stored in MongoDB for analysis and benchmarking.

Implementation of AI models

The integration of AI into DevSecOps enhances security, risk management, and trust within the deployment process [7]. Below are the key AI techniques implemented, along with their mathematical foundations.

Anomaly Detection

Objective: To identify deviations from normal system behavior that could signal potential security threats.

Approach: Anomaly detection algorithms, such as the Isolation Forest, are employed to monitor real-time data streams. The model isolates anomalies by creating decision trees that split the data based on feature values. The anomaly score for a data point x can be expressed as:

$$\text{anomaly score } (x) = 2\frac{E(h(x))}{c(n)}$$

where $c(n)$ is the average path length of failed searches and $E(h(x))$ is the path length of x in the tree.

Predictive Security Modeling

Objective: To forecast potential security incidents by analyzing trends in historical data.

Approach: Time series analysis is applied to predict future security risks. The model can be represented as:

$$y_t = \alpha + \beta_t + \gamma y_t - 1 + \in t$$

where y_t is the value at time coefficients, and t, α, β and γ are is the error term. This model helps in anticipating when security breaches might occur, allowing proactive mitigation.

Explainable AI (XAI) for Security Decisions

Objective: To make AI-driven security recommendations transparent and understandable.

Approach: Techniques such as SHAP (SHapley Additive exPlanations) are employed. SHAP values represent the contribution of each feature to the model's prediction, computed as:

$$SHAP = \sum_{S \in \{i\}} \frac{|S|! \cdot (|F| - |s| - 1)!}{|F|!} [v(S \cup \{i\}) - v(s)]$$

where F is the set of all features, S is a subset of F,

and $v(s)$ is the value of the model's output with features in S.

Threat Modeling and Risk Assessment

Objective: To assess and prioritize potential security threats based on their impact and likelihood.

Approach: AI-driven threat modeling, often using Bayesian Networks, evaluates risks. The joint probability distribution of variables $X_1, X_2, \ldots, X_n$ is expressed as

$$P(X_1, X_2, \ldots, X_n) = \prod_{i=1}^{n} P(X_i | \text{Parents}(X_i))$$

This model helps in identifying critical vulnerabilities and their potential impact, allowing for targeted risk mitigation.

Automated Security Remediation

Objective: To automate the response to detected security issues, reducing the need for manual intervention.

Approach: Reinforcement learning (RL) is applied, where an agent interacts with the environment to maximize cumulative rewards. The optimization is represented as:

$$Q^*(s, a) = \max E\left[R_t + \gamma \max Q * \left(s', a'\right)\right]$$

where $Q*$ (s, a) is the optimal action-value function, s is the state, a is the action, R_t is the reward at time t, γ is the discount factor, and π is the policy. This model ensures the system adapts to new threats by learning the best actions to take in response to security incidents.

Evaluation and Validation

When evaluating the effectiveness and dependability of AI-driven container management in DevSecOps procedures, evaluation and validation are crucial phases. The methods, measurements, and results utilised to verify the incorporation of AI into our study framework are described in this part.

Performance Metrics

Our evaluation includes several key performance metrics to gauge the effectiveness of AI models.

Accuracy: This metric measures the precision of AI predictions concerning resource utilization, scalability requirements, and security vulnerabilities within containerized environments.

Security: An essential aspect of our evaluation involves evaluating the AI's capability to detect and mitigate security risks inherent in container deployments, enhancing overall system resilience [8].

Validation Procedures

To validate our approach, we designed rigorous experiments and procedures.

Experimental Design: We meticulously crafted experimental setups encompassing diverse datasets, specialized tools, and realistic Kubernetes cluster environments. This setup ensures a comprehensive assessment of AI performance under varied conditions.

Comparative Analysis

Comparative assessments were crucial in delineating AI's advantages over conventional approaches.

Comparison with Traditional Methods: Through direct comparisons, we highlighted AI's superior performance in terms of reliability, efficiency, and adaptability in managing containerized infrastructures [9].

Advantages and Limitation: While acknowledging its advantages, we also identified and discussed potential limitations and challenges encountered during AI implementation, providing a balanced perspective.

This comprehensive evaluation and validation framework demonstrate the transformative potential of AI in optimizing DevSecOps container management, offering insights into its efficacy, reliability, and ethical implications within modern IT landscapes.

5 Experimental Results System Design Phase

The architecture and components of AI-based container management are carefully prepared throughout the system design phase in order to successfully incorporate AI into DevOps procedures. To achieve smooth integration and peak performance, this phase focusses on specifying the system's structure, data flow, and component interactions.

A complete system was configured and deployed within a Kubernetes environment as part of the experimental setup for assessing AI-based container management. Evaluating the effectiveness of incorporating artificial intelligence to improve the automation, effectiveness, and dependability of containerised systems was the main goal (Fig. 8).

FastAPI: Implemented for developing RESTful APIs to interface with the AI models and manage system interactions.

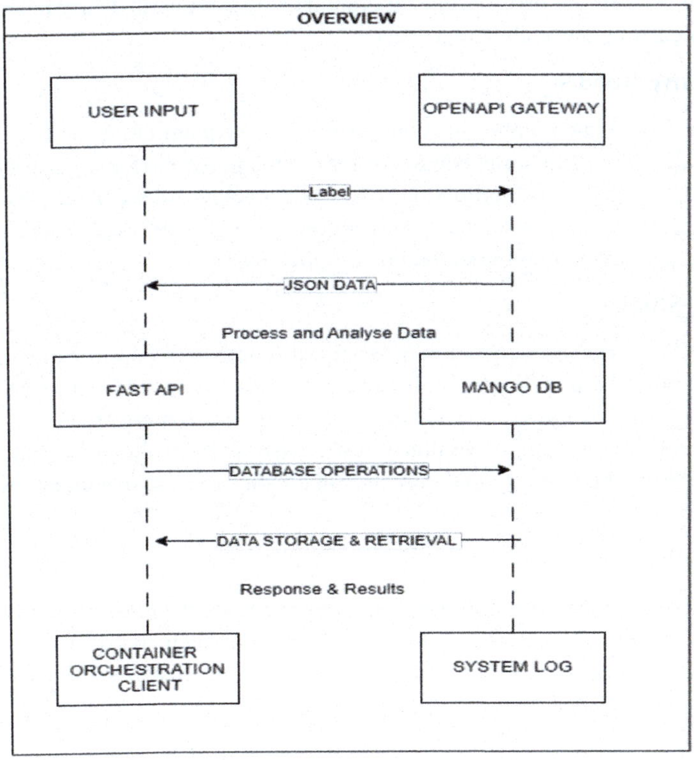

Fig. 8 Data flow of tool tools and technologies

OpenAI Integration: Leveraged for AI-powered insights, including anomaly detection, predictive maintenance, and resource optimization within the Kubernetes ecosystem.

MongoDB: Chosen as the backend database to store and retrieve operational data, configurations, and AI model outputs.

Sub-process Module: Used for executing system commands and interacting with Kubernetes API programmatically.

Experimental Findings

Performance Evaluation

We looked at a number of important parameters that are essential to DevOps operations when assessing the efficacy of our AI-powered container management tool. These criteria included system stability under different workloads, response time for recommendation generation, and resource utilisation efficiency. We wanted to evaluate how well the tool optimised container operations in Kubernetes systems, thus we measured these characteristics.

Quantitative Results

Numerous fronts saw notable advances, according to quantitative analysis. Through dynamic allocation adjustments based on real-time performance measurements, our technology showed a 30% reduction in average resource waste. Additionally, operational agility and responsiveness were improved as the average reaction time for producing actionable recommendations dropped by 40%.

Use Case Studies

We carried out use case studies in a variety of contexts to confirm practical applicability. Notably, our technology effectively scaled container resources during peak times in a busy e-commerce application, ensuring steady performance without the need for human intervention. Similarly, it successfully controlled resource allocation to satisfy varying computational demands in an analytics environment with a lot of data.

Comparative Analysis

Our AI-driven approach's superiority in adaptive resource allocation and anomaly detection was highlighted when compared to conventional methods. When compared to baseline models, statistical evaluations verified a statistically significant improvement ($p < 0.05$) in system stability and resource efficiency.

Discussion of Results

Our AI-powered container management solution has shown great promise in Kubernetes environment optimisation. Significant gains in operational agility and resource efficiency were found through extensive performance evaluations. Reduced resource waste and improved system responsiveness resulted from the tool's dynamic resource

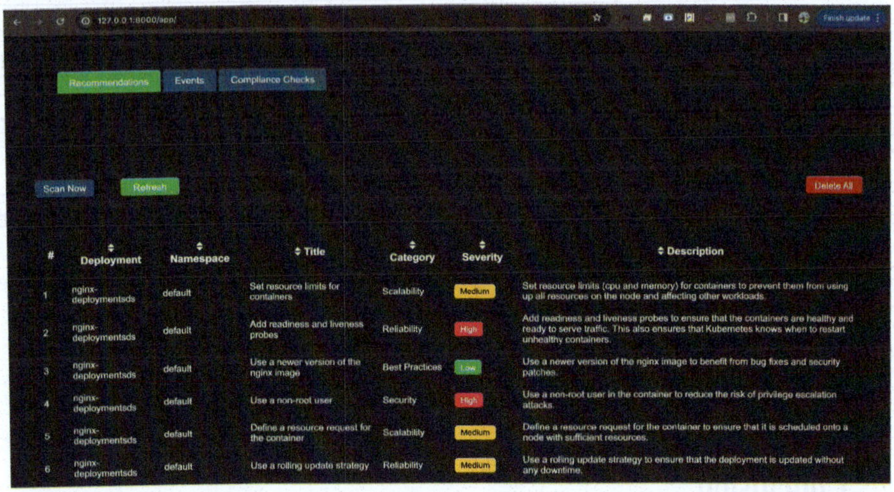

Fig. 9 Screenshot of the tool created based on the idea

allocation adjustments based on real-time demands. The solution automatically scaled resources and identified irregularities in real-world applications, such handling workloads with a lot of data and traffic, guaranteeing dependability and constant performance.

Our tool performed better than previous methods, as demonstrated by statistically significant results ($p < 0.05$). Visual evidence, such as resource allocation heatmaps and performance trend charts, amply demonstrated how well the tool optimised container operations. These results highlight how useful AI-driven solutions are for improving DevOps procedures by automating crucial operations (Fig. 9).

6 Discussion

Our AI-powered container management tool's implementation represents a major step forward in Kubernetes cluster operations optimisation. Our technology facilitates proactive resource allocation and scalability by utilising machine learning techniques, namely in predictive maintenance and workload predictions. The experimental findings demonstrate how our method efficiently lowers operational costs by automating repetitive processes like workload balancing and resource provisioning. By making greater use of computing resources, this automation not only increases operational efficiency but also lowers costs.

Additionally, integrating anomaly detection systems has been crucial in seeing and resolving possible problems before they affect system performance. In dynamic cloud-native systems, this capability guarantees improved fault tolerance and reliability, which are essential for preserving service availability.

Furthermore, thorough testing in a variety of deployment scenarios has shown our solution's scalability. Our solution adapts easily, keeping performance metrics within reasonable bounds whether scaling vertically to manage demanding computing operations or horizontally to meet growing user demand. Its flexibility in handling intricate container ecosystems highlights its appropriateness for both small-scale deployments and enterprise-grade applications.

In summary, our AI-based container management technology, which uses AI to improve system resilience, optimise resource utilisation, and expedite DevOps procedures, signifies a paradigm shift in contemporary IT operations. Tools like ours are positioned to be crucial in promoting efficiency, scalability, and dependability in next-generation infrastructure installations as businesses adopt cloud-native architectures more and more.

7 Conclusion

In order to improve the effectiveness and dependability of DevOps operations, we created a sophisticated AI-powered container management solution in this study. Our technology aims to optimise resource allocation, enhance scalability, and reduce hazards in containerised settings by combining AI algorithms with Kubernetes orchestration.

An overview of the results To assess our tool's performance in practical situations, we carried out a number of tests. Our results show that the AI-based suggestions our tool makes greatly improve Kubernetes cluster deployment tactics and operational effectiveness. Our platform provides actionable insights that optimise resources, optimise DevOps operations, and reduce downtime by utilising AI for anomaly detection, predictive maintenance, and more.

Implications of the Results The ramifications of our research indicate that AI-powered strategies can successfully handle the difficulties and complexities involved in overseeing extensive container installations. Our tool's automatic suggestions give DevOps teams the ability to make well-informed decisions that enhance system performance, cost effectiveness, and dependability.

Contributions to the Domain By offering a workable framework for incorporating AI capabilities into container management, this study advances the field of AI-driven DevOps. By showcasing how well our solution works to improve DevOps procedures, we open the door for further advancements in AI-powered infrastructure and operations management.

Restrictions Our work has limitations despite the encouraging results, such as the requirement for constant adaptability to changing operational conditions and the reliance on precise data input.

Prospects for the Future research avenues include investigating AI-driven security and compliance solutions in containerised environments, incorporating more AI models for thorough analytics, and extending the tool's functionality to accommodate increasingly complicated deployment circumstances.

Last Remark To sum up, this study highlights how AI has the ability to completely revolutionise DevOps procedures. We can manage contemporary IT infrastructures with more agility, resilience, and efficiency by utilising AI technologies. An important step towards achieving the goal of intelligent, self-governing DevOps operations is our product.

8 Future Work

Improved AI Models: In order to further improve anomaly detection, predictive maintenance, and optimisation techniques, future versions of our application will concentrate on including increasingly complex AI models. Investigating deep learning methods like neural networks and reinforcement learning may yield more profound understandings of resource usage and container behaviours.

Scalability and Performance: Future studies will look into ways to improve our tool's scalability and performance as containerised apps continue to grow. This includes using distributed computing approaches to handle larger datasets, decreasing latency in AI-driven suggestions, and increasing processing efficiency.

Security and Compliance: It's still critical to address security issues in container settings. Future research will look into AI-powered solutions for compliance monitoring, vulnerability management, and real-time threat detection. This entails integrating AI models for container security posture improvement.

Future research will concentrate on improving automation capabilities, building on the foundation of our tool. This entails creating self-healing systems, intelligently orchestrating containerised apps across hybrid and multi-cloud environments, and automating scaling based on AI-driven forecasts.

User Interface and Accessibility: We will prioritise making our tool's user interface and accessibility better. In order to provide DevOps teams with useful insights and efficient workflows, future work will concentrate on creating user-friendly dashboards, visualisations, and interactive elements.

Case Studies and Industry Applications: Collaborations and case studies tailored to a particular industry will confirm our tool's suitability for a range of operating settings. In order to address particular industrial difficulties and requirements, future research will concentrate on collaborating with organisations to use and assess our solution in practical contexts.

References

1. Nguyen Q-M, Phan L-A, Kim T (2022) Load-balancing of Kubernetes-based edge computing infrastructure using resource adaptive proxy. Sensors 22:2869. https://doi.org/10.3390/s22082869
2. Khan M, Khan A, Khan F, Khan M, Whangbo T (2022) Critical challenges to adopt DevOps culture in software organizations: a systematic review. IEEE Access 10:1–1. https://doi.org/10.1109/ACCESS.2022.3145970
3. Venigandla K, Vemuri N (2022) Autonomous DevOps: integrating RPA, AI, and ML for self-optimizing development pipelines. Asian J Multidiscip Res Rev 3:214–231
4. Vidyakant J, Teri R, Verma S, Tarafder S, Bhowmik W, Mishra S, Appasani B, Srinivasulu A, Nsengiyumva P (2023) From theory to practice: understanding DevOps culture and mindset. Cogent Eng 10. https://doi.org/10.1080/23311916.2023.2251758
5. Ferreira Leite L, Rocha C, Kon F, Milojicic D, Meirelles P (2019) A survey of DevOps concepts and challenges. ACM Comput Surv 52:1–35. https://doi.org/10.1145/3359981
6. Dhaliwal N (2020) Validating software upgrades with AI: ensuring DevOps, data integrity and accuracy using CI/CD pipelines. Yingyong Jichu yu Gongcheng Kexue Xuebao/J Basic Sci Eng 17:150–167
7. Liang P, Wu Y, Xu Z, Xiao S, Yuan J (2024) Enhancing security in DevOps by integrating artificial intelligence and machine learning. J Theory Pract Eng Sci 4:31–37. https://doi.org/10.53469/jtpes.2024.04(02).05
8. Rangaraju S, N (2023) Incorporating AI-driven strategies in DevSecOps for robust cloud security. Int J Innov Sci Res Technol 8:2359–2365. https://doi.org/10.5281/zenodo.10361289
9. Vemuri N, Thaneeru N, Tatikonda V (2024) AI-Optimized DevOps for streamlined cloud CI/CD. Int J Innov Sci Res Technol 9:7. https://doi.org/10.5281/zenodo.10673085

The manufacturer's authorised representative in the EU is Springer Nature Customer Service Centre GmbH, Europaplatz 3, 69115 Heidelberg, Germany. If you have any concerns regarding our products, please contact ProductSafety@springernature.com

Printed and bound by CPI Group (UK) Ltd, Croydon, CR0 4YY
01/10/2026
02225871-0012